REFERENCE MATERIAL FROM PLANE GEOMETRY

1. If a whole angle (a complete revolution) is divided into 360 equal parts, each of these is said to be an angle of *one degree,* denoted 1°.
2. Two angles are said to be *complementary* if their sum is 90°.
3. Two angles are said to be *supplementary* if their sum is 180°.
4. The sum of the three angles of a triangle is 180°.
5. In a right triangle, the sides that form the right angle are called the *legs;* the side opposite the right angle is called the *hypotenuse.*
6. **Pythagorean theorem:** *The square of the hypotenuse of a right triangle equals the sum of the squares of its legs.*
7. In a right triangle with angles 30°, 60°, 90°, the hypotenuse is twice the shorter leg.
8. An *isosceles* triangle has two equal sides, and hence two equal angles.
9. An *equilateral* triangle has three equal sides. Each angle is 60°.
10. A *tangent* line to a circle is perpendicular to the radius drawn to the point of tangency.
11. If two triangles have the three angles of one equal respectively to the three angles of the other, the triangles are *similar* and their corresponding sides are *proportional*.

Plane Trigonometry

SCHAUM'S SOLVED PROBLEMS BOOKS

Each title in this series is a complete and expert source of solved problems containing thousands of problems with worked out solutions.

Titles on the Current List Include:

3000 Solved Problems in Calculus
2500 Solved Problems in College Algebra and Trigonometry
2500 Solved Problems in Differential Equations
2000 Solved Problems in Discrete Mathematics
3000 Solved Problems in Linear Algebra
2000 Solved Problems in Numerical Analysis
3000 Solved Problems in Precalculus

Also Available from McGraw-Hill

SCHAUM'S OUTLINE SERIES IN MATHEMATICS & STATISTICS

Most outlines include basic theory, definitions, and hundreds of solved problems and supplementary problems with answers.

Titles on the Current List Include:

Advanced Calculus
Advanced Mathematics
Analytic Geometry
Beginning Calculus
Boolean Algebra
Calculus, 3d edition
Calculus for Business, Economics, & the Social Sciences
Calculus of Finite Differences & Difference Equations
College Algebra
College Mathematics, 2d edition
Complex Variables
Descriptive Geometry
Differential Equations
Differential Geometry
Discrete Math
Elementary Algebra, 2d edition
Essential Computer Math
Finite Mathematics
Fourier Analysis
General Topology
Geometry, 2d edition
Group Theory
Laplace Transforms
Linear Algebra, 2d edition
Mathematical Handbook of Formulas & Tables
Matrix Operations
Modern Abstract Algebra
Modern Elementary Algebra
Modern Introductory Differential Equations
Numerical Analysis, 2d edition
Partial Differential Equations
Probability
Probability & Statistics
Real Variables
Review of Elementary Mathematics
Set Theory & Related Topics
Statistics, 2d edition
Technical Mathematics
Tensor Calculus
Trigonometry, 2d edition
Vector Analysis

Plane Trigonometry

SEVENTH EDITION

E. Richard Heineman
Late Professor of Mathematics
Texas Tech University

J. Dalton Tarwater
Professor of Mathematics
Texas Tech University

McGraw-Hill, Inc.
New York St. Louis San Francisco Auckland Bogotá Caracas
Lisbon London Madrid Mexico Milan Montreal New Delhi
Paris San Juan Singapore Sydney Tokyo Toronto

Plane Trigonometry

1 2 3 4 5 6 7 8 9 0 DOC DOC 9 0 9 8 7 6 5 4 3 2

ISBN 0-07-028187-4

This book was set in Times Roman by Beacon Graphics Corporation.
The editors were Karen M. Minette, Michael Johnson, and James W. Bradley;
the design was done by Circa 86, Inc.
the production supervisor was Kathryn Porzio.
New drawings were done by Vantage Art.
R. R. Donnelley & Sons Company was printer and binder.

Library of Congress Cataloging-in-Publication Data

Heineman, E. Richard (Ellis Richard)
Plane trigonometry / E. Richard Heineman, J. Dalton Tarwater.—
7th ed.
p. cm.
Includes index.
ISBN 0-07-028187-4
1. Plane trigonometry. I. Tarwater, J. Dalton. II. Title.
QA533.H47 1993
516.24′2—dc20 92-17811

Dedication

E. Richard Heineman was born and raised in Wisconsin, where he taught high school geometry and chemistry. After earning his degrees from the University of Wisconsin, Madison, he taught at Michigan State University, before moving on to Texas Tech University. After a distinguished career at Texas Tech spanning 45 years, and almost two decades of retirement, he died in 1991.

Since its first edition, Professor Heineman's *Plane Trigonometry* has helped well over a million students learn trigonometry. It remains the standard for all other trigonometry texts.

This book is dedicated to his memory.

CONTENTS

PREFACE

The seventh edition of *Plane Trigonometry* is a significant revision of this classic text. Several chapters have been left basically intact, a few have been merged or revised slightly, while others have been completely rewritten. Two new chapters have been added. Every effort has been made to retain the brief expository style of the previous edition while updating the content to meet current standards.

Major changes include:

1. The introduction of radian measure in Chapter 4 and its extensive use afterward
2. An optional Section 8.6 on graphing calculators
3. A new Chapter 11 on polar coordinates and a new Chapter 13 on topics from analytic geometry
4. The assumption that the student uses a scientific calculator; however, the Appendix has a section on the use of tables for students without a calculator
5. The addition of a list of Key Terms and a set of Review Exercises at the end of each chapter
6. More emphasis on graphing throughout the text and the addition of graphs in the Answers section

Major features retained from the previous editions include:

1. Miscellaneous points include (*a*) a note to the student, (*b*) problems that are encountered in calculus, (*c*) a careful explanation of the concept of infinity, (*d*) memory schemes, (*e*) the uses of the sine and cosine curves, and (*f*) interesting applied problems.
2. The problems in each exercise are so arranged that by assigning numbers 1, 5, 9, etc., or similar sets beginning with 2, 3, or 4, the instructor can obtain balanced coverage of all points involved without undue emphasis on some principles at the expense of others.
3. Answers to three-fourths of the problems are given at the back of the book.

4. Many of the exercises contain true-false questions to test the student's ability to avoid pitfalls and to detect camouflaged truths. The duty of the instructor is not only to teach correct methods but also to convince the student of the error in the false methods.
5. All problem sets are carefully graded and contain an abundance of simple problems that involve nothing more than the principles being discussed. There is also an ample supply of problems of medium difficulty and some "head-scratchers."

A number of *supplements* have been provided for both the student and instructor. They include:

The Instructor's Resource Manual, prepared by Joan Van Glabek of Edison Community College, contains teaching suggestions, chapter exams and final exams, and solutions to every fourth problem from the text.

The Student's Solutions Manual, also prepared by Joan Van Glabek, contains the solutions to three-fourths of the problems from the text.

The Videotape Series, prepared by John Jobe of Oklahoma State University, consists of 11 tapes that present essential trigonometry topics for review.

The Professor's Assistant is a computerized test generator that allows the instructor to create tests using algorithmically generated test questions and those from a standard testbank. This testing system enables the instructor to choose questions either manually or randomly by section, question type, difficulty level, and other criteria. This system is available for IBM, IBM compatible, and Macintosh computers.

The Print Test Bank is a printed and bound copy of the questions found in the standard test bank.

For further information about these supplements, please contact your local College Division sales representative.

For over half of a century many people have contributed to the success of this text. We are grateful to the many students and instructors who have made suggestions and have offered criticisms. For this edition we are particularly indebted to Patrick Tarwater of Texas Tech University for the graphs in the Answers Section, and to Beth Martin, also of Texas Tech, for her excellent typing of the manuscript. Several reviewers have made substantial improvements in the presentation. Thanks to Sandra Beken, Horry-Georgetown Technical College; Steven Blasberg, West Valley College; Forrest G. Lowe, Longview Community College; Glenn T. Smith II, Santa Fe Community College; and Carol M. Walker, Hinds Community College.

Special thanks to Karen M. Minette and Michael Johnson of McGraw-Hill for their undying patience.

Finally, we acknowledge that the quotation attributed to Thomas Jefferson is taken from E. R. Hogan, The Beginnings of Mathematics in a Howling Wilderness, *Historia Mathematica,* v. 1, 1974, p. 156. Also, some exercises, examples, and formulas in Chapters 11 and 13 were borrowed from G. Fuller/D. Tarwater, *Analytic Geometry,* 7th ed., © 1992, Addison-Wesley Publishing Company, Inc. They are reprinted with permission of the publisher.

J. Dalton Tarwater

NOTE TO THE STUDENT

A mastery of the subject of trigonometry requires (1) a certain amount of memory work, (2) a great deal of practice and drill in order to acquire experience and skill in the application of the memory work, and (3) an insight and understanding of "what it is all about." Your instructor is a "troubleshooter" who attempts to prevent you from going astray, supplies missing links in your mathematical background, and tries to indicate the "common sense" approach to the problem.

The memory work in any course is one thing that students can and should perform by themselves. The least you can do for your instructor and yourself is to *commit to memory each definition and theorem as soon as you contact it.* This can be accomplished most rapidly not by reading, but by writing the definition or theorem until you can reproduce it without the aid of the text.

In working the problems, do not continually refer back to the illustrative examples. Study the examples so thoroughly (by writing them) that you can reproduce them with your text closed. Only after the examples are entirely clear and have been completely mastered should you attempt the unsolved problems. These problems should be worked *without referring to the text.*

Bear in mind, too, that *memory* and technical *skill* are aided by *understanding;* therefore, as the course develops you should review the definitions and theorems from time to time, always seeking a deeper insight into them.

Plane Trigonometry

The Trigonometric Functions

1.1 HISTORICAL OVERVIEW

Mathematics, like music and poetry, is a creation of the human mind. Just as we can't remember who wrote the first song or the first verse, the origins of mathematics are lost in antiquity. This is especially true of the branch of mathematics known as trigonometry, which means "triangle measurement." We have evidence of some aspects of trigonometry in Babylonia as early as 1600 B.C. Later, around 200 B.C., the Greek astronomers of Alexandria were well aware of the uses of trigonometry.

Since then, there have been many theoretical advances in trigonometry. There have been countless applications of trigonometry to astronomy, cartography, electronics, engineering, music theory, navigation, physics, surveying, and other endeavors. Indeed, in the early days of America, trigonometry was learned for its central role in surveying, navigation on the high seas, and astronomy. It was essential for a young country which was trying to explore, settle, and defend a continent. Thomas Jefferson, in 1799, wrote that

> trigonometry... is most valuable to every man. There is scarcely a day in which he will not resort to it for some of the purposes of common life...

For two centuries Jefferson's view of the utility of trigonometry has proved sound, if not in ways he meant or even in ways he could have imagined.

In the last half century, since this text was first published, the appearance of computers, hand calculators with trigonometric functions, and graphing calculators has removed the tedium of many trigonometric calculations, increased the accuracy of such calculations, and provided visual reiteration of the principles of the subject. While we embrace these technological advances, we caution the student that these devices do not lessen your need to know, or even to commit to memory, basic principles, definitions, and theorems of trigonometry.

We begin our study of trigonometry with a brief review of fundamental ideas of coordinate geometry.

1.2 THE RECTANGULAR COORDINATE SYSTEM

A directed line is a line on which one direction is considered positive and the other negative. Thus, in Figure 1.1, the arrowhead indicates that all segments measured from left to right are positive. Hence if $OA = 1$ unit of length, then $OB = 3$ and $BC = -5$. Observe that since the line is directed, CB is not equal to BC. However, $BC = -CB$; or $CB = -BC$. If A, B, and C are any three points on the line, then $AB + BC = AC$ and $AB + BC + CA = 0$.

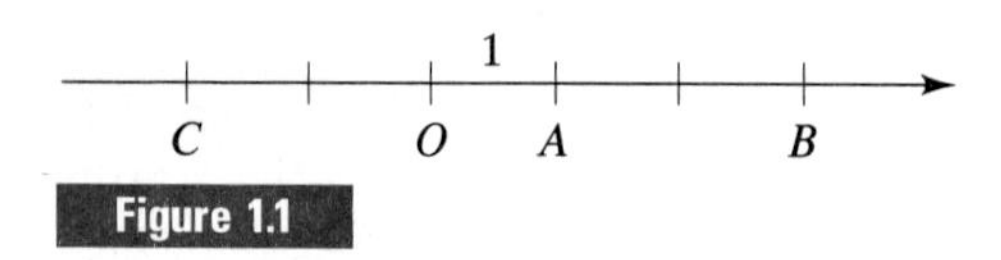

Figure 1.1

A rectangular (or cartesian) coordinate system consists of two perpendicular *directed* lines. It is conventional to draw and direct these lines as in Figure 1.2. The **x axis** and the **y axis** are called the **coordinate axes;** their intersection O is called the **origin.** The position of any point in the plane is fixed by its distances from the axes.

The x coordinate of point P is the length of the directed segment NP (or OM) measured from the y axis to point P. The y coordinate of point P is the length of the directed segment MP, measured from the x axis to point P.

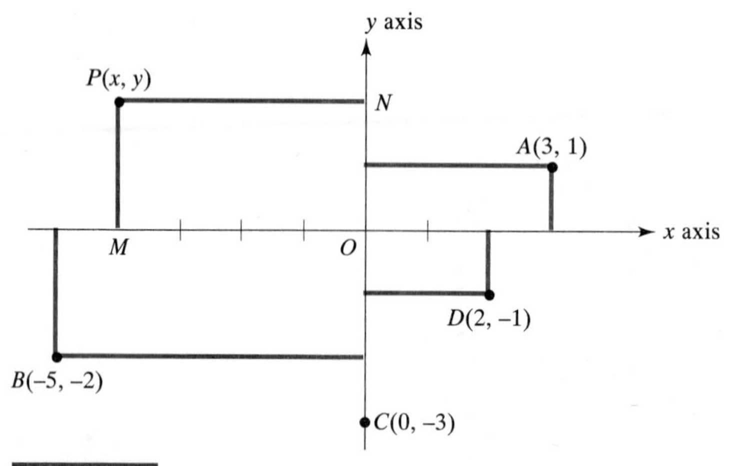

Figure 1.2

It is necessary to remember that each coordinate is measured *from axis to point*. Thus the x coordinate of P is NP (not PN); the y coordinate of P is MP (not PM). The point P, with x coordinate x and y coordinate y, is denoted by $P(x, y)$. It follows that the x coordinate of any point to the right of the y axis is positive; to the left it is negative. Also the y coordinate of any point above the x axis is positive; below, it is negative.

To **plot** a point means to locate and indicate its position on a coordinate system. Several points are plotted in Figure 1.2.

The distance r from the origin O to point P is called the **radius vector** of P. This distance r is not directed and *is always positive* by agreement. Hence with each point of the plane we can associate three numbers: x, y, and r. The radius vector r can be found by using the Pythagorean relation $x^2 + y^2 = r^2$ (see Figure 1.3).

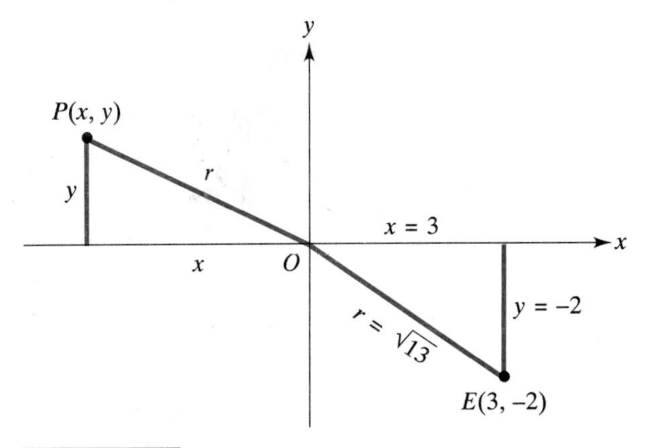

Figure 1.3

The coordinate axes divide the plane into four parts called **quadrants** as indicated in Figure 1.4. We shall sometimes denote these as Q I, Q II, Q III, and Q IV, respectively. We have abbreviated "$x < 0$" by "$x = -$" as an aid to memory.

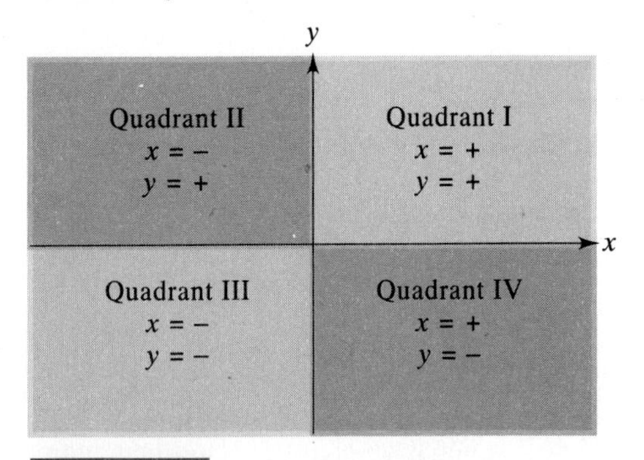

Figure 1.4

Illustration 1 To find the radius vector r for the point $(5, -12)$, use

$$r^2 = 5^2 + (-12)^2 = 169, \quad r = 13.$$

Illustration 2 If $x = 15$ and $r = 17$, we obtain y by using

$$x^2 + y^2 = r^2.$$

Hence $(15)^2 + y^2 = (17)^2$; $225 + y^2 = 289$; $y^2 = 64$; $y = \pm 8$. If the point is in quadrant I, $y = 8$; if the point is in quadrant IV, $y = -8$. (Since x is positive, the point cannot lie in either quadrant II or quadrant III.)

1.3 THE DISTANCE FORMULA

Let $P_1(x_1, y_1)$ and $P_2(x_2, y_2)$ be *any* two points in the xy plane (Figure 1.5). We shall use the Pythagorean theorem to express the distance P_1P_2 in terms of the coordinates of the points. Through P_1 draw a line parallel to the x axis. Through P_2 draw a line parallel to the y axis. These lines meet at $Q(x_2, y_1)$. The length of the positive segment P_1Q is $x_2 - x_1$ (that is, the *right* value of x minus the *left* value of x). More precisely, the x coordinate of the point on the right (the large x) minus the x coordinate of the point on the left (the small x). For $P_1(-2, 1)$ and $Q(5, 1)$, $P_1Q = 5 - (-2) = 7$. And the length of the positive segment QP_2 is $y_2 - y_1$ (that is, the *upper* y value minus the *lower* y). Since $(P_1P_2)^2 = (P_1Q)^2 + (QP_2)^2$,

$$P_1P_2 = \sqrt{(x_2 - x_1)^2 + (y_2 - y_1)^2}.$$

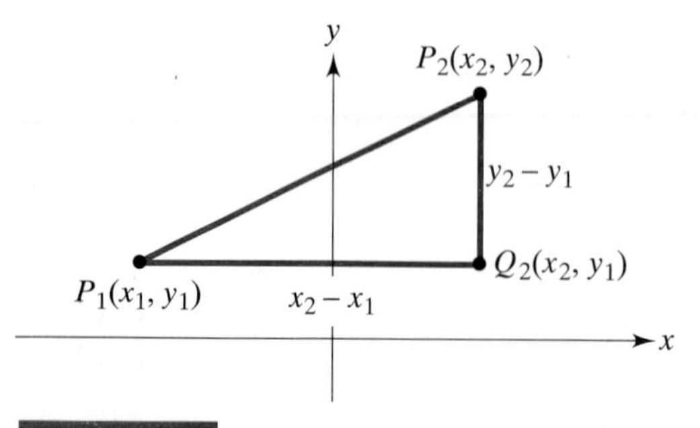

Figure 1.5

This formula holds for all positions of P_1 and P_2. If, for example, P_2 lies below P_1, then $QP_2 = y_2 - y_1$, which is a negative quantity. But the square of $(y_2 - y_1)$ is equal to the square of the positive quantity $(y_1 - y_2)$; that is, $(y_2 - y_1)^2 = (y_1 - y_2)^2$. Hence

the distance d between $P_1(x_1, y_1)$ and $P_2(x_2, y_2)$ is

$$d = \sqrt{(x_2 - x_1)^2 + (y_2 - y_1)^2}.$$

Illustration The distance between $A(-1,5)$ and $B(3,-2)$ is $d = \sqrt{(-1-3)^2 + (5+2)^2} = \sqrt{16+49} = \sqrt{65}$. Either A or B can be designated as P_1.

Exercise 1.1

1. Plot the following points on coordinate paper and then find the value of r for each: $(6,-8)$; $(-7,0)$; $(-5,-2)$; $(\sqrt{35},1)$.
2. Plot the following points on coordinate paper and then find the value of r for each: $(15,8)$; $(0,-1)$; $(-2,3)$; $(-\sqrt{7},-\sqrt{2})$.
3. Plot the following points on coordinate paper and then find the value of r for each: $(-24,7)$; $(0,4)$; $(3,-5)$; $(-6,-\sqrt{13})$.
4. Plot the following points on coordinate paper and then find the value of r for each: $(-5,-12)$; $(3,0)$; $(7,1)$; $(\sqrt{77},-2)$.
5. Use the Pythagorean theorem to find the missing coordinate and then plot the point:
   (a) $y = 15$, $r = 17$, point is in Q II.
   (b) $x = -2$, $r = 3$, point is in Q III.
   (c) $y = 0$, $r = 10$, x is negative.
6. Find the missing coordinate and then plot the point:
   (a) $x = 5$, $r = 13$, point is in Q IV.
   (b) $y = 6$, $r = \sqrt{61}$, point is in Q II.
   (c) $x = -1$, $r = 1$.
7. Find the missing coordinate and then plot the point:
   (a) $x = -3$, $r = 5$, point is in Q III.
   (b) $y = 2\sqrt{11}$, $r = 12$, point is in Q I.
   (c) $y = -6$, $r = 6$.
8. Find the missing coordinate and then plot the point:
   (a) $y = 24$, $r = 25$, point is in Q I.
   (b) $x = 4$, $r = 2\sqrt{13}$, point is in Q IV.
   (c) $x = 0$, $r = 9$, y is positive.
9. In which quadrants are the following ratios positive?
   (a) $\frac{y}{r}$ (b) $\frac{x}{r}$ (c) $\frac{y}{x}$

10. In which quadrants are the following ratios negative?

    (a) $\frac{y}{r}$ (b) $\frac{x}{r}$ (c) $\frac{y}{x}$

11. What is the y coordinate of all points on the x axis? What is the x coordinate of all points on the y axis?
12. Without plotting, identify the quadrant in which each of the following points lies if s is a negative number: $E(-2, s)$; $F(-s, s^2)$, $G(s^4, s^3)$; $H(-s^2, -s^3)$.
13. Use the distance formula to find the exact value of the distance between the points $(6, -7)$ and $(-2, -5)$.
14. Use the distance formula to find the distance between the points $(a, 1)$ and $(-3, 8)$.
15. Find a, if the distance from $(a, 1)$ to $(2, 3)$ is $\sqrt{8}$.

1.4 TRIGONOMETRIC ANGLES

There are several ways of measuring angles. One of the oldest and most widely used methods is to divide one revolution into 360 equal parts; each part is then called one **degree** (1°). Radian measure will be defined later.

A *ray,* or half-line, is the part of a line extending in one direction from a point on the line. In geometry, you may have thought of an angle as the "opening" between two rays that form the sides of the angle and that emerge from a point called the vertex of the angle.

A trigonometric angle is an amount of rotation used in moving a ray from one position to another.

A *positive* angle is generated by *counterclockwise* rotation; a *negative* angle, by *clockwise* rotation. Figure 1.6 illustrates the terms used and shows an angle of 200°.

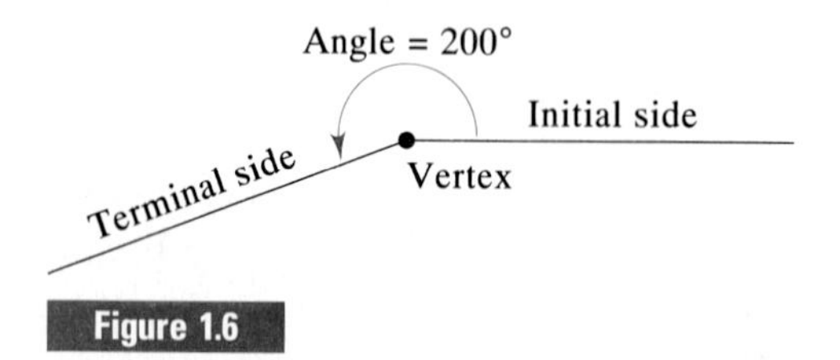

Figure 1.6

Figure 1.7 shows angles of 500° and −420°. The −420° angle may be thought of as the amount of rotation effected by the minute hand of a clock between 12:15 and 1:25. *To specify a trigonometric angle, we need,* in addition to its sides, *a curved arrow extending from its initial side to its terminal side.*

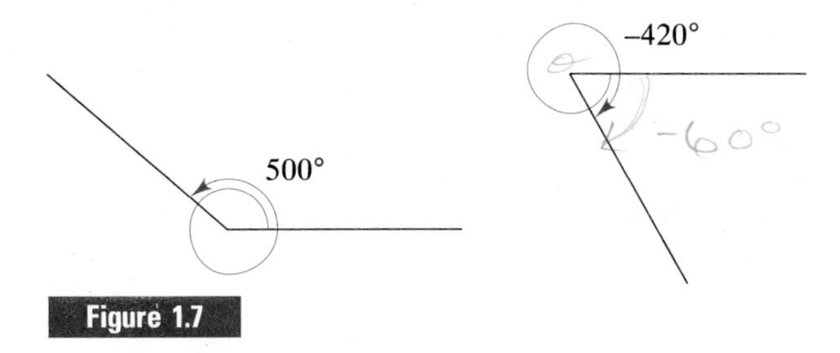

Figure 1.7

An angle is said to be in *standard position* if its vertex is at the origin and its initial side coincides with the positive x axis.

An angle is said to be in a certain quadrant if its terminal side lies in that quadrant *when the angle is in standard position.* For example, 600° is in the third quadrant; $-70°$ is a fourth-quadrant angle.

Angles are said to be **coterminal** if their terminal sides coincide when the angles are in standard position. For example, 200°, 560°, and $-160°$ are coterminal angles. From a trigonometric viewpoint these angles are not equal; they are merely coterminal.

Exercise 1.2

Place each of the following angles in standard position; draw a curved arrow to indicate the direction of rotation. Draw and find the size of two other angles, one positive and one negative, that are coterminal with the given angle.

1. 70°
2. 110°
3. 270°
4. 220°
5. 550°
6. 700°
7. 520°
8. 370°

Each of the following points is on the terminal side of a positive angle in standard position. Plot the point; draw the terminal side of the angle; indicate the angle by a curved arrow; use a protractor to find, to the nearest degree, the size of the angle.

9. $(5, 6)$
10. $(7, -1)$
11. $(-8, -5)$
12. $(-9, 4)$
13. $(-10, 7)$
14. $(-1, -4)$
15. $(7, 4)$
16. $(1, -\sqrt{3})$

17. A wheel makes 1600 revolutions per minute (rpm). Through how many degrees does it move in 1 second (s)?
18. Through how many degrees does the minute hand on a clock move in 17 min? In $37\frac{1}{2}$ min? In one day?

1.5 DEFINITIONS OF THE TRIGONOMETRIC FUNCTIONS OF A GENERAL ANGLE

Recall that a **relation** is a set $\{(u, v)\}$ of ordered pairs. The **domain** of the relation is the set of all first elements u of ordered pairs in the relation while the **range** of the relation is the set of all second elements v of ordered pairs in the relation.

A **function** is a relation in which no two ordered pairs have the same first element and distinct second elements. Thus, if a function has (u, v) and (u, w) in it, it must be that $v = w$. Otherwise, it is not a function. If a relation is a function, then for each member u in its domain, there is one and only one member v in its range. If f is the name of the function, we write $v = f(u)$, read "v equals f of u," to denote that the ordered pair (u, v) is in f. For most functions that the reader has encountered there is specified a rule or recipe to show how to find $f(u)$ if u is given.

Illustration $f(u) = u^2$ and $g(u) = (3u - 1)/(u^2 + 1)$ are examples of functional rules. In order to find the value of the function f when u is 3, we substitute in the recipe $f(3) = 3^2 = 9$. Likewise $g(3) = (3 \cdot 3 - 1)/(3^2 + 1) = 4/5$, and $g(-1) = -2$. The fact that $f(-1) = f(1) = 1$ does not destroy the functional nature of f. It very well may be that distinct members, 1 and -1 in this case, in the domain correspond to the same element in the range.

The whole subject of trigonometry is based on the six **trigonometric functions.** The names of these functions, with their abbreviations in parentheses, are: sine **(sin)**, cosine **(cos)**, tangent **(tan)**, cotangent **(cot)**, secant **(sec)**, and cosecant **(csc)**. In a certain sense, the following definitions are the most important in this book.

A Complete Definition of the Trigonometric Functions of any Angle θ

1. Place the angle θ in standard position (see Figure 1.8).
2. Choose a point $P(x, y)$, distinct from the origin, on the terminal side of θ.

3. Drop *a* perpendicular to the x axis, thus forming a right triangle of reference for θ.
4. The values of the six trigonometric functions of θ are defined in terms of the coordinates (x, y) of P and the radius vector r as follows:

$$\sin\theta = \frac{y}{r}, \qquad \csc\theta = \frac{r}{y},$$

$$\cos\theta = \frac{x}{r}, \qquad \sec\theta = \frac{r}{x},$$

$$\tan\theta = \frac{y}{x}, \qquad \cot\theta = \frac{x}{y}.$$

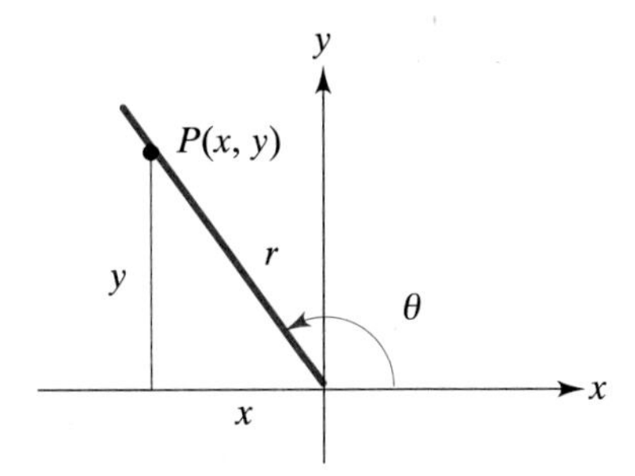

Figure 1.8

According to the functional notation mentioned above, we should write $\sin(\theta)$, $\tan(\theta)$, etc., to denote the value of the trigonometric functions at the angle θ. Traditionally, the parentheses are omitted and we simply write $\sin\theta$, $\tan\theta$, etc.

We need to prove that the procedure outlined in the definition does indeed define six functions. First note that the *domain* of each function consists of all θ for which the corresponding denominator is not 0; thus $\tan\theta$ and $\sec\theta$ are not defined when $x = 0$, and $\cot\theta$ and $\csc\theta$ are not defined when $y = 0$. You will recall that *division by zero is impossible.* The definition of division states that $a/b = c$ if and only if $bc = a$, provided that c is a unique number. If $\frac{1}{0} = a$, then $(0)(a)$ must equal 1. No such number a exists. Another explanation: When we write $\frac{12}{3}$, we are asking, "How many 3's add up to 12?" Consequently $\frac{1}{0}$ means "How many zeros will add up to 1?" Such a question is obviously absurd.

In order to prove that sine is a function, we must show that the value of $\sin\theta$, as computed by using the definition, would be the same if another point $P'(x', y')$ on the terminal side of θ were chosen.

Let $P'(x', y')$ be any other point on OP (see Figure 1.9). Then, using the coordinates of P', we have $\sin\theta = y'/r'$. Since triangles $OP'M'$ and OPM are similar, it follows that

$$\frac{y'}{r'} = \frac{y}{r}$$

and the value of $\sin\theta$ is the same whether it is obtained by using P or P'.

Since the value of $\sin\theta$ is determined solely by θ and is independent of the choice of the point P, we have verified that for any θ, there is one and only one value for $\sin\theta$. This establishes that sine is a function.

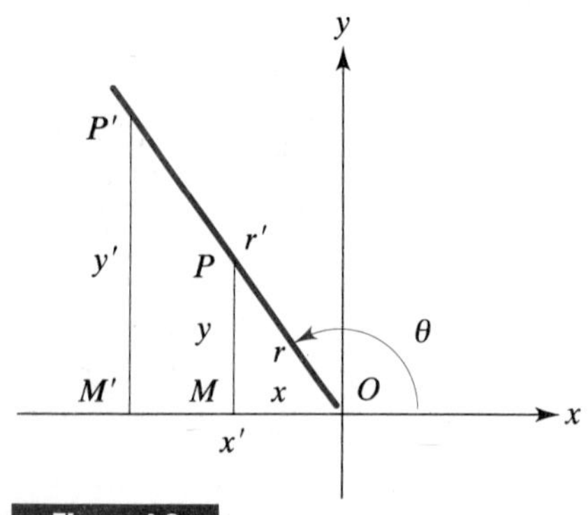

Figure 1.9

In a similar fashion, we may verify that the five other ratios in the definition define functions.

In order to examine the **range** of the sine and cosine functions, we observe that the hypotenuse of a right triangle is never shorter than either of the sides. Thus for any θ, and for any $P(x, y)$ on the terminal side of θ, we have $|x| \le r$ and $|y| \le r$. Hence

$$|\sin\theta| = \left|\frac{y}{r}\right| \le 1 \quad\text{and}\quad |\cos\theta| \le \left|\frac{x}{r}\right| \le 1\,.$$

These inequalities are the same as

$$-1 \le \sin\theta \le 1 \quad\text{and}\quad -1 \le \cos\theta \le 1\,,$$

which define the ranges of the two functions.

Since r is always positive, $\sin\theta$ is positive whenever y is positive, that is, when θ is in the upper quadrants, I and II. Similarly, $\sin\theta$ is negative in the lower quadrants, III and IV. Also $\sin\theta$ is 0 when $y = 0$, which occurs when θ is coterminal with 0° or 180° (see Figure 1.10).

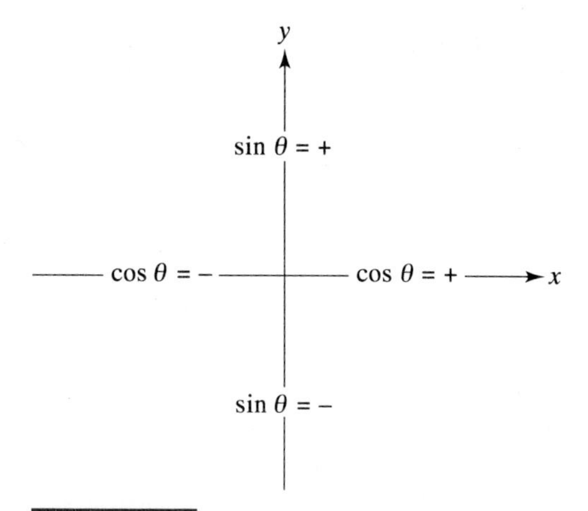

Figure 1.10

Likewise, $\cos \theta$ has the same sign as x. Hence $\cos \theta$ is positive in the right-hand quadrants, I and IV. Also $\cos \theta$ is negative in the left-hand quadrants, II and III. And $\cos \theta$ is 0 whenever $x = 0$, which occurs when θ is coterminal with 90° or 270°.

Moreover, $\tan \theta = y/x$ is positive when y and x have the same sign, namely, in quadrants I and III. Also $\tan \theta$ is negative when y and x have opposite signs, namely, in quadrants II and IV.

Example 1 Find the values of sin 160°, cos 160°, tan 160°.

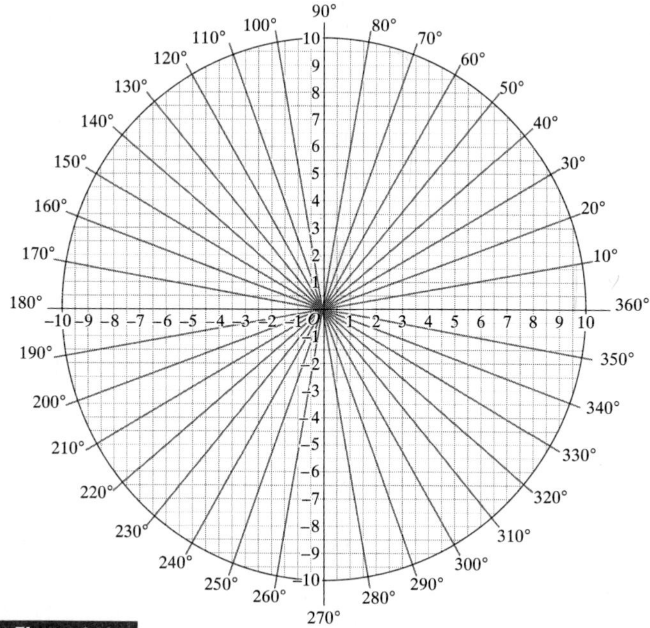

Figure 1.11

Solution In Figure 1.11, the angle 160° is in standard position. For convenience, on the terminal side of 160° choose point P so that $r = 10$. In forming a triangle of reference, we find that $x = -9.4$, $y = 3.4$. Since these numbers were obtained from a drawing, they are approximations. Equally acceptable would be 3.3 or 3.5 (see Figure 1.12). Hence

$$\sin 160° = \frac{y}{r} = \frac{3.4}{10} = 0.34,$$

$$\cos 160° = \frac{x}{r} = \frac{-9.4}{10} = -0.94,$$

$$\tan 160° = \frac{y}{x} = \frac{3.4}{-9.4} = -0.36.$$

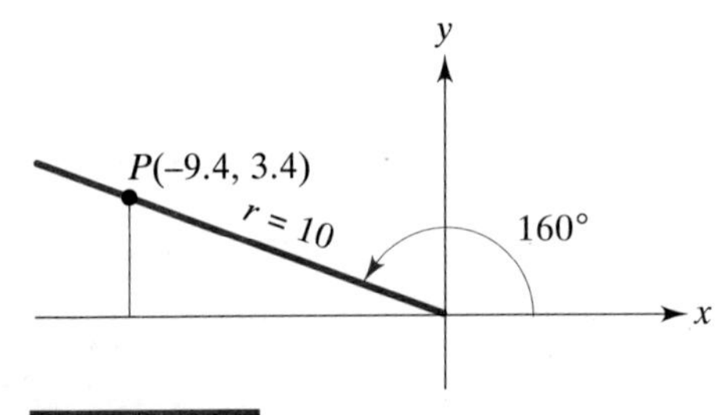

Figure 1.12

Example 2 Find the trigonometric functions of 180°.

Solution Place the angle in standard position. For point P let us choose $(-1, 0)$. Since r is always positive, $r = 1$. The triangle of reference has "collapsed," but P does have the coordinates x, y, r (see Figure 1.13).

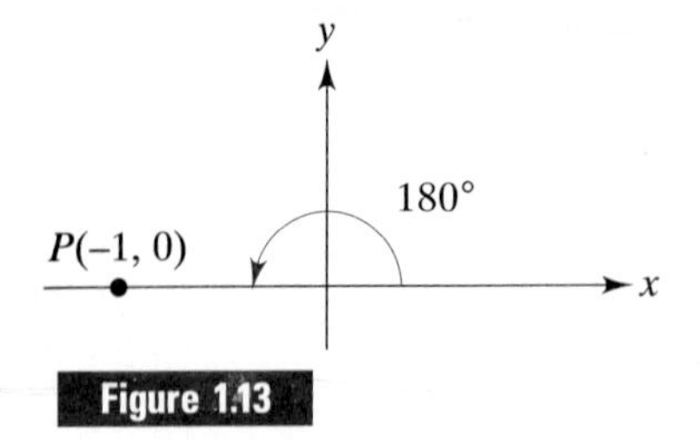

Figure 1.13

Then

$$\sin 180° = \frac{y}{r} = \frac{0}{1} = 0,$$

$$\cos 180° = \frac{x}{r} = \frac{-1}{1} = -1,$$

$$\tan 180° = \frac{y}{x} = \frac{0}{-1} = 0,$$

$$\cot 180° = \frac{x}{y} = \frac{-1}{0}, \text{ which does not exist,}$$

$$\sec 180° = \frac{r}{x} = \frac{1}{-1} = -1,$$

$$\csc 180° = \frac{r}{y} = \frac{1}{0}, \text{ which does not exist.}$$

Exercise 1.3

1. Verify in detail that the tangent is a function. What is its domain?
2. Verify in detail that secant is a function. What is its range?

In Problems 3 through 6, refer to Figure 1.14; then fill in each blank with first, second, third, or fourth. (Review the explanation in Section 1.4 of how to determine the quadrant in which an angle lies.)

3. α is a _____-quadrant angle
4. β is a _____-quadrant angle
5. γ is a _____-quadrant angle
6. δ is a _____-quadrant angle

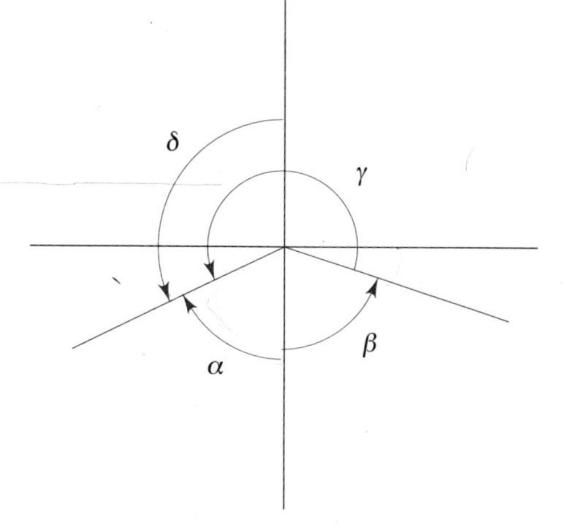

Figure 1.14

Place each of the following angles in standard position, using a curved arrow to indicate the rotation. Use Figure 1.11 to label the sides of the triangle of reference. Choose P so that $r = 10$. Then approximate the values of the sine, cosine, and tangent of each angle.

7. 80°
8. 110°
9. 290°
10. 160°
11. 220°
12. 310°
13. Using $r = 10$, read from Figure 1.11 the sine of each of the following angles: 0°, 10°, 20°, 30°, 40°, 50°, 60°, 70°, 80°, 90°.
14. Using $r = 10$, read from Figure 1.11 the cosine of each of the following angles: 0°, 10°, 20°, 30°, 40°, 50°, 60°, 70°, 80°, 90°.
15. Draw a figure and compute the trigonometric functions of 0°.
16. Draw a figure and compute the trigonometric functions of 90°.
17. Draw a figure and compute the trigonometric functions of 270°.
18. Draw a figure and compute the trigonometric functions of −180°.

1.6 CONSEQUENCES OF THE DEFINITIONS

(a) The following pairs of trigonometric functions are reciprocals of each other:

$$\{\sin \theta, \csc \theta\},$$

$$\{\cos \theta, \sec \theta\},$$

$$\{\tan \theta, \cot \theta\}.$$

The reciprocal of the number a is $1/a$. Hence the reciprocal of 3 is $\frac{1}{3}$; the reciprocal of $-\frac{1}{4}$ is -4; $\frac{2}{5}$ and $\frac{5}{2}$ are reciprocals. Since

$$\frac{x}{y} = \frac{1}{y/x}$$

we can say, for values of θ for which these functions are defined, that

$$\cot \theta = \frac{1}{\tan \theta}.$$

Similarly,

$$\sec \theta = \frac{1}{\cos \theta}$$

and
$$\csc \theta = \frac{1}{\sin \theta}.$$

Multiplying both sides of the last equation by $\sin \theta$, we get

$$\sin \theta \csc \theta = 1.$$

Dividing both sides of this equation by $\csc \theta$, we obtain

$$\sin \theta = \frac{1}{\csc \theta}.$$

Hence *$\sin \theta$ and $\csc \theta$ are reciprocals;* also *$\cos \theta$ and $\sec \theta$ are reciprocals;* and *$\tan \theta$ and $\cot \theta$ are reciprocals.* The following list indicates the reciprocal functions:

$$\left.\begin{array}{l} \sin \theta \\ \cos \theta \\ \tan \theta \\ \cot \theta \\ \sec \theta \\ \csc \theta \end{array}\right\} \text{Reciprocals}$$

Caution. The symbol *cos* in itself has no meaning standing alone. To have interpretation, it must be followed by some angle. Write *cos* θ, not *cos.* Notice that $\sin \theta \csc \theta = 1$ means that the sine of any angle times the cosecant of the *same* angle equals unity.

(b) Any trigonometric function of an angle is equal to the same function of all angles coterminal with it.

This is to say that if θ and φ are coterminal angles and if t is a trigonometric function, then $t(\theta) = t(\varphi)$. This follows directly from Section 1.5. Thus

$$\begin{aligned} \sin 370^\circ &= \sin(370^\circ - 360^\circ) = \sin 10^\circ, \\ \cos(-100^\circ) &= \cos(-100^\circ + 360^\circ) = \cos 260^\circ, \\ \tan 900^\circ &= \tan(900^\circ - 720^\circ) = \tan 180^\circ. \end{aligned}$$

The square of the number $\sin\theta$ is denoted $\sin^2\theta$ rather than $(\sin\theta)^2$. The same notation is employed for the other trigonometric functions and for the higher powers 3, 4, 5, and so forth. Thus $\sin^2 270° = 1$, $\cos^3 180° = -1$, and if $\tan\theta = -\sqrt{5}$ it must be true that $\tan^4\theta = 25$.

We can now consider the most basic relationship in trigonometry, which amounts to a simple restatement of the Pythagorean theorem.

(c) For any angle θ, $\sin^2\theta + \cos^2\theta = 1$.

To prove this very fundamental theorem, we let θ be any angle in standard position and choose (x, y), on the terminal side of θ, a positive distance r from the origin. This ensures that $r \neq 0$. By the Pythagorean theorem we have $x^2 + y^2 = r^2$ and by the definition of $\sin\theta$ and $\cos\theta$ we also have

$$\sin^2\theta = \left(\frac{y}{r}\right)^2 = \frac{y^2}{r^2},$$

and

$$\cos^2\theta = \left(\frac{x}{r}\right)^2 = \frac{x^2}{r^2}.$$

Thus $\sin^2\theta + \cos^2\theta = \dfrac{y^2 + x^2}{r^2} = \dfrac{r^2}{r^2} = 1$, which completes the proof.

Example 1 If θ is in Q IV and if $\cos\theta = \frac{12}{13}$, find $\sin\theta$.

Solution We have $\sin^2\theta + \cos^2\theta = 1$, so

$$\begin{aligned}\sin^2\theta &= 1 - \cos^2\theta \\ &= 1 - \left(\frac{12}{13}\right)^2 \\ &= 1 - \frac{144}{169} \\ &= \frac{25}{169}.\end{aligned}$$

Hence

$$\sin\theta = \pm\frac{5}{13}.$$

We choose the negative value for $\sin\theta$, since θ was given to be in Q IV, where $\sin\theta$ is negative. Therefore $\sin\theta = -\frac{5}{13}$.

Example 2 If $\sin\theta = \frac{9}{10}$, can $\cos\theta = \frac{4}{10}$?

Solution If we had an angle θ for which $\sin\theta = \frac{9}{10}$ and $\cos\theta = \frac{4}{10}$, then we would have

$$\begin{aligned}\sin^2\theta + \cos^2\theta &= \left(\frac{9}{10}\right)^2 + \left(\frac{4}{10}\right)^2 \\ &= \frac{81}{100} + \frac{16}{100} \\ &= \frac{97}{100} \\ &\neq 1.\end{aligned}$$

So there is no angle θ for which $\sin\theta = \frac{9}{10}$ and $\cos\theta = \frac{4}{10}$.

Students will use the identity

$$\sin^2\theta + \cos^2\theta = 1$$

repeatedly in this course and through their mathematical careers.

Example 3 Assuming that angle θ is in standard position, compute the trigonometric functions of θ if point $P(-2, -3)$ is on its terminal side.

Solution The Pythagorean theorem gives us $r = \sqrt{13}$. Angle θ and its triangle of reference are shown in Figure 1.15.

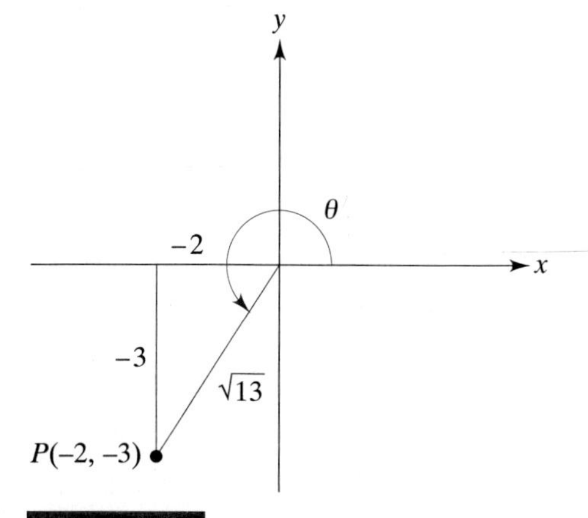

Figure 1.15

Then

$$\sin\theta = \frac{y}{r} = \frac{-3}{\sqrt{13}} = -\frac{3\sqrt{13}}{13},$$

$$\cos\theta = \frac{x}{r} = \frac{-2}{\sqrt{13}} = -\frac{2\sqrt{13}}{13},$$

$$\tan\theta = \frac{y}{x} = \frac{-3}{-2} = \frac{3}{2},$$

$$\cot\theta = \frac{x}{y} = \frac{-2}{-3} = \frac{2}{3},$$

$$\sec\theta = \frac{r}{x} = \frac{\sqrt{13}}{-2} = -\frac{\sqrt{13}}{2},$$

$$\csc\theta = \frac{r}{y} = \frac{\sqrt{13}}{-3} = -\frac{\sqrt{13}}{3}.$$

Exercise 1.4

Each of the following points is on the terminal side of an angle θ in standard position. Use a curved arrow to specify θ. Construct and label the sides of the triangle of reference as in Figure 1.15. Find the exact values of the six trigonometric functions of θ. Leave the results in fractional form. Do not use a calculator.

1. $(15, 8)$
2. $(-4, 3)$
3. $(5, -12)$
4. $(-7, -24)$
5. $(-3, -1)$
6. $(5, -8)$
7. $(9, 4)$
8. $(-6, 7)$
9. $(4, -\sqrt{65})$
10. $(-\sqrt{91}, -3)$
11. $(-\sqrt{13}, \sqrt{3})$
12. $(2\sqrt{6}, 5)$

Copy the following statements and identify the quadrant in which θ must be in order to satisfy each set of conditions.

13. $\cos\theta \geq 0$ and $\tan\theta \leq 0$
14. $\tan\theta \geq 0$ and $\csc\theta \leq 0$
15. $\sin\theta \leq 0$ and $\cot\theta \geq 0$
16. $\sec\theta \leq 0$ and $\csc\theta \geq 0$
17. $\cos\theta \leq 0$ and $\cot\theta \leq 0$
18. $\cos\theta \geq 0$ and $\csc\theta \leq 0$
19. $\sin\theta \geq 0$ and $\sec\theta \geq 0$
20. $\sin\theta \leq 0$ and $\sec\theta \geq 0$

Copy the following statements, identify each as possible or impossible, and explain why. Do not use a calculator.

21. $\cos\theta = 1.2$
22. $\sec\theta = -\frac{1}{2}$
23. $\sin\theta = 0.001$
24. $\sin\theta = 5$
25. $\tan\theta = 0$
26. $\tan\theta = 700$
27. $\cos\theta = \frac{1}{3}$ and $\sec\theta = -3$
28. $\cot\theta = 2$ and $\csc\theta = 2$

Assume θ is between 0° and 90° and in each case state whether θ is close to 0° or close to 90°. Do not use a calculator.

29. $\sin\theta = 0.01$
30. $\sin\theta = 0.999$
31. $\tan\theta = 0.01$
32. $\tan\theta = 1492$
33. $\cos\theta = 0.01$
34. $\cos\theta = 0.99$

1.7 GIVEN ONE TRIGONOMETRIC FUNCTION OF AN ANGLE, TO DRAW THE ANGLE AND FIND THE OTHER FUNCTIONS

When we know (1) the quadrant in which an angle lies and (2) the value of one trigonometric function of this angle, it is possible, by using the Pythagorean theorem and the general definition, to draw the angle and find its other five trigonometric functions.

Example Given $\cos\theta = \frac{2}{5}$ and θ is not in Q I, draw θ and find its other functions.

Solution Since $\cos\theta$ is positive in the two right-hand quadrants and since Q I is ruled out, θ must lie in Q IV. Remembering that for all angles $\cos\theta = x/r$ and that in this case $\cos\theta = \frac{2}{5}$, we can use $x = 2$ and $r = 5$. Equally correct but not so convenient would be to choose $x = 6$, $r = 15$ or $x = 1$, $r = \frac{5}{2}$. By means of $x^2 + y^2 = r^2$, we find $y = -\sqrt{21}$, the negative sign being chosen because $P(x, y)$ is in Q IV (see Figure 1.16 on the next page). Then

$$\sin\theta = \frac{y}{r} = \frac{-\sqrt{21}}{5} = -\frac{\sqrt{21}}{5},$$

$$\tan\theta = \frac{y}{x} = \frac{-\sqrt{21}}{2} = -\frac{\sqrt{21}}{2},$$

$$\cot\theta = \frac{x}{y} = \frac{2}{-\sqrt{21}} = -\frac{2\sqrt{21}}{21},$$

$$\sec \theta = \frac{r}{x} = \frac{5}{2},$$

$$\csc \theta = \frac{r}{y} = \frac{5}{-\sqrt{21}} = -\frac{5\sqrt{21}}{21}.$$

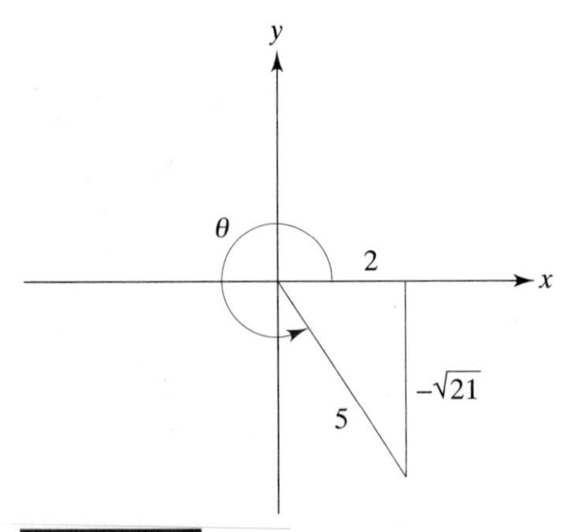

Figure 1.16

Exercise 1.5

Construct and label the sides of the triangle of reference. Use a curved arrow to indicate θ. Find the exact values of the remaining trigonometric functions. Verify that $\sin^2 \theta + \cos^2 \theta = 1$ in each case.

1. $\tan \theta = -\frac{2}{3}$, θ not in Q IV
2. $\sin \theta = -\frac{1}{6}$, θ not in Q IV
3. $\cos \theta = \frac{3}{8}$, θ not in Q I
4. $\tan \theta = \frac{7}{5}$, θ not in Q III
5. $\sin \theta = \frac{24}{25}$, $\cos \theta > 0$
6. $\cos \theta = -\frac{3}{5}$, $\tan \theta < 0$
7. $\tan \theta = \frac{5}{12}$, $\sin \theta < 0$
8. $\sin \theta = -\frac{8}{17}$, $\cos \theta > 0$
9. $\cos \theta = -\dfrac{2\sqrt{29}}{29}$, θ not in Q II

   $\left(\textit{Hint:} \quad -\dfrac{2\sqrt{29}}{29} = \dfrac{-2}{\sqrt{29}} \right)$

10. $\tan\theta = -\dfrac{\sqrt{2}}{4}$, θ not in Q II

$\left(\textit{Hint:} \quad -\dfrac{\sqrt{2}}{4} = \dfrac{-1}{2\sqrt{2}}\right)$

11. $\sin\theta = \dfrac{\sqrt{15}}{5}$, θ not in Q I

$\left(\textit{Hint:} \quad \dfrac{\sqrt{15}}{5} = \dfrac{\sqrt{3}\sqrt{5}}{\sqrt{5}\sqrt{5}} = \dfrac{\sqrt{3}}{\sqrt{5}}\right)$

12. $\cos\theta = \dfrac{4\sqrt{2}}{9}$, θ not in Q IV

13. $\sec\theta = \frac{41}{40}$, θ not in Q I
14. $\cot\theta = 7$, θ not in Q III
15. $\tan\theta = a$, θ *is* in Q I
16. $\csc\theta = b$, θ *is* in Q II

1.8 TRIGONOMETRIC FUNCTIONS OF AN ACUTE ANGLE

Let θ be an acute angle of an arbitrary right triangle as shown at the top of Figure 1.17. The arbitrary triangle may be rotated, as indicated by the arrow, so that θ is in standard position. It may even be necessary to turn the triangle over to get θ into standard position so that the bottom triangle becomes its triangle of reference. Once in standard position, the hypotenuse (*hyp*) is the radius vector r of the point P at the vertex opposite θ; the side

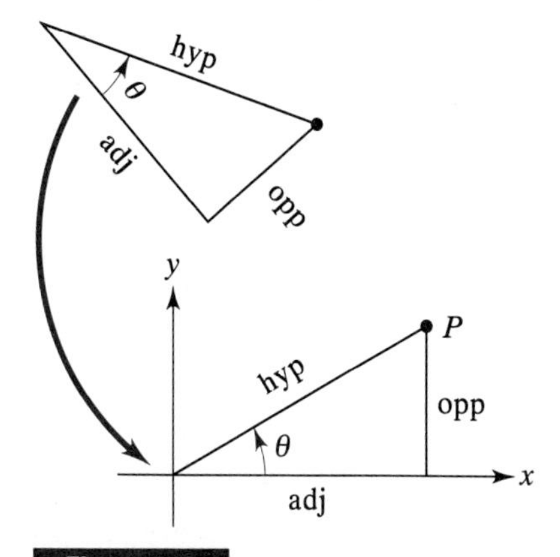

Figure 1.17

opposite θ (*opp*) becomes the y coordinate of P; and the side adjacent to θ (*adj*) would be the x coordinate of the point P on the terminal side of θ. Then the general definitions (Section 1.5) involving x, y, and r would become special definitions involving *adj*, *opp*, and *hyp*.

We conclude that *for any* **acute** *angle θ in a right triangle,*

$$\sin \theta = \frac{\text{opp}}{\text{hyp}},$$

$$\cos \theta = \frac{\text{adj}}{\text{hyp}},$$

$$\tan \theta = \frac{\text{opp}}{\text{adj}}.$$

The other three functions can be obtained through their reciprocals.

Exercise 1.6

For each of the following right triangles, write the sine, cosine, and tangent of each of the acute angles. Leave results in fractional form.

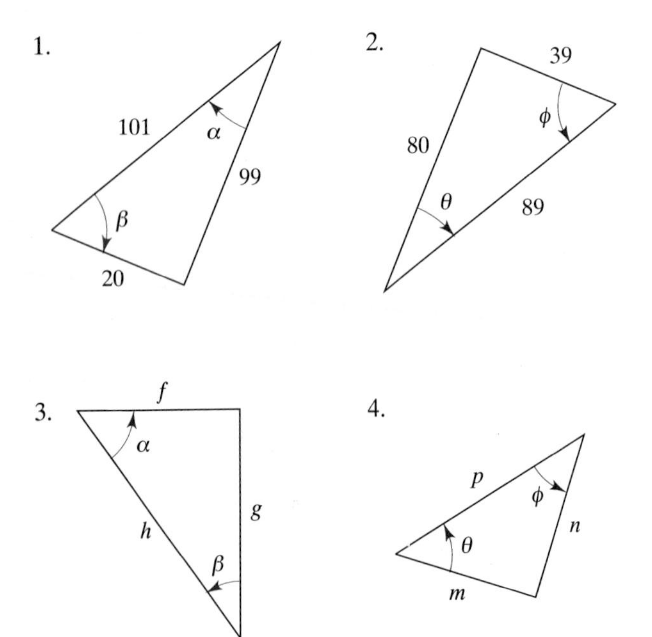

1.9 COFUNCTIONS

The sine and cosine are said to be *cofunctions;* that is, the cosine is the cofunction of the sine, and the sine is the cofunction of the cosine. Similarly, the tangent and cotangent are cofunctions, and the secant and cosecant are cofunctions.

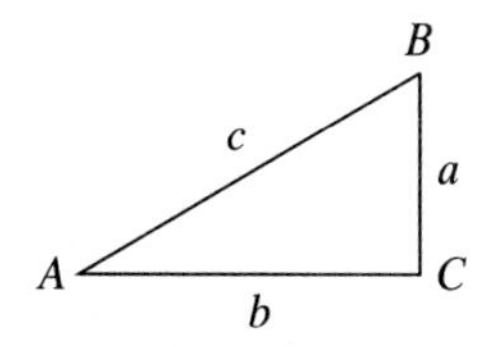

Figure 1.18

If A and B are the acute angles in right triangle ABC (Figure 1.18), then they are *complementary* because their sum is 90°. Moreover,

$$\sin A = \frac{a}{c} = \cos B,$$

$$\cos A = \frac{b}{c} = \sin B.$$

Similarly,

$$\tan A = \cot B, \qquad \cot A = \tan B,$$

$$\sec A = \csc B, \qquad \csc A = \sec B.$$

Hence we have the following:

Theorem **Any trigonometric function of an acute angle is equal to the cofunction of its complementary angle.**

Thus $\sin 70° = \cos 20°$, $\cos 80° = 10°$, and $\tan 50° = \cot 40°$.

Look again at Figure 1.11 with r fixed at 10. Now as θ increases from 0° to 90° the x coordinate of a point $P(x, y)$ on the circle is decreasing from 10 to 0, so $\cos \theta = x/10$ decreases from 1 to 0. Likewise, the y coordinate of P increases from 0 to 10, so $\sin \theta = y/10$ increases from 0 to 1. Furthermore $\tan \theta = y/x$ increases without bound from 0 to surpass any positive number N. Of course $\tan 90°$ is undefined, but as θ "gets close" to 90°, $\tan \theta$ achieves very large positive numbers.

It follows that as θ increases from 0° to 90°, $\sin \theta$, $\tan \theta$ and $\sec \theta$ increase while $\cos \theta$, $\cot \theta$, and $\csc \theta$ decrease.

1.10 THE TRIGONOMETRIC FUNCTIONS OF 30°, 45°, 60°

We now know that sin 60° = cos 30° and tan 70° = cot 20°, yet we do not know the values of these numbers. In this section we will find the value of cos 30° = sin 60°, but the value of cot 20° will not be determined until Chapter 2.

Consider an equilateral triangle of side 2. The bisector of one of the 60° angles also is the bisector of the opposite side (Figure 1.19). By the Pythagorean theorem, the length of the bisector is $\sqrt{3}$. Using Section 1.8, we find

$$\sin 30^\circ = \frac{1}{2}, \qquad \sin 60^\circ = \frac{\sqrt{3}}{2},$$

$$\cos 30^\circ = \frac{\sqrt{3}}{2}, \qquad \cos 60^\circ = \frac{1}{2},$$

$$\tan 30^\circ = \frac{1}{\sqrt{3}} = \frac{\sqrt{3}}{3}, \qquad \tan 60^\circ = \sqrt{3}.$$

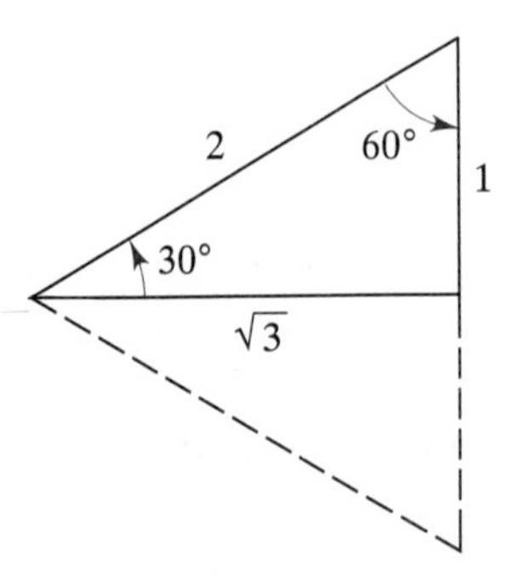

The 30°-60°-90° triangle can be easily remembered if we note that the longest side (the hypotenuse) is twice the shortest side.

To compute the functions of 45°, draw an isosceles right triangle of leg 1 (Figure 1.20). The hypotenuse, by the Pythagorean theorem, must be $\sqrt{2}$. Then, by Section 1.8,

$$\sin 45^\circ = \frac{1}{\sqrt{2}} = \frac{\sqrt{2}}{2},$$

$$\cos 45^\circ = \frac{1}{\sqrt{2}} = \frac{\sqrt{2}}{2},$$

$$\tan 45^\circ = \frac{1}{1} = 1.$$

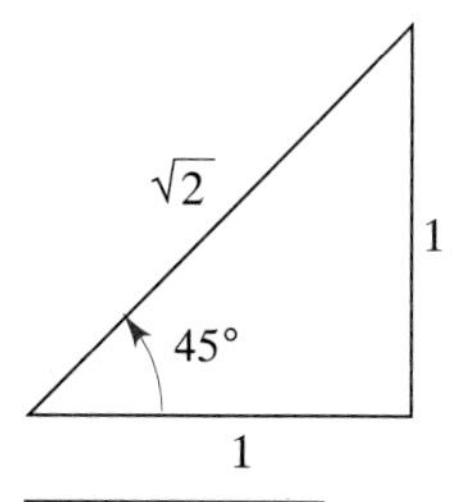

Figure 1.20

It is very important that the student learn these values, together with those displayed below. They occur repeatedly throughout the remainder of this course as well as in all future mathematics courses. If necessary, memorize the values and the figures from which they are derived.

$$\sin 30° = \cos 60° = \frac{1}{2},$$

$$\sin 45° = \cos 45° = \frac{\sqrt{2}}{2},$$

$$\sin 60° = \cos 30° = \frac{\sqrt{3}}{2}.$$

As we shall see later, the tangent of any angle can be obtained by dividing its sine by its cosine. The three remaining functions can be obtained through their reciprocals.

If we recall (Problem 15, Exercise 1.3) that $\sin 0° = 0$, $\cos 0° = 1$, $\sin 90° = 1$, $\cos 90° = 0$, we can easily remember the sine and cosine of special first-quadrant angles by forming a mental picture of Figure 1.21.

	0°	30°	45°	60°	90°
sin	$\frac{\sqrt{0}}{2}$	$\frac{\sqrt{1}}{2}$	$\frac{\sqrt{2}}{2}$	$\frac{\sqrt{3}}{2}$	$\frac{\sqrt{4}}{2}$
cos	$\frac{\sqrt{4}}{2}$	$\frac{\sqrt{3}}{2}$	$\frac{\sqrt{2}}{2}$	$\frac{\sqrt{1}}{2}$	$\frac{\sqrt{0}}{2}$

Figure 1.21

Observe that the radicands change by 1 as we change from angle to angle. Use this if needed to remember these important values. Naturally, after using this device to recall one of the numbers, one should convert the member to its more familiar form, e.g., $\sqrt{4}/2 = 1$ or $\sqrt{0}/2 = 0$.

Exercise 1.7

Compute the exact value of each of the following expressions. Leave the results in fractional form. Do not use a calculator.

1. $5 \sin^4 45° - \sin 60° \cos 30° + \cos^6 90°$
2. $6 \cos^2 45° - 5 \sin^3 30° - \sqrt{3} \cos 30°$
3. $4 \cos^3 60° + 3\sqrt{2} \sin 45° - 5 \cos^2 0° + 2 \sin 30°$
4. $\sqrt{3} \sin 60° - \cos^2 60° - 2 \cos^6 45°$
5. $2\sqrt{2} \cos 45° + 6 \sin^5 30° - 9 \cos^2 60°$
6. $5\sqrt{6} \cos^2 60° + 3 \sin 45° \sin 60° + \cos^4 30° \sin^5 0°$
7. $\sin^4 60° + 2 \cos^2 45° - \cos^2 30°$
8. $3 \cos^4 30° - 5 \sin^3 30° - \sin^2 45° - \sin 90°$

Identify as true or false and give reasons. Do not use a calculator.

9. $\tan 8° > \tan 5°$
10. $\cos 40° > \cos 20°$
11. $\cos(40° - \theta) = \sin(50° + \theta)$
12. $\tan 56° = \cot 34°$
13. $\sec 44° = \dfrac{1}{\sin 44°}$
14. $\csc 15° = \dfrac{1}{\cos 75°}$
15. $\sin^2 80° = \dfrac{1}{\sec^2 10°}$
16. $\cos 30° + \cos 60° = \cos 90°$
17. $\sin 60° = 2 \sin 30°$
18. $\sin 0° = \sin 180°$
19. $1492 \cot 90° + 1776 \tan 180° + 1992 \cos 270° = 0$

KEY TERMS

Rectangular coordinates, quadrant, distance formula, standard position of an angle, relation, function, trigonometric function, range of trigonometric functions, coterminal angles, complementary angles

REVIEW EXERCISES

1. State the definition of the six trigonometric functions of an angle θ.
2. How far is $(-3, -5)$, from the x axis? From the y axis? From the origin?

3. If the distance from $(a, 2)$ to $(4, 5)$ is $\sqrt{34}$, find a.
4. If $\sin\theta = 0.001$, find $\csc\theta$. If in addition, $\cos\theta < 0$, in what quadrant is θ?
5. Find the exact values of the trigonometric functions of θ if $\tan\theta = \frac{5}{6}$ and $\sin\theta < 0$. In what quadrant is θ?
6. Through how many degrees does the hour hand on a clock move in 1 h? In 3 h and 20 min?
7. Verify that cotangent is a function. What is its domain?
8. Fill in the following table from memory:

θ	0°	30°	45°	60°	90°
$\sin\theta$					
$\cos\theta$					
$\tan\theta$					

Compute the exact value of:

9. $\sin^2 45° + \cos^2 45° - \tan^2 45°$
10. $\sin 30° + \sin 60° - \sin 90°$

2

Trigonometric Functions of Any Angle

2.1 HAND CALCULATORS

We have seen how to find the values of the trigonometric functions of several special angles, including 0°, 30°, 45°, 60°, and 90°. It would be a very difficult task to find the values of those functions at say 23° by geometric arguments such as we were able to use in Chapter 1. Today there is no need to resort to lengthy calculations to find these values. The widespread availability of hand calculators with the trigonometric functions sin, cos, and tan makes the findings of sin 23°, cos 23°, and tan 23° a quick and easy matter. Furthermore, one can readily perform complex calculations with these values. Thus one can quickly use a scientific calculator to compute the value of an expression such as

$$\frac{123.7945 \sin 23.78^\circ}{\cos 67.49^\circ}.$$

Furthermore, this can be found with four-figure accuracy! Twenty years ago hand calculators were just being introduced for sale to the public. Before their introduction it would have been a slow and tedious process using tables, like those in the back of this text, to compute the value of the above expression to even three-figure accuracy.

It is assumed that the student has access to a scientific calculator with the three trigonometric functions, sin, cos, and tan, their inverses, and with a key ln x. Sadly, since there are several manufacturers of scientific calculators, the labeling of the keys and the order of pressing the keys to perform a given operation is not standardized. The reader is strongly urged to consult the user's manual for the calculator to be employed. We will indicate one or two keying sequences for algebraic and/or reverse-Polish notation calculators. Still, we do not claim to include all possible keying sequences for all calculators on the market.

Students who do not have access to a scientific calculator should consult Sections A.4 and A.5 in the Appendix for instruction in the use of trigonometric tables.

2.2 APPROXIMATIONS AND SIGNIFICANT FIGURES

If a given distance is *measured* and if its length is expressed in decimal form, it is conventional to write no more digits than are correct (or probably correct). Thus if we say that the measured distance between points A and B is 17 meters (m), we mean that the result is given to the nearest meter; that is, the true distance is closer to 17 m than it is to 16 m or 18 m. This is an example of two-figure accuracy. If we say that the measured distance AB is 17.0 m, we mean that the true distance is given to three significant figures; that is, it is closer to 17.0 m than it is to 16.9 or 17.1 m. This implies that the true distance is somewhere between 16.95 and 17.05 m. Notice that 17 and 17.0 do not mean the same thing when they represent approximate values.

The number of significant digits in a number is obtained by counting the digits from left to right, beginning with the first nonzero digit and ending with the rightmost digit. Ambiguity may result if the number in question is an integer that ends in one or more 0's. For example, if the radius of the earth is given as 4000 mi, we may not know how many 0's are significant. If, however, the number 4000 was obtained from 3960 by rounding it off to the nearest multiple of 100 mi, then the first 0 is significant; the other two are not. In a case of this kind we usually use **scientific notation.** In this notation the number is expressed as a product. The first factor is the number formed by the significant digits with a decimal point placed after the first digit; the second factor is an integral power of 10. Thus, $4.0(10^3)$ indicates the number 4000, in which only the first 0 is significant. Also, $0.0123 = 1.23(10^{-2})$. We see that 0.078060 has five significant digits, 70.00 has four, and 0.790 has only three. Notice that the number of significant digits does not depend on the position of the decimal point.

If calculations are made on approximate data, the result may have digits that are not significant. These digits should be rejected. The process of discarding these extra digits is called *rounding off* the number.

Illustration 1 When 3.141592 is rounded off to five (significant) digits, we obtain 3.1416. In this case, 1 must be added to the fifth digit of the given number to obtain the best five-figure approximation. For four-figure and three-figure approximations, 3.141592 becomes 3.142 and 3.14, respectively.

Illustration 2 In rounding off 78.25 to three-figure accuracy, we could logically get 78.2 or 78.3. In cases of this kind, it is conventional to choose the larger number. Thus, 78.25 rounds off to 78.3.

We shall adhere to the following two rules for rounding off the numbers that result when calculations are performed on approximate data.

1. When multiplication (or division) is performed on approximate data, *round off the result so that it will have as many significant figures as there are in the least accurate number in the data.*
2. When addition (or subtraction) is to be performed on approximate data, write one number beneath the other and add (or subtract); then round off the result so that the last digit retained is in the rightmost column in which both of the given numbers have significant digits.

If a small field is measured and found to be 11.3 m long and 10.7 m wide, we would be very tempted to say that its area is $(11.3)(10.7) = 120.91$ m^2 (square meters). To do so would be false accuracy. The result should be rounded off to three significant figures (the same as in the given data) to obtain 121 m^2. The first two figures in this result are correct, but the third is only a good approximation, because the true area is somewhere between $(11.25)(10.65) = 119.8125$ m^2 and $(11.35)(10.75) = 122.0125$ m^2.

Since nearly all the angles listed in the table starting on page 293 have trigonometric functions that are nonending decimals, the numbers appearing in this table are merely four-figure approximations. Hence most of the results obtained using this table will be approximations and should be considered as such. On a calculator there may be 8-, 10-, or 12-figure approximations, depending on the type of calculator. However, if you are using a calculator to perform computations on three-figure data, then the results should be rounded off to only three-figure accuracy.

In solving triangles, we agree to set up the following correspondence between accuracy in sides and angles:

Accuracy in Sides	Accuracy in Angles
Two-figure	Nearest degree
Three-figure	Nearest tenth of a degree
Four-figure	Nearest hundredth of a degree

Hence within each of the following sets of data, the same degree of accuracy prevails:

23, 42, 62°
0.0461, 61.2°, 44.8°, 74.0°
8.624, 55.78°, 82.00°

If the data include a side with two-figure accuracy and another side with three-figure accuracy, then the computed parts should be written with only two-figure accuracy, which means that computed angles should be taken to the nearest degree. In general, *our results can be no more accurate than the least accurate item of the data.* If the given data include a number whose degree of accuracy is doubtful, we shall (in this book) *assume the maximum degree of accuracy.* For example, with no information to the contrary, the side 700 km (kilometers) will be treated as a three-figure number.

Exercise 2.1

Round off the following numbers and angles to (*a*) four-figure accuracy, (*b*) three-figure accuracy, and (*c*) two-figure accuracy.

1. 0.71819
2. 8.9165
3. 45.7350
4. 0.067382
5. 30.625°
6. 22.947°
7. 9.6491°
8. 14.575°

The number 77.4 is the best three-figure approximation for all numbers from 77.35 inclusive to 77.45 exclusive. In other words, 77.4 is the best three-figure approximation for all numbers x such that $77.35 \leq x < 77.45$; that is, x is greater than or equal to 77.35 and less than 77.45. What range of numbers is covered by each of the following approximations?

9. 42.7
10. 262
11. 0.0758
12. 9.03
13. 6000
14. 0.1945
15. 2.817
16. 0.003486

2.3 TWO PROBLEMS

In Section 1.5 we saw that the trigonometric ratios did indeed define functions. In this section we want to examine two problems:

1. Given any θ, with $0° \leq \theta \leq 90°$, to find the trigonometric functions of θ,
2. To find θ if we are given the value of a trigonometric function of θ.

We begin with the first problem.

Example 1 Find sin 23° to four-figure accuracy.

Solution First be certain that the calculator is in the "degree" mode by pressing the [DRG] key until 'DEG' shows on the display. Then press the following keys:

[2] [3] [sin].

The displayed result is 0.390731128, which gives, to four-figure accuracy,

$$\sin 23° = 0.3907\,.$$

Example 2 Find cot 72.1° to four-figure accuracy.

Solution Be sure the calculator is set for degrees. Inasmuch as most calculators have no cotangent function, we observe that

$$\cot 72.1° = \frac{1}{\tan 72.1°}\,.$$

Press the following keys:

[7] [2] [.] [1] [tan] [1/x].

This reciprocal key finds the reciprocal of the displayed number, in this case, tan 72.1°.

An alternative procedure (leading to the same final result) is to press

[1] [÷] [7] [2] [.] [1] [tan] [=]

The result 0.3229911993 is now displayed. Therefore, to four-figure accuracy,

$$\cot 72.1° = 0.3230\,.$$

Throughout the remainder of the text we will measure degrees in decimal degrees, such as 23.75°. We do this because scientific calculators accept only decimal degrees and not the older degree-minute-second form, in which 60 minutes is 1 degree and 60 seconds is 1 minute. In the classical system 23°45′ represents 23 degrees and 45 minutes. In decimal degrees it is 23.75° since $45/60 = 0.75$. Many scientific calculators have a key to allow the user to convert from degree-minute-second form to decimal degrees. This key is usually denoted [DMS-DD], [DMS] or [DMS-DEG] and it may, in fact, be located on the "upper" or "second" register, which is addressed by first pressing the key [2nd].

Example 3 Find tan 27° 35′ 41″ to five decimal places.

Solution First we convert 27° 35′ 41″ to decimal degrees and then find tangent of the result. Enter 27.35 41 to represent 27° 35′ 41″:

[2] [7] [·] [3] [5] [4] [1]
[DMS-DD] [tan].

Note that the decimal equivalent is 27.5947222 and that the final answer is 0.522670082. We have, to five decimal places,

$$\tan 27^\circ\ 35'\ 41'' = 0.52267\,.$$

For those whose calculators do not have the [DMS] key, it is easy enough to convert from DMS to DD by simply using fractions. 27° 35′ 41″ represents

$$27 + \frac{35}{60} + \frac{41}{3600} = 27 + 0.583333 + 0.0113888$$

degrees. Hence

$$27^\circ\ 35'\ 41'' = 27.594722^\circ.$$

We turn now to the second problem, which in functional terms is this: If we know the function f and if we know the value of y in the range of f, can we find an x in the domain of f such that $f(x) = y$? If, for each y there is a unique such x then we say "x is f-inverse of y" and write $x = f^{-1}(y)$. Thus $f(f^{-1}(y)) = y$. In a later chapter we will explore this concept in more detail. For now, we will use the following rather evident facts:

1. For every y with $0 \le y \le 1$, there is one and only one acute angle θ such that $\sin \theta = y$. This θ is denoted $\sin^{-1} y$ and we say "θ equals inverse sine of y."
2. For every y with $0 \le y \le 1$, there is one and only one acute angle θ such that $\cos \theta = y$. We write $\theta = \cos^{-1} y$.
3. For every y with $y \ge 0$ there is one and only one acute angle θ such that $\tan \theta = y$. In this case, $\theta = \tan^{-1} y$.

We see then that $\sin^{-1} \sqrt{3}/2 = 60°$, $\cos^{-1} \frac{1}{2} = 60°$, and $\tan^{-1} 1 = 45°$.

In order to use a scientific calculator to evaluate the inverse of a trigonometric function, one must invariably press two keys. On many calculators the inverse function such as $\sin^{-1}$ is on the upper register over the function. In this event it is addressed by pressing [2nd] before pressing [sin]. On the other hand, if the inverse does not appear over the trigonometric function, there is probably an [INV] key to be pressed before the trigonometric function key.

Example 4 If $\cos \theta = 0.2889$, find θ to four-figure accuracy.

Solution If $\cos \theta = 0.2889$, then $\theta = \cos^{-1} 0.2889$. Being sure the calculator is set for degrees, press

The result is 73.20788, which to the required accuracy yields

$$\theta = \cos^{-1} 0.2889 = 73.21°.$$

Example 5 If $\cot \theta = 1.4014$, find θ to five-figure accuracy.

Solution Since the cotangent function does not appear on the calculator, we recall that if $\cot \theta = 1.4014$, then

$$\tan \theta = \frac{1}{1.4014}.$$

Hence, in the degree mode, we press

[1] [·] [4] [0] [1] [4] [1/x] [INV] [tan].

The result to five-figure accuracy is

$$\theta = \cot^{-1} 1.4014 = 35.511°.$$

Exercise 2.2

1. Why does a calculator have keys for sin, cos, and tan but not for csc, sec, and cot?

Convert the following to decimal degrees.

2. $78° 13' 58''$
3. $19° 19' 19''$
4. $123° 48' 51''$
5. $213° 31' 09''$

Find the value of each of the following to four decimal places if the number is less than 1 and to four significant digits if the number is greater than 1.

6. $\sin 73.2°$
7. $\cot 56.7°$
8. $\cos 22.6°$
9. $\tan 7.92°$
10. $\cot 18.56°$
11. $\sin 40.33°$
12. $\tan 83.0°$
13. $\cos 61.42°$
14. $\cos 49.88°$
15. $\sec 66.1°$
16. $\csc 37.82°$
17. $\csc 18.1°$

Find the acute angle θ to the nearest tenth of a degree from each of the following functions of θ

18. $\tan \theta = 8.6720$
19. $\cos \theta = 0.7181$
20. $\cot \theta = 0.5658$
21. $\sin \theta = 0.9191$
22. $\cos \theta = 0.1925$
23. $\tan \theta = 1.414$
24. $\sin \theta = 0.0019$
25. $\cot \theta = 3.876$

Calculate the value of each of the following to five significant figures.

26. $\dfrac{123.7945 \sin 23.78°}{\cos 67.49°}$
27. $2.41476 \tan 14° 48' 13''$
28. $\sin 56.94° \cos 33.06°$
29. $\sqrt{2} \csc 18° 36' 51''$
30. $\sin^2 48.96° + \cos^2 48.96°$

2.4 RELATED ANGLES

We explore ways to find the values of the trigonometric functions at angles which are not acute. As we will see, one can express the functional value of an arbitrary angle in terms of functions of an acute angle. Since coterminal angles have the same trigonometric functions, we shall consider only those angles between 0° and 360°. In order to determine the functions of an angle larger than 90°, we introduce the concept of the related angle, which some texts call the reference angle.

The related angle of a given angle θ is the positive acute angle between the x axis and the terminal side of θ, when θ is in standard position.

Hence (see Figure 2.1), *to find the related angle of* θ

If θ *is in* Q II, *subtract* θ *from* 180°.
If θ *is in* Q III, *subtract* 180° *from* θ.
If θ *is in* Q IV, *subtract* θ *from* 360°.

Note that, in finding the related angle, we always work to or from 180° or 360°, never 90° or 270°. Thus:

The related angle of 160° is 20°.
The related angle of 260° is 80°.
The related angle of 310° is 50°.
The related angle of 500° is the related angle of (500° − 360°) = 140°, which is 40°.

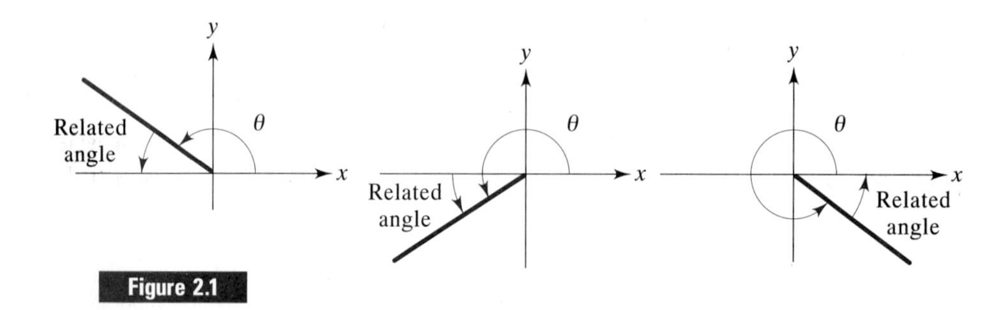

Figure 2.1

2.5 REDUCTION TO FUNCTIONS OF AN ACUTE ANGLE

Two numbers are said to be numerically equal if they are equal, except perhaps for sign. If a and b are numerically equal, then $a = \pm b$, or $|a| = |b|$.

Related-Angle Theorem

Any trigonometric function of an angle is *numerically* equal to the same function of its related angle.

To prove this, let θ_1 be any positive acute angle, and let θ_2, θ_3, θ_4 be positive angles, one in each of the other three quadrants, such that their common related angle is θ_1 (see Figure 2.2). Choose points P_1, P_2, P_3, P_4 on the terminal sides of these angles so that all four points have the same radius vector r. The four triangles of reference are congruent. Therefore, the corresponding sides are numerically equal. Hence

$$x_2 = -x_1, \qquad x_3 = -x_1, \qquad x_4 = x_1,$$

and

$$y_2 = y_1, \qquad y_3 = -y_1, \qquad y_4 = -y_1.$$

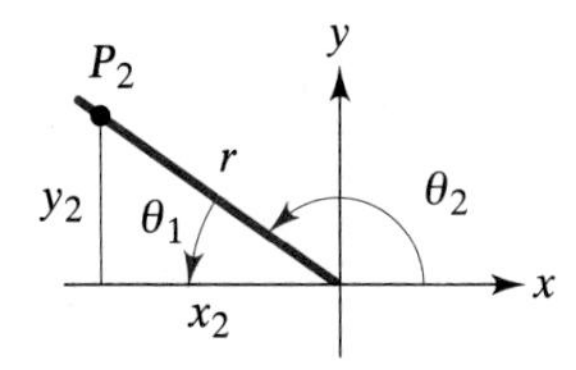

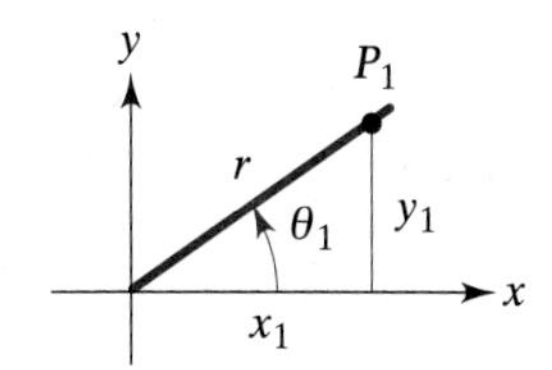

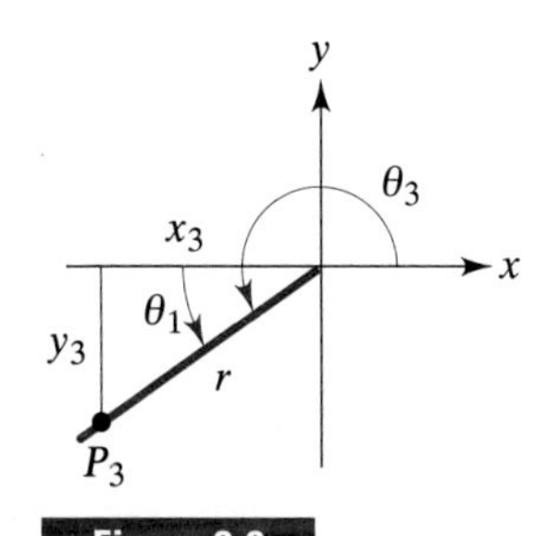

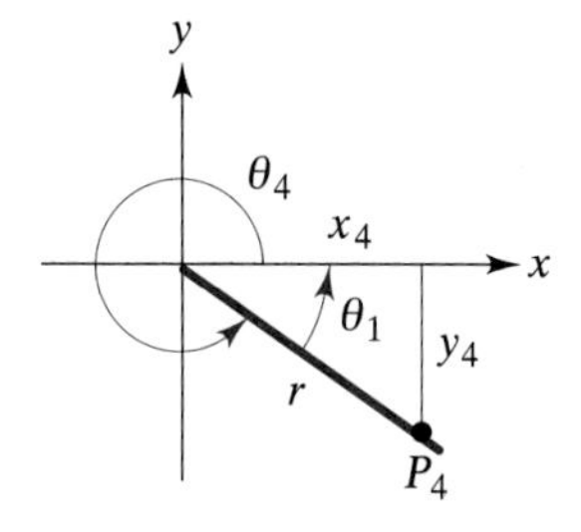

Figure 2.2

Then

$$\sin\theta_2 = \frac{y_2}{r} = \frac{y_1}{r} = \sin\theta_1,$$

$$\sin\theta_3 = \frac{y_3}{r} = \frac{-y_1}{r} = -\frac{y_1}{r} = -\sin\theta_1,$$

$$\sin\theta_4 = \frac{y_4}{r} = \frac{-y_1}{r} = -\frac{y_1}{r} = -\sin\theta_1,$$

and

$$\cos\theta_2 = \frac{x_2}{r} = \frac{-x_1}{r} = -\frac{x_1}{r} = -\cos\theta_1,$$

$$\cos\theta_3 = \frac{x_3}{r} = \frac{-x_1}{r} = -\frac{x_1}{r} = -\cos\theta_1,$$

$$\cos\theta_4 = \frac{x_4}{r} = \frac{x_1}{r} = \cos\theta_1.$$

Similarly, each of the other functions of θ_2, θ_3, and θ_4 is *numerically* equal to the same function of θ_1, the common related angle. The proper sign, + or −, is determined by the quadrant in which the given angle lies.

Example 1 Use the related-angle theorem to find sin 120° and cos 120° without a calculator.

Solution The related angle of 120° is 180° − 120° = 60°. Since 120° is in Q II, its sine is positive and its cosine is negative. Hence

$$\sin 120^\circ = +\sin 60^\circ = \frac{\sqrt{3}}{2},$$

$$\cos 120^\circ = -\cos 60^\circ = -\tfrac{1}{2}.$$

The remaining functions can be found by using the fundamental identities:

$$\tan 120^\circ = \frac{\sin 120^\circ}{\cos 120^\circ} = \frac{\sqrt{3}/2}{-\frac{1}{2}} = -\sqrt{3},$$

$$\cot 120^\circ = \frac{1}{\tan 120^\circ} = -\frac{1}{\sqrt{3}} = \frac{-\sqrt{3}}{3}, \text{ etc.}$$

Example 2 Use the related-angle theorem to find sin 255° and tan 255°.

Solution The related angle of 255° is 255° − 180° = 75°. Hence

$$\sin 255^\circ = -\sin 75^\circ = -0.9659,$$

$$\tan 255^\circ = +\tan 75^\circ = 3.732.$$

Exercise 2.3

Use the related-angle theorem to find the exact values of the sine and cosine of the following angles *without* using a calculator or tables.

1. 240°	2. 315°	3. 150°	4. 300°
5. 330°	6. 210°	7. 225°	8. 135°
9. 495°	10. 480°	11. 1020°	12. 930°

Prove or disprove the following statements *without* using a calculator or tables.

13. $\sec 111° = -\sec 21°$
14. $\csc 200° = -\sec 70°$
15. $\cos 222° > \cos 200°$
16. $\sin 144° < \sin 155°$
17. $\cos^2 310° + \cos^2 40° = 1$
18. $\tan^2 36° = \sec^2 306° - 1$
19. $\dfrac{\cos 340°}{\sin 160°} = -\cot 20°$
20. $\dfrac{\sin 234°}{\sin 324°} = \tan 54°$

Name one angle in each of the other three quadrants whose trigonometric functions are numerically equal to those of the given angle.

21. 333°
22. 100°
23. 44°
24. 246°

Use the related-angle theorem to find the values of the following functions to five significant figures.

25. cot 319°
26. cos 254°
27. tan 117°
28. cos 357°
29. sin 253.6°
30. cot 339.7°
31. sin 342.1°
32. tan 125.8°
33. cos 141.58°
34. sin 98.93°
35. cos 229.85°
36. sin 249.69°

2.6 TRIGONOMETRIC FUNCTIONS OF THE NEGATIVE OF AN ANGLE

Let θ be any angle. Then $-\theta$ indicates the same amount of rotation but in the *opposite* direction. Place both angles in standard position on the same coordinate system. (See Figure 2.3.) Choose any point $P(x, y)$ on the terminal side of θ. Drop a perpendicular from P to the x axis and extend it until it strikes the terminal side of $-\theta$ at $P'(x, y')$. The triangles of reference, OPM and $OP'M$, are congruent. (Why?) Hence $OP = OP' = r$. But y and y' are only numerically equal. Since $y' = -y$, we conclude that

$$\sin(-\theta) = \frac{y'}{r} = \frac{-y}{r} = -\frac{y}{r} = -\sin\theta,$$

$$\cos(-\theta) = \frac{x}{r} = \cos\theta,$$

$$\tan(-\theta) = \frac{y'}{x} = \frac{-y}{x} = -\frac{y}{x} = -\tan\theta.$$

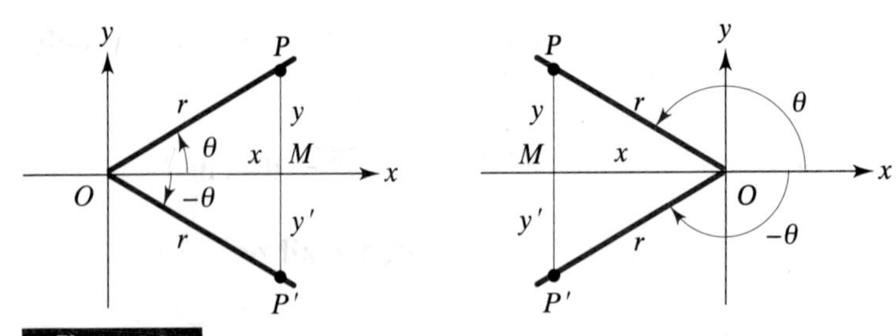

Figure 2.3

Similarly,

$$\cot(-\theta) = -\cot\theta,$$

$$\sec(-\theta) = \sec\theta,$$

$$\csc(-\theta) = -\csc\theta.$$

The student should draw figures for θ in the other quadrants and also for θ a negative angle. For all possible positions of θ, the following is true:

Theorem

For any angle θ,

$$\sin(-\theta) = -\sin\theta,$$

$$\cos(-\theta) = \cos\theta,$$

$$\tan(-\theta) = -\tan\theta.$$

The other three functions behave as do their reciprocals.

Example Compute $\sin(-225°)$, $\cos(-225°)$, and $\tan(-225°)$.

Solution Using the preceding theorem and the related-angle theorem, we obtain

$$\sin(-225°) = -\sin 225° = -(-\sin 45°) = \frac{\sqrt{2}}{2},$$

$$\cos(-225°) = \cos 225° = -\cos 45° = -\frac{\sqrt{2}}{2},$$

$$\tan(-225°) = -\tan 225° = -\tan 45° = -1.$$

A function $f(\theta)$ is said to be an *even* function if, for all permissible values of θ, $f(-\theta) = f(\theta)$.

Examples of even functions of θ are: $7\theta^2$, $\cos\theta$, $5\theta^6 - \theta^2$, and 1.

A function $f(\theta)$ is called an *odd* function if, for all permissible values of θ, $f(-\theta) = -f(\theta)$.

Examples of odd functions of θ are $8\theta^3$, $\sin\theta$, $\tan\theta$, and $-\theta^5$. Some functions, such as $\theta^5 + \theta^4$, are neither even nor odd; 0 is both odd and even.

Exercise 2.4

Use the theorem in Section 2.6 to compute the exact values of the sine, cosine, and tangent of each of the following angles. Do not use a calculator or tables.

1. $-30°$
2. $-45°$
3. $-90°$
4. $-60°$

Prove or disprove statements 5 through 15 without using a calculator or tables.

5. $\cos(-1000°)\sec 1000° = -1$
6. $\tan(-160°) = -\tan 20°$
7. $\tan(-100°)\cot 100° = -1$
8. $\sin(-200°)\csc 20° = -1$
9. $\dfrac{\sin(-48°)}{\sin(-42°)} = \cot 42°$
10. $\dfrac{\sin(-\theta)}{\cos(-\theta)} = -\tan\theta$
11. $\sin^2(-\theta) - \cos^2\theta = -1$
12. $\cot^2(-\theta) + 1 = \csc^2\theta$
13. $\cos(-130°)$ is a positive number.
14. $\sin(-260°)$ is a negative number.
15. $\tan(-880°)$ is a positive number.
16. Draw a figure and prove that, if θ is a positive angle in Q III, then (a) $\sin(-\theta) = -\sin\theta$, (b) $\cos(-\theta) = \cos\theta$.
17. Draw a figure and prove that, if θ is a negative angle in Q I, then (a) $\sin(-\theta) = -\sin\theta$, (b) $\cos(-\theta) = \cos\theta$.
18. Identify each of the following functions as an even function, an odd function, or neither.
    (a) $\sin^3\theta$ (b) $\tan^8\theta$ (c) $7\theta^5 + 8$ (d) $2\theta^3 + 4\theta$
    (e) $\cos(-3\theta)$ (f) $\tan 4\theta$ (g) $3\theta^6 - 4\theta^2$ (h) $\cos\theta - \sin\theta$

KEY TERMS

Significant digit, degree-minute-second, decimal degree, inverse of a function, related angle, odd function, even function

REVIEW EXERCISES

1. Find tan 78.321°, sec 78.321°, and csc 78.321° to five-figure accuracy.
2. If $\theta = 156° \, 35' \, 45''$, find
   (a) θ in decimal degrees.
   (b) $\sin \theta$, $\cos \theta$, and $\tan \theta$ to five-figure accuracy.
   (c) the related angle of θ.
3. Find the acute angle θ, to the nearest hundredth of a degree, for which
   (a) $\cot \theta = 2.1034$
   (b) $\cos \theta = 0.99931$
   (c) $\tan \theta = 1.0101$
4. Find the related angle of 217.983°, of 1050°, and of −92.5°.
5. Without using a calculator find $\sin \theta$ and $\cos \theta$, if
   (a) $\theta = -150°$
   (b) $\theta = 420°$
   (c) $\theta = 120°$
6. If $f(\theta)$ is an odd function and $g(\theta)$ is an even function, state whether the following functions are odd, even, or neither.
   (a) $\dfrac{1}{g(\theta)}$, provided $g(\theta) \neq 0$
   (b) $\dfrac{f(\theta)}{g(\theta)}$, provided $g(\theta) \neq 0$
   (c) $f(\theta) + g(\theta)$
   (d) $f(\theta)g(\theta)$

3

Solving Triangles and Applications

3.1 THE SOLUTION OF RIGHT TRIANGLES

To solve a triangle means to find from the given parts the values of the remaining parts. A right triangle is determined by

1. Two of its sides, or
2. One side and an acute angle.

In either case it is possible to find the remaining parts by using the special definitions in Section 1.8 together with the fact that the acute angles of a right triangle are complementary. For convenience we list here again these special definitions.

For any acute angle θ lying in a right triangle:

$$\sin \theta = \frac{\text{opp}}{\text{hyp}},$$

$$\cos \theta = \frac{\text{adj}}{\text{hyp}},$$

$$\tan \theta = \frac{\text{opp}}{\text{adj}}.$$

For any triangle we shall use the small letters a, b, and c to denote the lengths of the sides that are opposite the angles A, B, and C, respectively. In a right triangle we shall always reserve the letter c for the hypotenuse. Hence $C = 90°$.

Example 1 Solve the right triangle having an acute angle of 38.7°, the side adjacent to the angle being 311.

Solution Plan We first draw the triangle and label numerically the parts that are known (Figure 3.1).

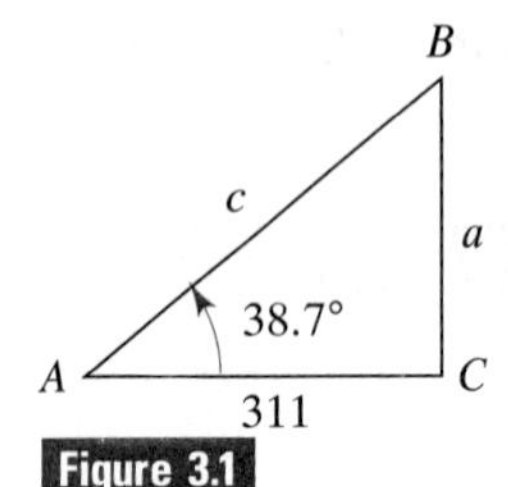

Figure 3.1

Then

(1) $$B = 90° - 38.7°$$
$$= 51.3°.$$

(2) To find a, we observe that the given side and the required side are related to the given angle by the equation

$$\tan 38.7° = \frac{a}{311}.$$

Multiply both sides of the equation by 311:

$$311 \tan 38.7° = a.$$

Hence $$a = 311 \tan 38.7°.$$

(3) To find c, we notice that the given parts, 38.7° and 311, are related to the required part through the cosine of the angle:

$$\cos 38.7° = \frac{311}{c}.$$

Multiply both sides by c:

$$c \cos 38.7° = 311.$$

Divide both sides by cos 38.7°:

$$c = \frac{311}{\cos 38.7°}.$$

Solution (2) $$a = 311 \tan 38.7°.$$

Using an algebraic operating system (AOS) with the switch set for degrees, we press

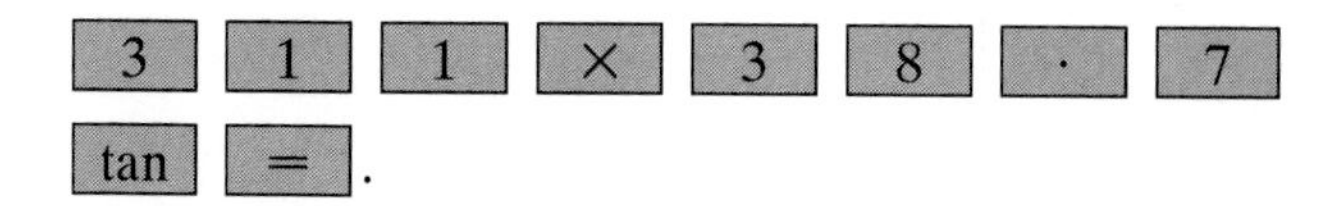

Using Reverse Polish Notation (RPN) with the switch set for degrees, we press

[3] [1] [1] [ENTER] [3] [8] [·] [7]
[tan] [×].

In each case the displayed result is 249.1579829. Rounding off to three figures, we get

$$a = 249\,.$$

(3) $$c = \frac{311}{\cos 38.7°}$$

We press:

Algebraic logic [311] [÷] [38.7] [cos] [=],

RPN [311] [ENTER] [38.7] [cos] [÷].

In both cases the displayed result is 398.4980558, which, rounded off to three significant figures, gives

$$c = 398\,.$$

This problem illustrates three-figure accuracy in the data and the computed results.

Example 2 Solve the right triangle whose hypotenuse is 40.50 and one of whose legs is 34.56.

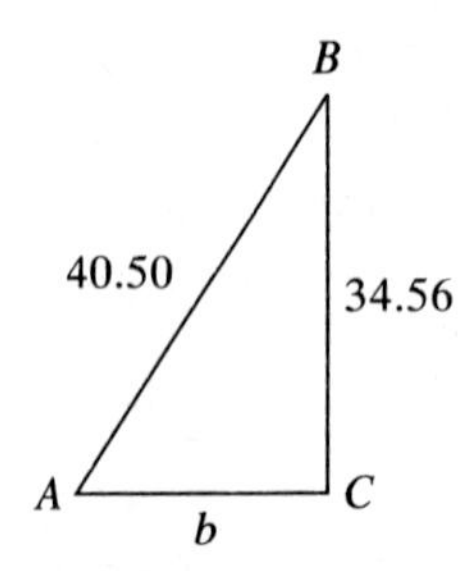

Figure 3.2

Solution Plan Draw the triangle (Figure 3.2) and label numerically the given parts.

(1) Since the hypotenuse and the side opposite A are given,

$$\sin A = \frac{34.56}{40.50},$$

or

$$A = \sin^{-1} \frac{34.56}{40.50}.$$

(2)

$$B = 90° - A.$$

(3) To find b, use

$$\cos A = \frac{b}{40.50},$$

which is equivalent to

$$b = 40.50 \cos A.$$

Solution (1)

$$A = \sin^{-1} \frac{34.56}{40.50}.$$

Using an algebraic logic calculator set for degrees, we press

[34.56] [÷] [40.50] [=] [inv] [sin].

Using RPN with the calculator set for degrees, we press

[34.56] [ENTER] [40.50] [÷] [inv] [sin].

In each case the displayed result is 58.57609351, which we should jot down, or store in the calculator, for further use in finding b. Four-place accuracy in the given sides implies that the angles should be found to the nearest hundredth of a degree. Hence

$$A = 58.58°.$$

(2)
$$B = 90° - 58.58° = 31.42°.$$

(3)
$$b = 40.50 \cos A.$$

Assuming that the "unrounded off" value of A is still displayed on the calculator, we press

Algebraic logic [cos] [×] [40.50] [=],

RPN [cos] [ENTER] [40.50] [×].

The displayed result is 21.11531198. Hence, to four-place accuracy,

$$b = 21.12.$$

Check: Since $a^2 + b^2 = c^2$, then $b = \sqrt{c^2 - a^2}$. If the calculator has a *square key* and a *square root key,* for an algebraic logic instrument we press

[40.50] [x^2] [−] [34.56] [x^2] [=] [$\sqrt{x}$].

The result on display is 21.11531198, which agrees with our value for b. This problem illustrates four-figure accuracy in the data and the results.

Exercise 3.1

Solve the following right triangles.

1. $B = 18°$, $a = 55$
2. $A = 69°$, $a = 3.8$
3. $A = 72.5°$, $b = 20.1$
4. $B = 38.0°$, $b = 567$
5. $A = 52.25°$, $c = 99.44$
6. $B = 40.8°$, $c = 0.0666$

7. $a = 351$, $b = 890$
8. $a = 1620$, $b = 1776$
9. $a = 40.0$, $c = 60.5$
10. $b = 7531$, $c = 9843$
11. $b = 50.01$, $c = 83.38$
12. $a = 3.9$, $c = 4.6$
13. $a = 2783$, $b = 1988$
14. $a = 60$, $b = 73$
15. $B = 23°$, $c = 45$
16. $A = 78.78°$, $c = 586.2$
17. $b = 0.456$, $c = 0.803$
18. $A = 80.00°$, $b = 1492$
19. $B = 36.12°$, $a = 5916$
20. $b = 708$, $c = 900$

3.2 ANGLES OF ELEVATION AND DEPRESSION; BEARING OF A LINE

The *angle of* $\left\{\begin{matrix} \textit{elevation} \\ \textit{depression} \end{matrix}\right\}$ of a point P as seen by an observer O is the angle measured from the horizontal line through O $\left\{\begin{matrix} \textit{upward} \\ \textit{downward} \end{matrix}\right\}$ to the line of sight OP (see Figure 3.3).

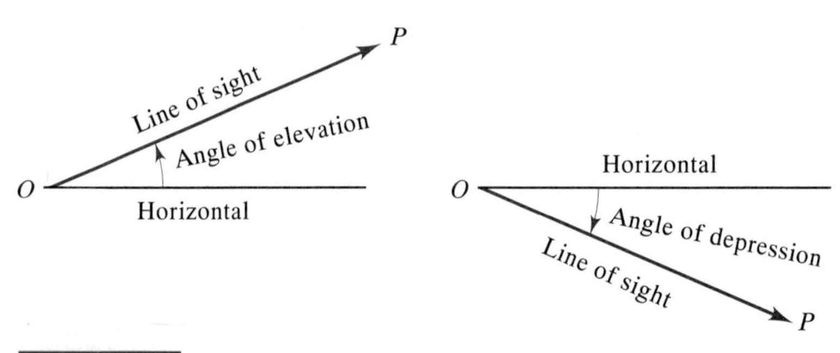

Figure 3.3

Example 1 A vertical stake 20.0 centimeters (cm) high casts a horizontal shadow 12.5 cm long. What time is it if the sun rose at 6:00 A.M. and will be directly overhead at noon?

Solution The angle of elevation of the sun (Figure 3.4) is found by

$$\tan\theta = \frac{20.0}{12.5} = 1.600,$$

$$\theta = 58.0°.$$

It takes the earth 6 hours (h) to rotate through 90°. Since this rotation is uniform, each degree of elevation of the sun will correspond to $\frac{6}{90}$ of an hour, or 4 minutes (min). Consequently a rotation through 58.0° will require (58) (4 min) = 232 min = 3 h and 52 min. Hence the time is 9:52 A.M.

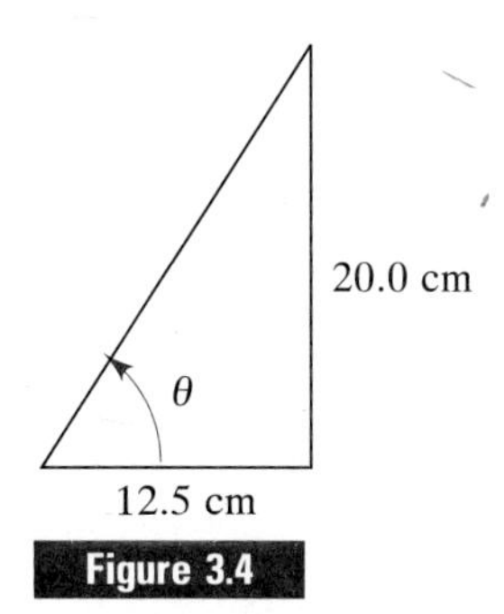

Figure 3.4

In surveying and some kinds of navigation, the *bearing* of a line in a horizontal plane is the *acute* angle made by this line with a north-south line. In giving the bearing of a line, write first the letter *N* or *S*, then the angle of deviation from north or south, then the letter *E* or *W*. Thus, in Figure 3.5 the bearing of line *OA* is N 70° E, or the bearing of point *A* from point *O* is N 70° E.

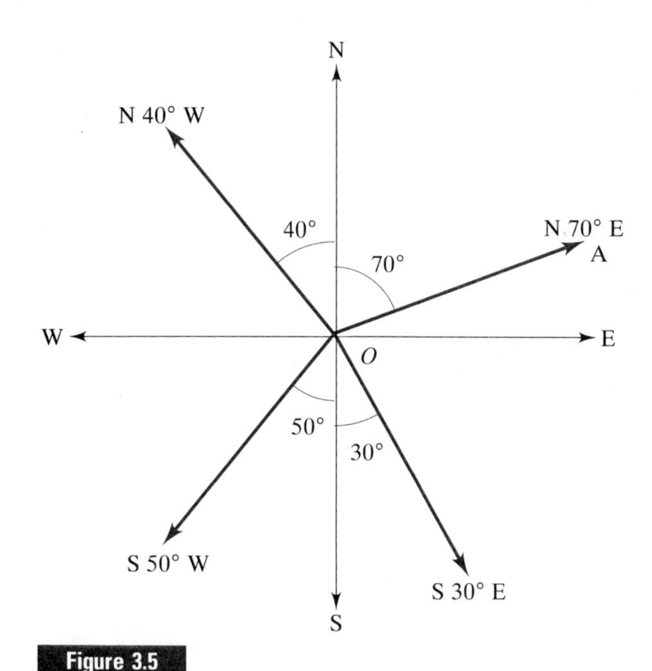

Figure 3.5

Example 2 From a lookout tower A a column of smoke is sighted due south. From a second tower B, 5.00 miles (mi) west of A, the smoke is observed in the direction S 63.0° E. How far is the fire from B? From A? (See Figure 3.6.)

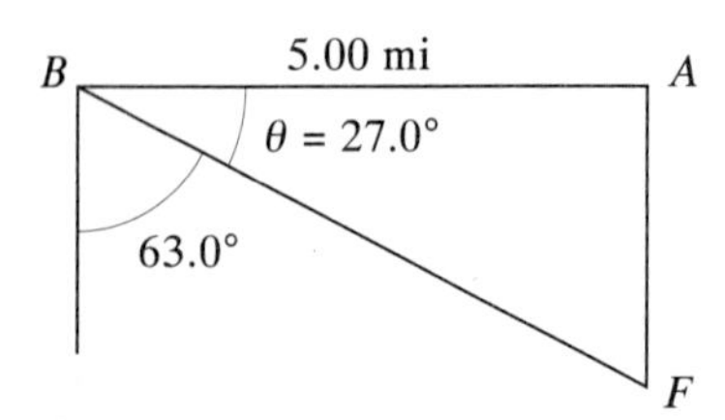

Figure 3.6

Solution Angle $FBA = \theta = 90° - 63.0° = 27.0°$. To get BF, use

$$\cos 27.0° = \frac{5.00}{BF}.$$

Hence
$$BF = \frac{5.00}{\cos 27.0°} = \frac{5.00}{0.8910}$$
$$= 5.61 \text{ mi}.$$

To obtain AF, use

$$\tan 27.0° = \frac{AF}{5.00}.$$

Hence
$$AF = 5.00 \tan 27.0° = (5.00)(0.5095)$$
$$= 2.55 \text{ mi}.$$

In some countries on the metric system, the grad is the preferred unit of angular measurement for surveyors. By definition, 100 grads equals one right angle. Thus 400 grads equals 360° and 50 grads equals 45°.

Example 3 A French surveyor must measure the distance across a small lake as shown in Figure 3.7. If $C = 100$ grads, $A = 37$ grads, and $b = 1020$ m, find the distance BC across the lake.

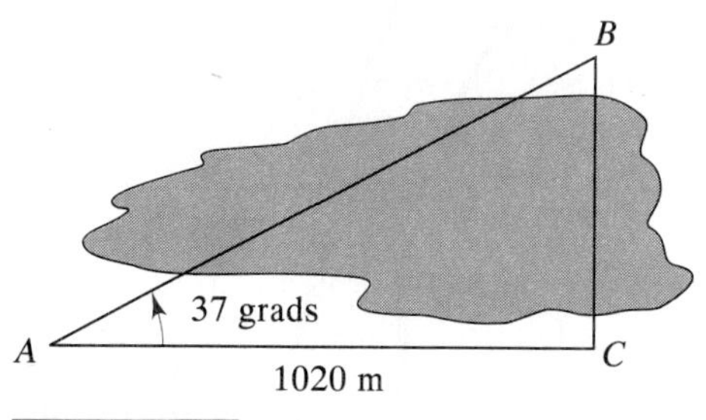

Figure 3.7

Solution Since C is a right angle we know that $a = 1020 \tan A$. With the calculator set in the GRAD mode, we press the keys

to produce the result 670.0147668. Thus the distance from B to C is 670 m.

Exercise 3.2

In Exercises 9 through 12, ignore the curvature of the earth and assume only two-place accuracy. Get angles to the nearest degree and distances to the nearest multiple of 10 mi.

1. A balloon hovers 307 feet (ft) above one end of the Rip Van Winkle Bridge, which spans the Hudson River at Catskill, New York. The angle of depression of the other end of the bridge from the balloon is 21.0°. How long is the bridge?
2. The angle of elevation of the top of the Sphinx in Gizeh, Egypt, from a point on level ground 12 m from its base is 59°. What is the height of the Sphinx?
3. From a point on the ground 366 ft away from the foot of the Washington Monument, the angle of elevation of the top of the monument is 56.6°. How high is the monument?
4. From a point A on the top of a cliff that is 115 m above sea level, the angle of depression of an anchored rowboat B is 22.0°. Find the length of the airline distance AB from the cliff to the boat.
5. A flagpole 8.20 m high casts a horizontal shadow 4.00 m long. What time is it if the sun was directly overhead at 12:10 P.M. and will set at 6:10 P.M.?

6. The Royal Gorge Bridge crosses the Arkansas River near Canon City, Colorado. From a point 750.0 ft from, and in the same horizontal plane with, the bridge, the angle of depression of the river flowing beneath the bridge is 54.54°. Find the distance from the bridge down to the river.
7. A French surveyor notes that the angle of elevation of the Eiffel Tower, 300 m tall, is 16.52 grads. How far is it to the tower?
8. The angle of elevation of a treetop from a point on the ground 10 m from the base of the tree is 67°. What is the height of the tree?
9. Harrisburg, Pennsylvania, is due west of Trenton, New Jersey. Elmira, New York, is due north of Harrisburg and is 170 mi N 42° W from Trenton. How far is Harrisburg from Trenton? From Elmira?
10. Schenectady, New York, is 150 mi due west of Haverhill, Massachusetts. Poughkeepsie, New York, is due south of Schenectady and is 160 mi from Haverhill. What is the bearing of Haverhill from Poughkeepsie? Poughkeepsie from Haverhill?
11. Tulsa, Oklahoma, is 350 mi due south of Omaha, Nebraska. Nashville, Tennessee, is 500 mi due east of Tulsa. What is the bearing of Nashville from Omaha? Omaha from Nashville?
12. Glen Ellyn, Illinois, is 280 mi due east of Cedar Rapids, Iowa. Rolla, Missouri, is S 57° W from Glen Ellyn and due south of Cedar Rapids. How far is Rolla from Glen Ellyn? From Cedar Rapids?
13. A boat leaves a certain port at 11:40 A.M. and moves S 60° W at 9 km/h. At noon another boat leaves the same port and travels N 30° W at 10 km/h. Find the bearing (to the nearest degree) of the second boat from the first boat at 2 P.M.
14. The tallest structure in the continental United States is the Blanchard TV Tower in North Dakota. Find the height of the tower if the angle of elevation of its top from a point on the ground 400.2 ft from its base is 79.02°.
15. Find the perimeter of a regular polygon of 180 sides inscribed in a circle of diameter 1.00000 m. Compare this number with the circumference of the circle.
16. A ship moving due east at 10.0 km/h is 3.00 km due south of a lighthouse at 10:00 A.M. When (to the nearest minute) will the lighthouse have a bearing of N 77.0° W from the ship?
17. An airplane flying 278 km/h is coming down at an angle of 12.9° with the horizontal. What is its rate of descent?
18. Tangents AB and AC are drawn to a circle through a point A on the outside of the circle, which has center O. If $AO = 36.72$ and the radius of the circle is 15.84, find the angle BAC made by the two tangents.

19. Two boats leave the same port at noon. The first boat moves in the direction S 31.6° W; the second travels N 58.4° W. If the first boat travels twice as fast as the second boat, find the bearing of the second boat from the first boat at 12:45 P.M.
20. In order to measure the width CB of a river, a surveyor laid off a distance of 20.0 m on the bank CA at right angles to CB. Using a transit, he found angle BAC to be 76.4°. How wide is the river?
21. Let ABC be any triangle with acute angles A and B and included side c. If CD is the perpendicular from C to AB, show that

$$CD = \frac{c}{\cot A + \cot B}.$$

(*Hint:* Show that $AD = CD \cot A$ and $DB = CD \cot B$. Then add these two equations.)
22. From the top of a building a meters tall, on level ground, the angle of depression of the bottom of a television tower is θ. From the bottom of the building the angle of elevation of the top of the tower is ϕ. Show that the height of the tower is ($a \cot \theta \tan \phi$) meters.
23. From the top of a lighthouse h meters above sea level at high tide, the angle of depression of a gull standing on a buoy is θ at high tide and ϕ at low tide. Show that the height of the tide is $h(\cot \theta \tan \phi - 1)$ meters.
24. Prove that the perpendicular bisector of the hypotenuse of a 30°-60°-90° triangle divides the larger leg into two parts having the ratio 2 to 1. (*Hint:* Extend the perpendicular bisector and the smaller leg until they intersect. Prove that the large right triangle thus formed and the original triangle are congruent.)
25. A ladder 10.00 m long stands on level ground and leans against a vertical wall. The angle of elevation of the ladder is 77.0°.
    (a) Find the distance from the foot of the ladder to the wall.
    (b) How far is the top of the ladder above the ground?
    (c) If the foot of the ladder is moved 3.00 m farther from the wall, how much does the top of the ladder fall?
26. What is the angle made by a diagonal of a cube and a touching diagonal of a face of the cube?

3.3 VECTORS

A *vector quantity* is a quantity that has both magnitude and direction. Examples of vector quantities are forces, velocities, accelerations, and displacements. A *vector* is a directed line segment. A vector quantity can be represented by means of a vector if (1) the direction of the vector is the same as that of the vector quantity and (2) the length of the vector represents, to some convenient scale, the magnitude of the vector quantity. For example, a velocity of 30 mi/h in a northerly direction can be represented by a 3-inch (in.) line segment pointed north.

The *resultant* (or vector sum) of two vectors is the diagonal of a parallelogram having the two given vectors as adjacent sides. In physics it is shown that like vector quantities, such as forces, are combined according to this vector law of addition. In Figure 3.8, vector *OR* is the resultant of vectors *OA* and *OB*. If the lengths and directions of *OA* and *OB* are known, then the length and direction of *OR* can be found by solving triangle *OAR*.

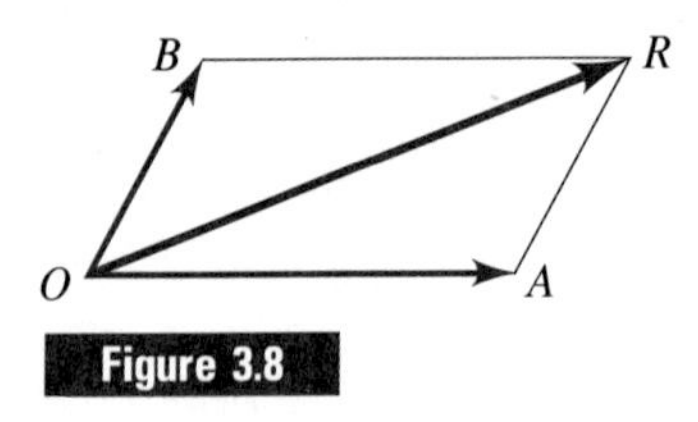

Figure 3.8

Two vectors that have a certain resultant are said to be *components* of that resultant. If the two vectors are at right angles to each other, they are called *rectangular components.* It is possible to resolve a given vector into components along any two different specified directions. For example (Figure 3.9), a force of 70 lb acting in the direction N 62° E can be resolved into an easterly force and a northerly force by solving the right triangle *ORN*.

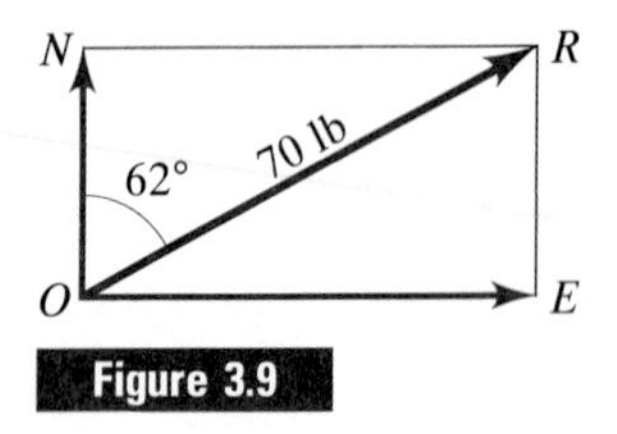

Figure 3.9

In air navigation, the bearing of a line *OC* is the angle measured *clockwise from the north* to the line *OC*. Referring to Figure 3.10, in aviation the bearing of line *OC* is 110°, or the direction *OC* is 110°; the bearing of *OD* is 260°.

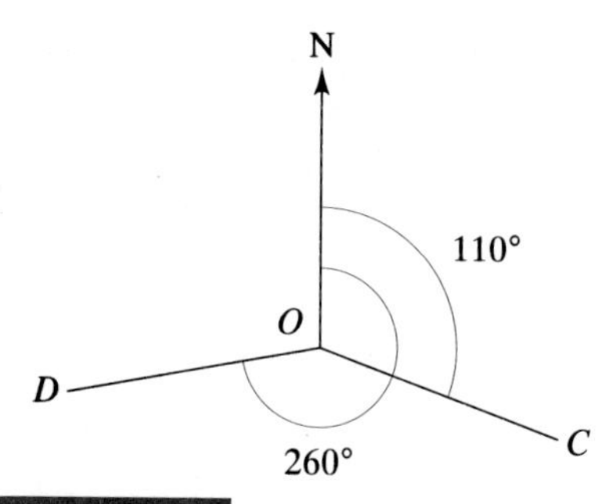

Figure 3.10

Example 1 An airplane with an *air speed* (speed in still air) of 178 mi/h is headed due south. (Its *heading* is 180.0°.) If a west wind of 27.5 mi/h is blowing, find the *course* of the plane (the direction it travels) and its *ground speed* (actual speed with respect to the ground).

Solution To find θ (see Figure 3.11), use

$$\tan \theta = \frac{178}{27.5}$$

or

$$\theta = 81.22° \rightarrow 81.2°.$$

To find OR, use

$$\cos \theta = \frac{27.5}{OR}.$$

Hence

$$OR = \frac{27.5}{\cos \theta} = 180.1 \rightarrow 180.$$

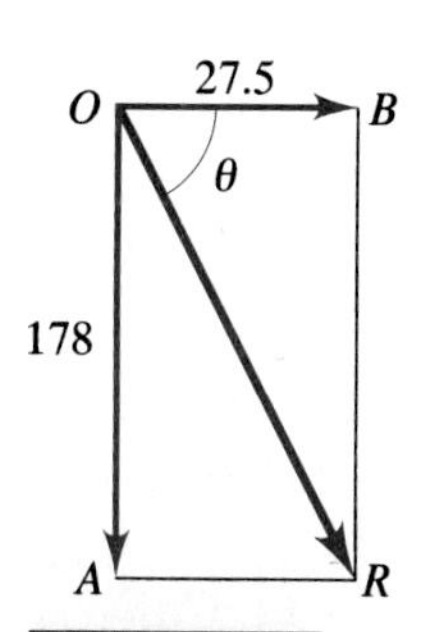

Figure 3.11

Inasmuch as the original data indicate only three-figure accuracy, the results should be rounded off to three significant digits. Hence the course of the plane is $90.0° + 81.2° = 171.2°$, and its ground speed is 180 mi/h.

Example 2 What is the minimum force required to prevent a 90-pound (lb) barrel from rolling down a plane that makes an angle of 17° with the horizontal? Find the force of the barrel against the plane.

Solution The weight or force of 90 lb, *OR*, acting vertically downward can be resolved into two rectangular components, one parallel to the plane and the other perpendicular to it (see Figure 3.12). Hence

$$OA = OR \sin \angle ORA = 90 \sin 17° = 26.3\,.$$

The force required to prevent the barrel from rolling is 27 lb (to the nearest pound).

The force of the barrel against the plane is

$$OB = OR \cos \angle ROB = 90 \cos 17° = 86 \text{ lb}\,.$$

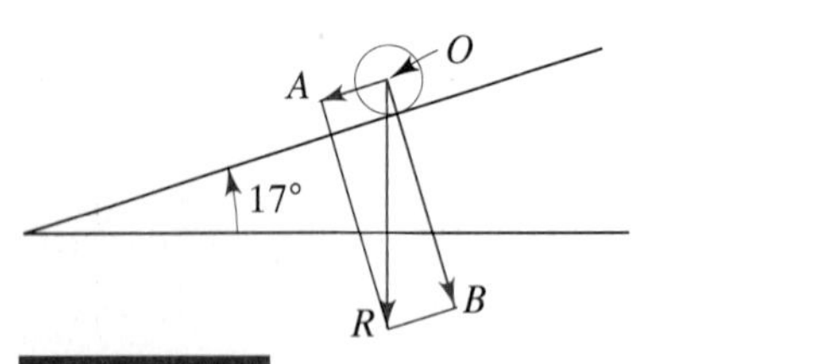

Figure 3.12

Exercise 3.3

1. West Point, New York, is 87 mi due east of Scranton, Pennsylvania. An airplane leaves West Point at 10:00 A.M. and flies 250 mi/h in the direction 327°. When (to the nearest minute) will the plane be due north of Scranton?
2. Ypsilanti, Michigan, is 60 mi due north of Bowling Green, Ohio. A helicopter leaves Bowling Green at 3:00 P.M. and travels 40 mi/h in

the direction 338°. When (to the nearest minute) will it be closest to Ypsilanti?

3. At noon a plane leaves *A* and flies east at 300 km/h. At 12:30 P.M. the plane changes its heading to 206.0° and continues until it is due south of *A*. Find the total distance traveled by the plane since it left *A*.
4. A plane flies 63.50 km due west of *B* and then turns to the south and flies 49.00 km. Find the bearing of the plane from *B*.
5. A bus weighing 7113 lb exerts a force of 6922 lb on an inclined ramp. (a) What angle does the ramp make with the horizontal? (b) What force must be exerted by the brakes to prevent the bus from rolling down the ramp?
6. A cable that can withstand a tension of 2305 lb is used to pull vehicles up an inclined ramp to the second floor of a storage garage. If the inclination of the ramp is 17.75°, find the weight of the heaviest truck that can be pulled with the cable.
7. A barrel resting on an inclined ramp exerts a force of 85 lb against the ramp, which makes an angle of 20° with the horizontal. (a) Find the weight of the barrel. (b) What force is required to prevent the barrel from rolling down the ramp?
8. A force of 518 lb is required to keep a 4230-lb automobile from rolling down a hill. What angle does the hill make with the horizontal?
9. The wind is from the south. If a small plane's heading is 260.0°, it will travel due west at 230 km/h. Find the wind speed and the plane's air speed.
10. A wind of 35 km/h is blowing in the direction 110°. A pilot wants her helicopter to move in the direction 200° with a ground speed of 85 km/h. Find the proper heading and the air speed.
11. The pilot of an airplane wishes to fly due west. His plane has a cruising speed of 150 mi/h in still air. In what direction should he head his plane if a north wind of 38.0 mi/h is blowing? Find the ground speed of the plane.
12. An airplane with a cruising speed of 220 km/h in still air moves with a heading of 72°. Find the speed of a north wind that makes the plane travel due east.
13. A river flows due south at 30.0 meters per minute (m/min). By heading her motorboat in the direction N 75.0° E, a young woman is able to make her boat move due east across the river. How long will it take to cross the river if it is 280 m wide?
14. A launch capable of a speed of 500 feet per minute (ft/min) in still water is at the north bank of a river that flows east. Find the proper heading for the launch if it is to travel due south at 400 ft/min. What is the speed of the current?

15. A baseball is thrown southwest at 50 mi/h from a train traveling southeast (S 45° E) at 20 mi/h. Find the speed of the baseball and the direction in which it moves.
16. The cruising speed of an airplane in still air is 305 km/h. The plane is headed due south. Because of an east wind, the plane's ground speed is 310 km/h. Find the wind speed and the plane's course.
17. An 8000-lb truck traveling north collided with a 2000-lb car going east at 20 mi/h. After the impact, the wreckage moved in the direction N 10° E. Was the truck exceeding the 30-mi/h speed limit? [*Hint:* The momentum of a moving object is the product of its mass (i.e., its weight) and its velocity; that is, momentum = (weight) (velocity). Consider each vehicle's momentum as a vector quantity.]
18. A balloon is rising at 120 ft/min and is carried eastward by a west wind. If the sun is directly overhead and the balloon's shadow is moving at 365 ft/min on level ground, what is the angle made by the path of the balloon and the horizontal?
19. Find the direction and magnitude of the resultant of forces of 23.75 lb acting S 56.14° W and 48.61 lb acting S 33.86° E.
20. An observer at P looks due north and sees a meteor with an angle of elevation of $\theta°$. At the same instant, another observer, a miles east of P, sees the same meteor and approximates its position as $N\ \phi°$ W but fails to note its angle of elevation. Show that the height of the meteor is $a \tan\theta/\tan\phi$ miles and its distance from P is $a/\tan\phi\cos\theta$ miles.

3.4 OBLIQUE TRIANGLES

A triangle that does not contain a right angle is called an oblique triangle. Since the ratio of two sides of an *oblique* triangle does not represent a function of an angle of the triangle, additional formulas are needed for solving oblique triangles. These formulas are the law of sines and the law of cosines.

The six parts of a triangle are its three sides a, b, c, and the opposite angles A, B, C, respectively. If three parts, at least one of which is a side, are given, the remaining parts can be determined. For convenience, we shall divide the possibilities into the following four cases:

***SAA*: Given one side and two angles.**
***SSA*: Given two sides and the angle opposite one of them.**
***SAS*: Given two sides and the included angle.**
***SSS*: Given three sides.**

3.5 THE LAW OF SINES

We will need the following theorem.

The Law of Sines

In any triangle the sides are proportional to the sines of the opposite angles:

$$\frac{a}{\sin A} = \frac{b}{\sin B} = \frac{c}{\sin C}.$$

Proof Consider the two cases: all angles acute (Figure 3.13) and one angle obtuse (Figure 3.14). Let h be the perpendicular from C to AB (or AB extended). In either case,

$$\sin B = \frac{h}{a}, \qquad \text{hence} \quad h = a \sin B.$$

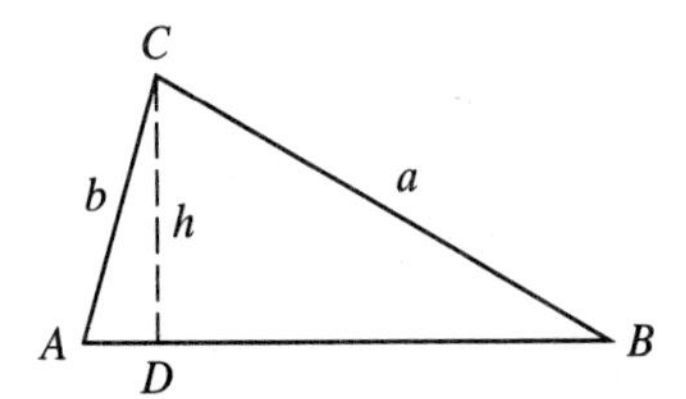

Figure 3.13

Also in Figure 3.14, $\sin A = \sin \angle CAD = h/b$,

$$\sin A = \frac{h}{b}, \qquad \text{hence} \quad h = b \sin A.$$

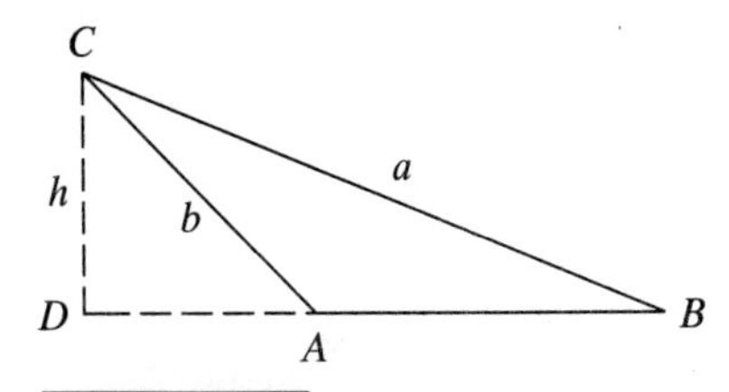

Figure 3.14

Equate the two values of h: $a \sin B = b \sin A$.

Divide by $\sin A \sin B$: $\dfrac{a}{\sin A} = \dfrac{b}{\sin B}$.

Similarly, $\dfrac{a}{\sin A} = \dfrac{c}{\sin C}$.

Therefore $\dfrac{a}{\sin A} = \dfrac{b}{\sin B} = \dfrac{c}{\sin C}$.

The law of sines is equivalent to the following three equations:

$$\frac{a}{\sin A} = \frac{b}{\sin B}, \qquad \frac{a}{\sin A} = \frac{c}{\sin C}, \qquad \frac{b}{\sin B} = \frac{c}{\sin C}.$$

If the three given parts of the triangle include one side and the opposite angle, then one of these equations will involve three known quantities and one unknown. We can solve for this unknown by using ordinary algebraic methods. Thus, if a, c, and C are the known parts, we can find A by solving the equation

$$\frac{\sin A}{a} = \frac{\sin C}{c},$$

$$\sin A = \frac{a \sin C}{c}.$$

Whenever four parts of a triangle are known (the fourth part may have been obtained from the other three), *we can use the law of sines to find the remaining two parts.* Why?

3.6 APPLICATIONS OF THE LAW OF SINES: *SAA*

When one side and two angles of a triangle are known, we can immediately find the third angle from the relation $A + B + C = 180°$. The remaining two sides can then be found by using the law of sines.

Example 1 Solve the triangle ABC, given $a = 20$, $A = 30°$, $B = 40°$.

Solution (See Figure 3.15.)

(1) $$C = 180° - (30° + 40°)$$
$$= 110°.$$

(2) To find b, use

$$b = \frac{a \sin B}{\sin A} = \frac{20 \sin 40°}{\sin 30°} = \frac{20(0.6428)}{0.5000} \doteq 26.$$

(3) To find c, use

$$c = \frac{a \sin C}{\sin A} = \frac{20 \sin 110°}{\sin 30°} = \frac{20(0.9397)}{0.5000} \doteq 38.$$

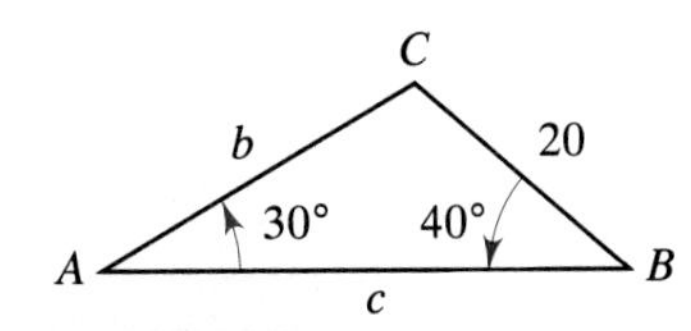

Figure 3.15

The values of b and c are rounded off to two-figure accuracy. The results can be checked by the law of cosines (Section 3.8) or by finding c with the formula

$$c = \frac{b \sin C}{\sin B}.$$

Example 2 Solve the triangle ABC, given $b = 191, A = 55.7°, C = 81.5°$ (see Figure 3.16).

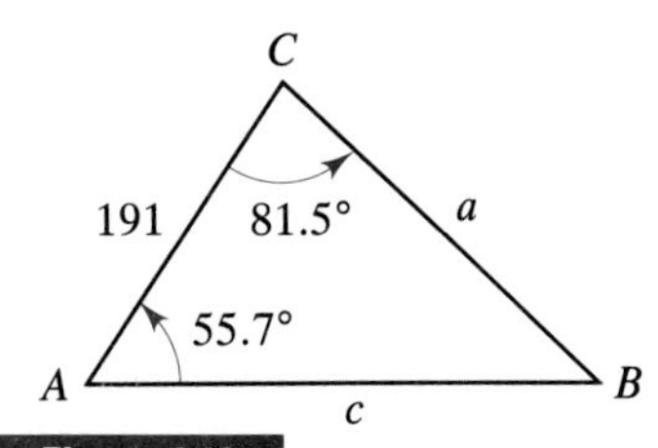

Figure 3.16

Solution Plan (1) $$B = 180° - (55.7° + 81.5°)$$
$$= 42.8°.$$

(2) $$a = \frac{b \sin A}{\sin B} = \frac{191 \sin 55.7^\circ}{\sin 42.8^\circ}.$$

(3) $$c = \frac{b \sin C}{\sin B} = \frac{191 \sin 81.5^\circ}{\sin 42.8^\circ}.$$

Solution (2) $$a = \frac{191 \sin 55.7^\circ}{\sin 42.8^\circ}.$$

Using an algebraic calculator with the switch set for degrees, we press

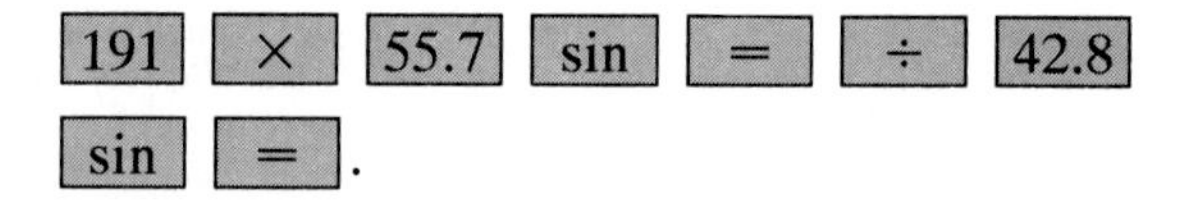

The displayed result is 232.2272333. Rounding off to three-figure accuracy, we obtain

$$a = 232.$$

Using an RPN calculator, in degree mode, we press

[191] [ENTER] [55.7] [sin] [×] [42.8]
[sin] [÷].

In this case the displayed result is 232.2272332, which differs slightly from the algebraic calculator answer. (Observe that two different calculators may produce results that are not quite identical.) Once again, however, the answer to three significant figures is

$$a = 232.$$

(3) Using the same procedure, we find $c = 278.0255317$ or 278.0255318, each of which, after rounding off, becomes

$$c = 278.$$

Exercise 3.4

Solve the following triangles in Exercises 1 through 12.

1. $b = 200$, $A = 56.7°$, $C = 40.0°$
2. $c = 456$, $A = 49.7°$, $B = 100.0°$
3. $a = 6.0$, $B = 38°$, $C = 67°$
4. $b = 12$, $A = 46°$, $C = 79°$
5. $c = 0.36$, $A = 99°$, $C = 22°$
6. $a = 5.6$, $A = 38°$, $B = 82°$
7. $b = 48.5$, $B = 39.2°$, $C = 40.8°$
8. $c = 946$, $A = 18.0°$, $C = 111.0°$
9. $a = 30.15$, $B = 62.58°$, $C = 75.25°$
10. $b = 1812$, $A = 68.20°$, $B = 74.57°$
11. $c = 5665$, $A = 44.37°$, $B = 80.50°$
12. $a = 0.02736$, $A = 28.28°$, $C = 81.70°$
13. Two angles of a triangle are 58.11° and 57.89°. If the smallest side is 197.3, find the longest side.
14. Bogota, Colombia, is 770 km S 50° E from Panama. The bearing of San Cristobal, Venezuela, from Panama is S 81° E. The bearing of Bogota from San Cristobal is S 30° W. How far is San Cristobal from Panama? From Bogota? Ignore the curvature of the earth and assume only two-place accuracy. Get distances to the nearest multiple of 10 km.
15. The heading of an airplane is 250°. A 40-km/h wind, blowing toward 320°, causes the plane to move in the direction 255°. Find the air speed and the ground speed of the plane.
16. A train is moving at 60 mi/h on a straight track in the direction N 37° W. The sun, setting in the direction N 80° W, casts the train's shadow onto the front of a building that faces due west. How fast is the train's shadow moving?
17. Show that if $C = 90°$, the law of sines gives the definition of the sine of an acute angle.
18. Why does the law of sines fail to solve the *SAS* and *SSS* cases?

3.7 THE AMBIGUOUS CASE: *SSA*

If two sides and the angle opposite one of them are given, the triangle is not always uniquely determined. With the given parts, we may be able to construct two triangles, or only one triangle, or no triangle at all. Because the possibility of two triangles exists, this is usually called the ambiguous case.

To avoid unneccessary confusion, we shall use A to designate the given angle, a to represent the opposite side, and b to indicate the other given side. Construct angle A and lay off b as an adjacent side, thus fixing the vertex C. With C as center and a as radius, strike an arc cutting the other side adjacent to A. Figure 3.17 illustrates the various possibilities.

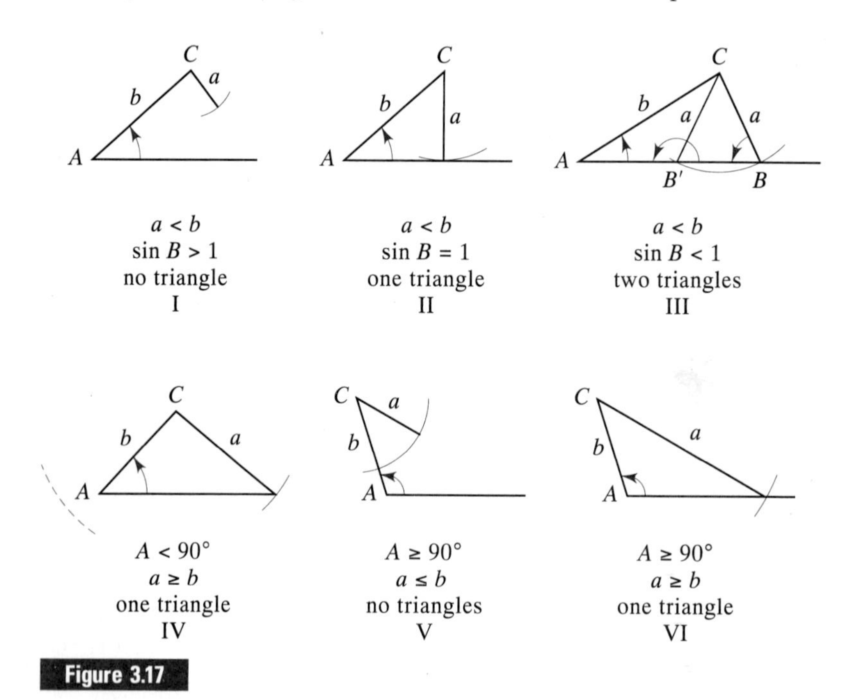

Figure 3.17

The last three cases (diagrams IV, V, and VI) can be quickly identified by merely noting the relative sizes of a and b. If A is acute and $a < b$, it is necessary to begin the solution before we can state how many triangles are possible. To do this, use the law of sines:

$$\frac{\sin B}{b} = \frac{\sin A}{a}, \qquad \sin B = \frac{b \sin A}{a}.$$

After determining the value of sin B, we can definitely classify our problem as one of the various types. It is frequently possible to determine the number of solutions by merely constructing a figure to scale.

Illustration 1 Given $a = 100$, $b = 70$, $A = 80°$. This is case IV because $a > b$. There is one triangle.

Illustration 2 Given $a = 40$, $b = 42$, $A = 110°$. This is case V because $A > 90°$ and $a < b$. There is no triangle.

Illustration 3 Given $a = 80$, $b = 100$, $A = 54°$. Since A is acute and $a < b$, this is case I, II, or III. A carefully constructed figure leaves some doubt as to whether a is long enough to reach the horizontal side of angle A. Using the law of sines, we find

$$\sin B = \frac{b \sin A}{a} = \frac{100(0.8090)}{80} = 1.0112.$$

Since no angle can have a sine greater than 1, B is impossible and there is no triangle. This is case I.

Example Solve the triangle ABC, given $a = 48.8$, $b = 69.2$, $A = 37.2°$.

Solution A scale drawing (Figure 3.18) indicates that there are two triangles. Using the law of sines, we get

$$\frac{\sin B}{b} = \frac{\sin A}{a};$$

$$\sin B = \frac{b \sin A}{a} = \frac{69.2 \sin 37.2°}{48.8}.$$

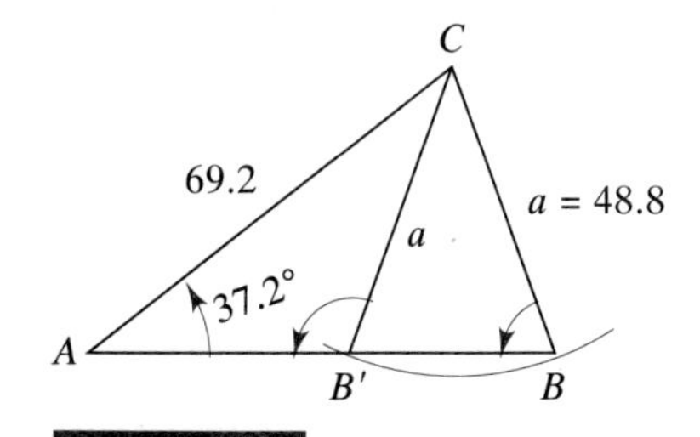

Figure 3.18

Using a calculator, we find $B = 59.0°$.

Continuing with the solution of large triangle ABC (Figure 3.19), we find

$$C = 180° - (37.2° + 59.0°)$$
$$= 83.8°.$$

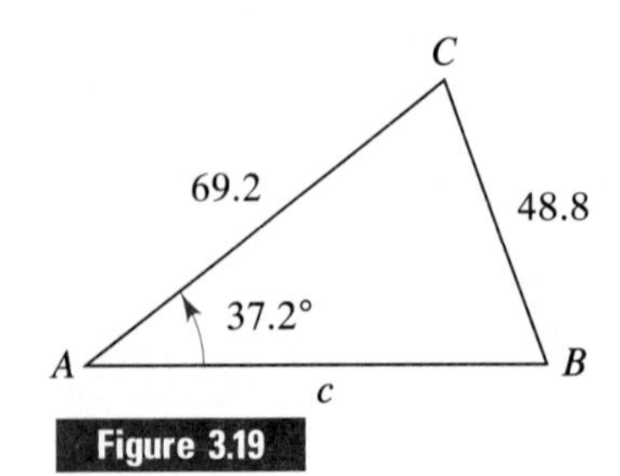

Figure 3.19

To find c, use

$$\frac{c}{\sin C} = \frac{a}{\sin A}; \qquad c = \frac{a \sin C}{\sin A} = \frac{48.8 \sin 83.8^\circ}{\sin 37.2^\circ}.$$

Hence the computed parts of triangle ABC are $B = 59.0^\circ$, $C = 83.8^\circ$, $c = 80.2$.

We shall now solve the small triangle $AB'C'$. Since triangle BCB' (Figure 3.18) is isosceles, $\angle BB'C = B$. Hence

$$B' = 180^\circ - B = 180^\circ - 59.0^\circ = 121.0^\circ.$$

Also (see Figure 3.20),

$$C' = 180^\circ - (37.2 + 121.0^\circ) = 21.8^\circ.$$

Using the law of sines, we find $c' = 30.0$.

Hence the computed parts of triangle $AB'C'$ are $B' = 121.0^\circ$, $C' = 21.8^\circ$, $c' = 30.0$.

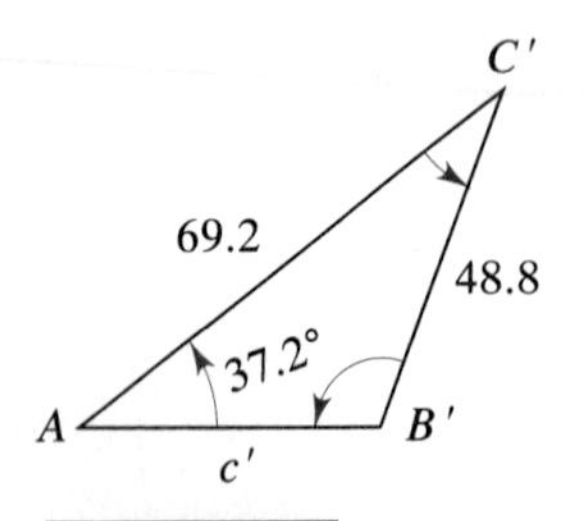

Figure 3.20

Exercise 3.5

Draw a figure and solve all possible triangles in Exercises 1 through 12.

1. $a = 4.5, b = 6.0, A = 29°$
2. $a = 58, b = 48, A = 76°$
3. $a = 66, b = 77, A = 59°$
4. $a = 3040, b = 7486, A = 17.56°$
5. $a = 99.44, b = 70.80, A = 103.55°$
6. $a = 0.08216, b = 0.09123, A = 97.42°$
7. $a = 37.1, b = 44.6, A = 53.2°$
8. $a = 0.379, b = 0.758, A = 30.0°$
9. $a = 192, b = 291, A = 41.3°$
10. $a = 150, b = 200, A = 33.3°$
11. $a = 61.20, b = 61.20, A = 61.20°$
12. $a = 7.01, c = 7.25, A = 126.0°$
13. A motorboat is due east of a freighter moving along the shoreline in the direction N 21° E. If the freighter travels 10 km/h and the motorboat can do 20 km/h, in what direction should the motorboat travel in order to reach the freighter as soon as possible?
14. Little Rock, Arkansas, is 380 mi from Vincennes, Indiana, and 200 mi from Jackson. The bearing of Little Rock from Vincennes is S 44° W. The bearing of Jackson from Vincennes is S 17° W. How far is Jackson from Vincennes? Ignore the curvature of the earth and assume only two-place accuracy. Get distances to the nearest multiple of 10 miles.
15. The earth is 93 million mi from the sun. The planet Mercury is 36 million mi from the sun. How far is Mercury from the earth when the sun is about to rise in the east and Mercury is 15° above the horizon in the east? (Get the result to the nearest multiple of a million miles.)
16. A 10-ft pole placed vertically on a hillside casts a shadow of 12 ft straight *down* the slope. At the tip of the shadow, the angle subtended by the pole is 35°. Find the angle made by the hillside with the horizontal. Find the elevation of the sun.

3.8 THE LAW OF COSINES

When three sides, or two sides and the included angle, are known, we turn to the law of cosines to solve the triangle.

The Law of Cosines

The square of any side of a triangle is equal to the sum of the squares of the other two sides minus twice their product times the cosine of the included angle:

$$a^2 = b^2 + c^2 - 2bc \cos A, \qquad \cos A = \frac{b^2 + c^2 - a^2}{2bc}.$$

$$b^2 = c^2 + a^2 - 2ca \cos B, \qquad \cos B = \frac{c^2 + a^2 - b^2}{2ca}.$$

$$c^2 = a^2 + b^2 - 2ab \cos C, \qquad \cos C = \frac{a^2 + b^2 - c^2}{2ab}.$$

Proof (1) If all the angles are acute: Draw CD perpendicular to AB. Then (see Figure 3.21)

$$\begin{aligned} a^2 &= h^2 + \overline{DB}^2 \\ &= h^2 + (c - AD)^2 \\ &= \underbrace{h^2 + \overline{AD}^2}_{} + c^2 - 2cAD \\ &= \quad b^2 \quad + c^2 - 2c(b \cos A) \\ &= b^2 + c^2 - 2bc \cos A. \end{aligned}$$

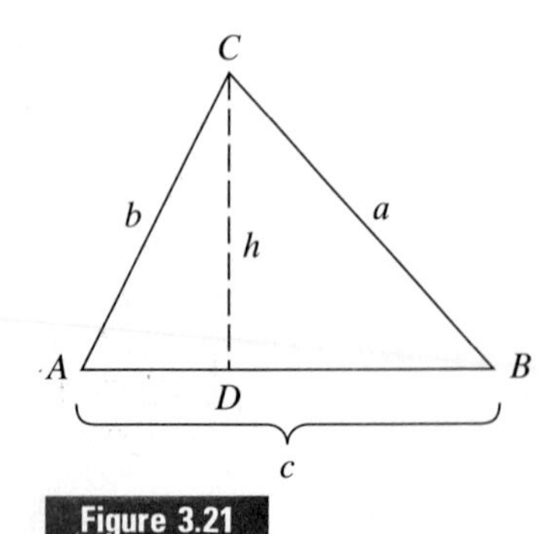

Figure 3.21

(2) If one angle is obtuse: Draw CD perpendicular to AB extended. Notice that if AB is considered a positive directed segment, then AD is negative in Figure 3.22 while DB is

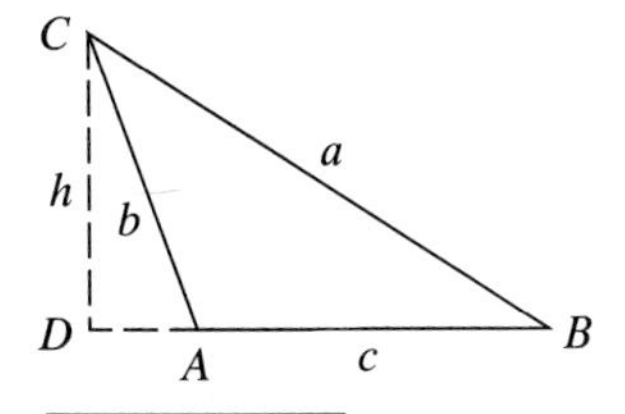

Figure 3.22

negative in Figure 3.23. With this understanding, the proofs for Figures 3.22 and 3.23 are exactly the same as that for Figure 3.21. The student should verify this.

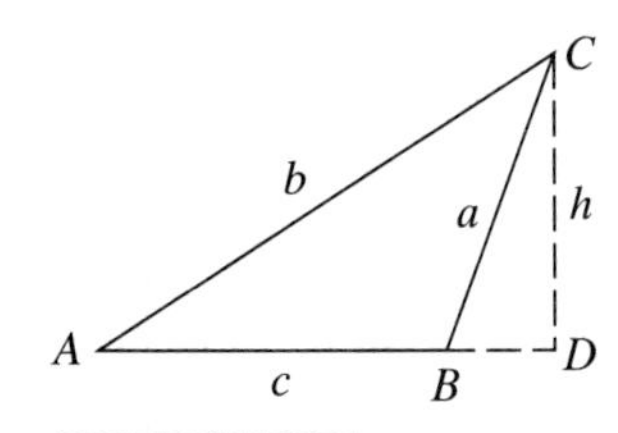

Figure 3.23

This is a general proof, because a may be used to represent any side of the given triangle. The other two forms of the law of cosines can be obtained from the first form by the method of *cyclic permutation,* in which

a is changed to b,	A is changed to B,
b is changed to c,	B is changed to C,
c is changed to a,	C is changed to A.

If $A = 90°$, $\cos A = 0$, and $a^2 = b^2 + c^2 - 2bc \cos A$ reduces to the form $a^2 = b^2 + c^2$. We conclude that the Pythagorean theorem is a special case of the law of cosines; that is, the law of cosines is a generalization of the Pythagorean theorem.

3.9 APPLICATIONS OF THE LAW OF COSINES: *SAS* AND *SSS*

The law of cosines is used in many geometric problems, one of which is the solution of triangles. Since three sides and one angle are involved in every form of the law of cosines, it can be used to solve the *SAS* and *SSS* cases described in Section 3.4.

Example 1 Solve the triangle ABC, given $a = 50.0$, $c = 60.0$, $B = 100.0°$ (see Figure 3.24).

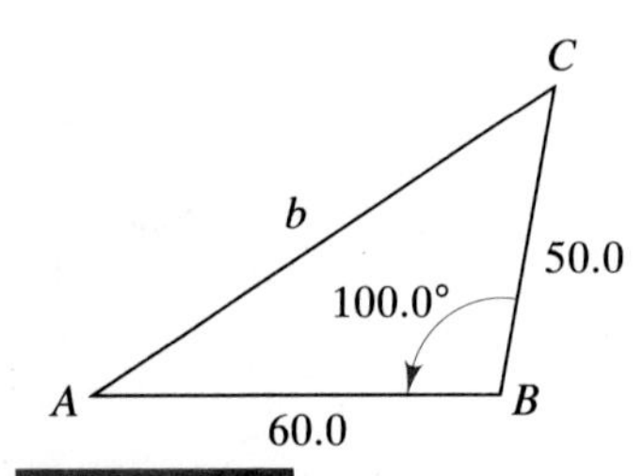

Figure 3.24

Solution This is the two sides and included angle case, *SAS*. Hence

(1)
$$\begin{aligned} b^2 &= c^2 + a^2 - 2ca \cos B \\ &= 60^2 + 50^2 - 2(60)(50)\cos 100° \\ &= 3600 + 2500 - 6000(-0.1736) \\ &= 7141.6. \end{aligned}$$

$$b = 84.50969806 \rightarrow 84.5.$$

Notice that $\cos 100° = -\cos 80° = -0.1736$.

(2) To find A, use the law of sines:

$$\sin A = \frac{a \sin B}{b} = \frac{50 \sin 100°}{b}.$$

$$\begin{aligned} A &= \sin^{-1}\left(\frac{50 \sin 100°}{84.5}\right) \\ &= 35.6°. \end{aligned}$$

(3)
$$\begin{aligned} C &= \sin^{-1}\left(\frac{60 \sin 100°}{84.5}\right) \\ &= 44.4°. \end{aligned}$$

Check: $A + B + C = 35.6 + 100.0° + 44.4° = 180.0°$.

Comments While you are using a calculator to find A and C, be sure to use the displayed value of b (84.50969806) rather than the rounded-off value (84.5).

This example illustrates three-figure accuracy in the given data and the computed results.

In the *SAS* case, if the given angle is acute, one of the other angles may be obtuse. If c is the longest side of a triangle and if $c^2 > a^2 + b^2$, then C must be an obtuse angle.

Example 2 Given $a = 8.20$, $b = 5.10$, $c = 4.10$, find the largest and smallest angles (see Figure 3.25).

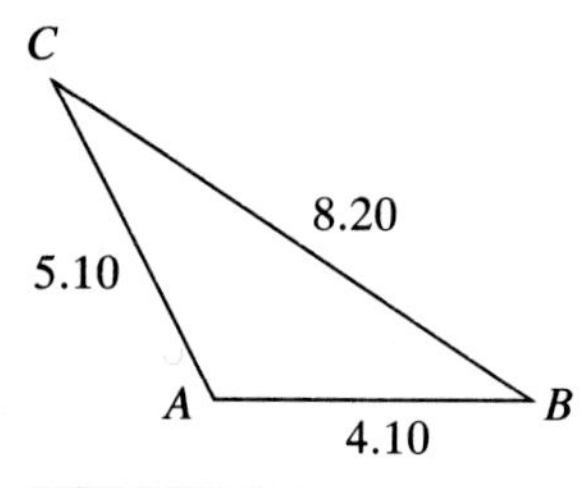

Figure 3.25

Solution The largest and smallest angles lie opposite the largest and smallest sides, respectively.

(1) To find A, use

$$\cos A = \frac{b^2 + c^2 - a^2}{2bc}$$

$$= \frac{5.1^2 + 4.1^2 - 8.2^2}{2(5.1)(4.1)}$$

$$A = \cos^{-1}\left[\frac{5.1^2 + 4.1^2 - 8.2^2}{2(5.1)(4.1)}\right].$$

Using an algebraic logic calculator, in degree mode, we press

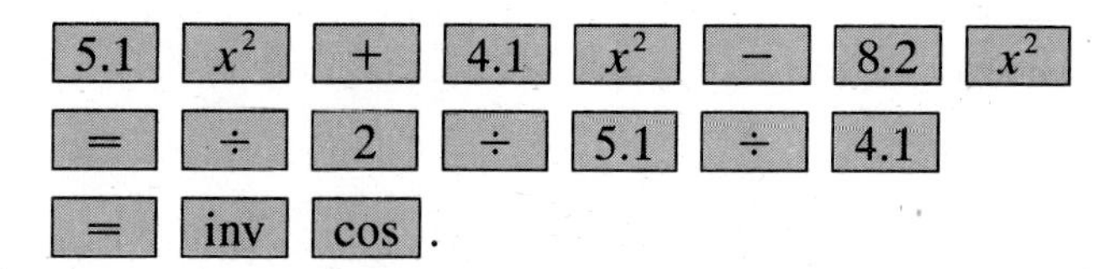

.

When the displayed result is rounded off to the nearest tenth of a degree, we get

$$A = 125.7°.$$

(If $-1 \leq u < 0$, then the calculator is programmed to give the angle in Q II in evaluating $\cos^{-1} u$. See Section 9.2.)

With an RPN calculator in degree mode, press

[5.1] [ENTER] [5.1] [×] [4.1] [ENTER]

[4.1] [×] [+] [8.2]

[ENTER] [8.2] [×] [−]

[2] [ENTER] [5.1] [×] [4.1] [×] [÷]

[inv] [cos].

Rounding off the displayed result to three-figure accuracy, we obtain $A = 125.7°$.

(2) To find C, use

$$\cos C = \frac{a^2 + b^2 - c^2}{2ab} = \frac{8.2^2 + 5.1^2 - 4.1^2}{2(8.2)(5.1)}$$

$$\doteq 0.91392.$$

$$C = 23.9°.$$

Exercise 3.6

Solve the triangles in Exercises 1 through 12.

1. $a = 18$, $c = 23$, $B = 106°$
2. $b = 8.64$, $c = 5.79$, $A = 48.3°$
3. $a = 380$, $b = 460$, $C = 71.4°$
4. $b = 4.4$, $c = 5.5$, $A = 98°$
5. $a = 401$, $b = 372$, $c = 293$
6. $a = 40$, $b = 70$, $c = 50$
7. $a = 9.6$, $b = 6.2$, $c = 4.3$

8. $a = 19.4$, $b = 28.5$, $c = 33.6$
9. $a = 8.310$, $b = 3.770$, $C = 58.10°$
10. $a = 77.88$, $c = 33.55$, $B = 44.99°$
11. $a = 3702$, $b = 3015$, $c = 1122$
12. $a = 417.3$, $b = 913.6$, $c = 748.5$
13. Jerusalem is 22 km S 65° W from Jericho. Bethlehem is 29 km S 51° W from Jericho. How far is Bethlehem from Jerusalem?
14. Copenhagen, Denmark, is 480 km S 13° E from Oslo, Norway. Stockholm, Sweden, is 510 km from Copenhagen and 410 km from Oslo. What is the bearing of Stockholm from Oslo? From Copenhagen? (Stockholm, of course, lies north and east of the Oslo–Copenhagen line.) Ignore the curvature of the earth and assume only two-place accuracy. Get angles to the nearest degree. Get distances to the nearest multiple of 10 km.
15. An outboard-motor race is held over a triangular course, the first leg of which is 700 m in an easterly direction. The second and third legs, of 800 m and 900 m, respectively, lie to the south of the first leg. In what direction do the boats move on the second leg?
16. Discuss the equation $c^2 = a^2 + b^2 - 2ab \cos C$ (a) when C approaches 180°, (b) when C approaches 0°.
17. An airplane is moving with an air speed of 100 mi/h. A north wind (direction 180°), blowing at 20 mi/h, causes the plane to travel with a ground speed of only 90 mi/h. Find the plane's heading and its course.
18. A force of 10 lb acts due south on point A. A force of 9 lb acts in the direction N 60° E on point A. Find the magnitude of the resultant force. (Consider all numbers as exact.)

3.10 THE AREA OF A TRIANGLE

We conclude this chapter by developing three formulas for the area of an arbitrary triangle.

I. The area of a triangle is equal to one-half the product of any two sides times the sine of the included angle:

$$K = \tfrac{1}{2} bc \sin A = \tfrac{1}{2} ca \sin B = \tfrac{1}{2} ab \sin C.$$

Proof Let h be the altitude from C to AB (Figure 3.26).

$$\text{Area } \triangle ABC = K = \tfrac{1}{2} ch = \tfrac{1}{2} bc \sin A,$$

since $h = b \sin A$. The other forms can be obtained by cyclic permutation.

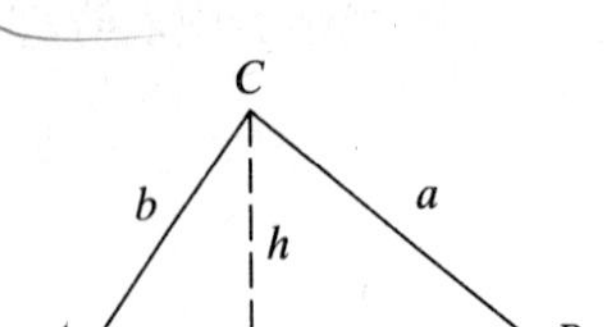

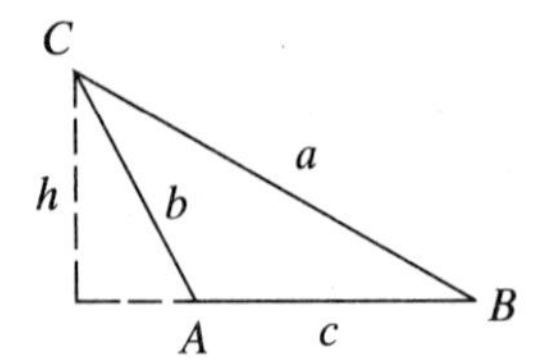

Figure 3.26

II. The area of triangle ABC is

$$K = \sqrt{s(s-a)(s-b)(s-c)},$$

where $s = \frac{1}{2}(a+b+c)$.

Proof Since $K = \frac{1}{2}bc \sin A$,

$$K^2 = \frac{1}{4}b^2c^2 \sin^2 A = \frac{b^2c^2}{4}(1 - \cos^2 A)$$

$$= \frac{bc}{2}(1 + \cos A)\frac{bc}{2}(1 - \cos A).$$

Using the law of cosines, we get

$$K^2 = \frac{bc}{2}\left(1 + \frac{b^2 + c^2 - a^2}{2bc}\right)\frac{bc}{2}\left(1 - \frac{b^2 + c^2 - a^2}{2bc}\right)$$

$$= \frac{2bc + b^2 + c^2 - a^2}{4} \cdot \frac{2bc - b^2 - c^2 + a^2}{4}$$

$$= \frac{(b+c)^2 - a^2}{4} \cdot \frac{a^2 - (b-c)^2}{4}$$

$$= \frac{b+c+a}{2} \cdot \frac{b+c-a}{2} \cdot \frac{a-b+c}{2} \cdot \frac{a+b-c}{2}.$$

Let $$s = \tfrac{1}{2}(a+b+c),$$

then $$\frac{b+c-a}{2} = s - a, \text{ etc.}$$

Hence

$$K^2 = s(s-a)(s-b)(s-c).$$

$$K = \sqrt{s(s-a)(s-b)(s-c)}.$$

III. The area of triangle ABC is

$$K = \frac{a^2 \sin B \sin C}{2 \sin A} = \frac{b^2 \sin C \sin A}{2 \sin B} = \frac{c^2 \sin A \sin B}{2 \sin C}.$$

Proof We start with $K = \frac{1}{2}ab \sin C$ and use the law of sines to replace b with

$$b = \frac{a \sin B}{\sin A}.$$

This gives

$$K = \frac{a^2 \sin B \sin C}{2 \sin A}.$$

This formula should be used when one side and two angles are given.

Exercise 3.7

Find the areas of the triangles in Exercises 1 through 12.

1. $a = 12$, $b = 11$, $C = 30°$
2. $b = 15$, $c = 40$, $A = 161°$
3. $a = 10$, $A = 53°$, $B = 74°$
4. $b = 3.3$, $A = 47°$, $C = 48°$
5. $a = 40$, $b = 85$, $c = 117$ (consider the numbers as exact)
6. $a = 23$, $b = 14$, $c = 13$ (consider the numbers as exact)
7. $a = 20.0$, $c = 30.0$, $B = 40.0°$
8. $a = 64.63$, $b = 87.19$, $C = 107.37°$
9. $c = 1.27$, $B = 27.8°$, $C = 19.3°$
10. $a = 17.33$, $A = 52.68°$, $C = 63.15°$
11. $a = 6.006$, $b = 8.008$, $c = 9.998$
12. $a = 46.2$, $b = 75.8$, $c = 93.6$
13. The area of a triangle is 88.66 m^2. Two of its sides are 32.10 m and 17.82 m. Find the included angle.
14. Determine the number of acres in a triangular field whose sides are 700 ft, 800 ft, and 900 ft. (One acre contains 43,560 ft^2.)
15. Use area formula I to derive an expression for the area of an equilateral triangle. Check your result by using (a) formula II, (b) formula III.

16. Use area formula II to show that the area of an isosceles triangle with sides a, a, and b is $(b/4)\sqrt{4a^2 - b^2}$.
17. Prove that the area of any quadrilateral is equal to one-half the product of its diagonals times the sine of the included angle.

KEY TERMS

Angle of elevation, angle of depression, bearing of a line, grad, vector, resultant, heading, law of sines, law of cosines, area formulas

REVIEW EXERCISES

For each of the eight following problems, (a) draw the triangle, (b) write the equation that should be used in solving for the required part, and (c) find the required part.

1. $b = 7.0$, $c = 8.5$, $A = 34°$; find a.
2. $c = 20$, $A = 75°$, $B = 56°$; find b.
3. $a = 49$, $b = 52$, $c = 38$; find B.
4. $a = 1.7$, $c = 1.9$, $A = 22°$; find C.
5. $b = 436$, $c = 371$, $C = 94.7°$; find B.
6. $a = 88.4$, $b = 70.6$, $c = 42.5$; find A.
7. $a = 6.66$, $A = 57.0°$, $C = 20.7°$; find c.
8. $a = 0.305$, $c = 0.856$, $B = 130.0°$; find b.
9. In a triangle, $A = 78.75°$, $b = 5432$, and the length of the median from C to the midpoint of c is 6040. Find c.
10. In a quadrilateral $PQRS$, $PQ = 20$, $QR = 10$, $RS = 11$, angle $Q = 164°$, and angle $R = 63°$. Find PS.
11. The Leaning Tower of Pisa makes an angle of 5.3° with the vertical. When the angle of elevation of the sun is 22.8°, the shadow of the tower (falling on a horizontal plane) is 409 ft long and appears on the side *from which* the tower leans. Find the distance from the bottom to the top of the tower.
12. A ladder leaning against a building has an angle of elevation of α. After the foot of the ladder has been moved k meters closer to the building, the angle of elevation is β. Show that the number of meters in the length of the ladder is

$$\frac{k}{\cos \alpha - \cos \beta}.$$

13. Monterrey, Mexico, is 710 km N 9° W of Mexico, D.F. Guadalajara, Mexico, is 640 km from Monterrey and 460 km from Mexico, D.F. What is the bearing of Guadalajara from Mexico, D.F.? From Monterrey? (Guadalajara lies west of the line connecting the other two cities.) Get the angles to the nearest degree.
14. Tucson, Arizona, is 260 mi N 83° W from El Paso, Texas. Albuquerque, New Mexico, is due north of El Paso. The bearing of Tucson from Albuquerque is S 52° W. How far is Albuquerque from El Paso? From Tucson? Ignore the curvature of the earth and assume only two-place accuracy. Get distances to the nearest multiple of 10 mi.
15. Amsterdam is 350 km N 74° E from London. Paris is 340 km S 29° E from London. Find the bearing and the distance of Amsterdam from Paris. Ignore the curvature of the earth and assume only two-place accuracy. Get angle to the nearest degree. Get distance to the nearest multiple of 10 km.
16. State College, Pennsylvania, is 210 mi due east of Wooster, Ohio. The bearing of Washington from Wooster is S 64° E. State College is 130 miles from Washington. How far is Washington from Wooster? Ignore the curvature of the earth and assume only two-place accuracy. Get distance to the nearest multiple of 10 mi.
17. Two boys 1085 ft apart on a level plain are gazing northward, watching a drifting cloud that is in the same vertical plane as the boys. The angles of elevation of the cloud as observed by the two boys are 57.50° and 66.50°, respectively. How high is the cloud above the ground?
18. On a hillside that makes an angle of 17° with the horizontal, a 30-ft tree leans downhill. When the elevation of the sun is 51°, the tree's shadow is 15 ft long and falls straight *up* the slope. Find the angle made by the tree with the vertical.
19. In Figure 3.27, express x in terms of a, α, δ, γ, and ϕ.
20. In Figure 3.27, show that

$$x = \csc \alpha \sqrt{a^2 \sin^2 \delta + b^2 \sin^2 \alpha - 2ab \sin \alpha \sin \delta \cos \gamma}.$$

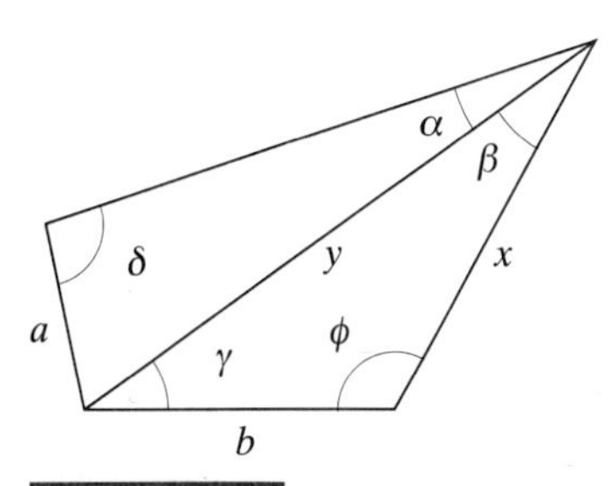

Figure 3.27

21. An island, known to be 3.00 km wide, subtends an angle of 24.8° from point A, 4.00 km from one end of the island. Find the distance of A from the other end of the island.
22. If R is the radius of the circle circumscribed about triangle ABC, prove that

$$2R = \frac{a}{\sin A} = \frac{b}{\sin B} = \frac{c}{\sin C}.$$

(*Hint:* Let A be any acute angle of the triangle and let O be the center of the circle. Connect O with the midpoint M of BC. Then, by geometry, angle MOB is equal to A.)
23. A 50-ft flagpole stands on the top of a 25-ft building. How far from the base of the building should a person stand if the flagpole and the building are to subtend equal angles at her eye, which is 5 ft above the ground? (Consider the numbers as exact.)
24. Prove or disprove: If the sides of a triangle are rational numbers, then the cosines of the angles of the triangle are also rational numbers.
25. The Great Pyramid of Gizeh, Egypt, has a square base. Its faces are congruent isosceles triangles that intersect in its four edges. Before vandals removed the outer limestone casing and the top 31.00 ft of the pyramid, each edge was 719.2 ft long and made an angle of 42.01° with the horizontal. Find the pyramid's original height and the length of a side of its base. Also find the area of the base in acres.
26. In Figure 3.28, prove that

$$\tan \gamma = \frac{\sin \alpha \sin(\alpha + \beta)}{\sin \beta - \sin \alpha \cos(\alpha + \beta)}.$$

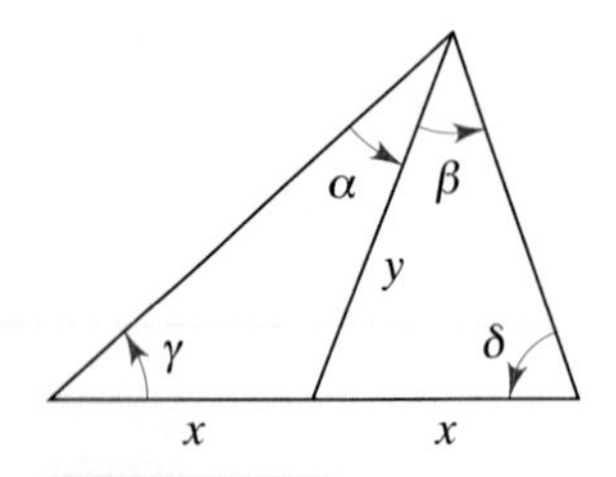

Figure 3.28

4

Radian Measure

4.1 THE RADIAN

Thus far we have employed the *degree* as the unit of measure for angles. It may be thought of as $\frac{1}{360}$ of the angular magnitude about a point. For many practical purposes, the degree is a convenient unit, but most of the applications of the trigonometric functions that require higher mathematics, particularly calculus, are simplified if another unit, the radian, is used. For this reason, students planning to study calculus should make a special effort to learn radian measure thoroughly and to use it, so that it will be familiar to them when they need it.

A radian is an angle that, if its vertex is placed at the center of a circle, intercepts an arc equal in length to the radius of the circle (Figure 4.1).

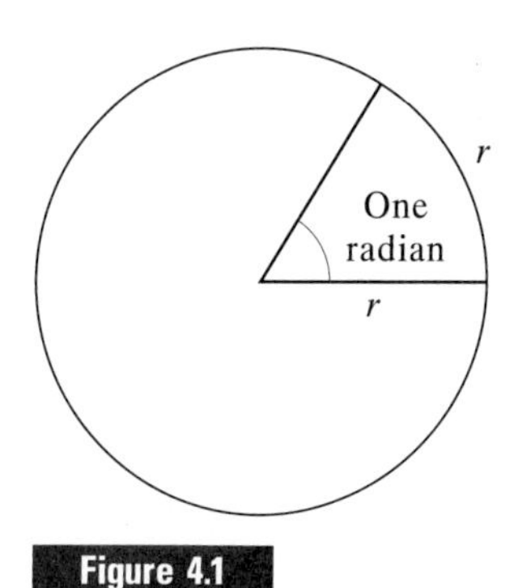

Figure 4.1

According to the definition of a radian, the number of radians in a circle is equal to the number of times the radius can be laid off along the circumference (Figure 4.2). Since $c = 2\pi r$, the number of radians in a circle is $2\pi r/r = 2\pi$. But the number of degrees in a circle is 360°. Hence

$$2\pi \text{ radians} = 360°$$

or

$$\boldsymbol{\pi} \textbf{ radians} = \textbf{180°}. \tag{1}$$

Thus

$$1 \text{ radian} = \frac{180°}{\pi} \doteq \frac{180°}{3.14159} \tag{2}$$

$$\doteq 57.2958° \doteq 57°\ 17.75',$$

where $60' = 60 \text{ minutes} = 1°$.

The symbol $\doteq$ means "is approximately equal to." It is worth observing that 1 radian is a little less than 60°.

Also,

$$1° = \frac{\pi}{180} \text{ radians} \doteq 0.017453 \text{ radian}. \tag{3}$$

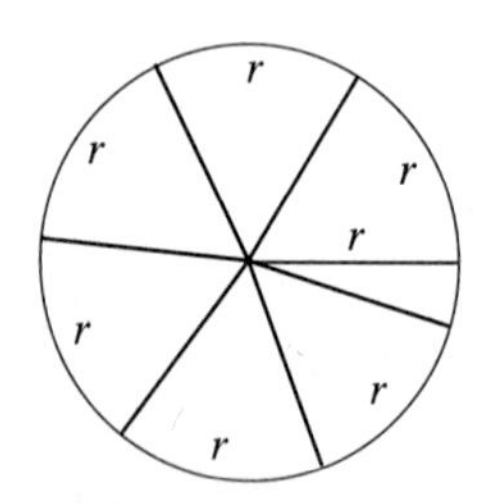

Figure 4.2

It is better not to try to learn Equations (2) and (3). Memorize (1) and derive your results from it. In expressing the more common angles in terms of radians, we usually leave the result in terms of π. Thus $90° = \pi/2$. *When no unit of measure is indicated, it is understood that an angle is expressed in radians.* Thus if we read $\theta = \pi/6$, we understand that $\theta = \pi/6$ radian $= 30°$.

The student should become quite familiar with the radian measure of 30°, 45°, 60°, and 90°, and the angles that are related to them, namely, 120°, 135°, 150°, 180°, 210°, 225°, 240°, 270°, 300°, 315°, and 330°.

Example 1 Express $7\pi/9$ in terms of degrees.

Solution Since

$$\pi = 180°,$$

$$\frac{7\pi}{9} = \frac{7(180°)}{9} = 140°.$$

Example 2 Express 6.75° in terms of radians.

Solution 1 Since

$$180° = \pi,$$

$$1° = \frac{\pi}{180},$$

$$6.75° = 6.75\left(\frac{\pi}{180}\right) = \frac{675\pi}{18{,}000} = \frac{3\pi}{80}.$$

This is the exact result. If π is replaced by 3.1416, we get

$$6.75° = 0.1178,$$

correct to four decimal places.

Solution 2 Some calculators have [DRG ▸] key which converts from degrees to radians to grads. With the calculator in the degree mode, enter 6.75 and then press the [DRG ▸] key. The result is 0.1178097 radians. Pressing [DRG ▸] again yields 7.5 grads.

With a [DRG ▸] key one can easily convert between the three measurement systems. On some calculators, one may have to press the [2nd] key to activate the [DRG ▸] conversion.

Example 3 Express 2 radians in terms of degrees.

Solution 1 Since

$$\pi \text{ radians} = 180°,$$

$$1 \text{ radian} = \frac{180°}{\pi},$$

and

$$2 \text{ radians} = \frac{360°}{\pi}.$$

This exact result may be approximated by the decimal form

$$\frac{360°}{3.1416} \doteq 114.59°,$$

which is correct to a hundredth of a degree.

Solution 2 Since

$$\begin{aligned} 1 \text{ radian} &= 57.2958°, \\ 2 \text{ radians} &= 2(57.2958°) \\ &= 114.5916° \to 114.59°. \end{aligned}$$

Example 4 Evaluate csc $7\pi/6$.

Solution

$$\begin{aligned} \csc \frac{7\pi}{6} &= \csc \frac{7(\overset{30°}{\cancel{180°}})}{\cancel{6}} \\ &= \csc 210° = \frac{1}{\sin 210°} = \frac{1}{-\sin 30°} \\ &= \frac{1}{-\frac{1}{2}} = -2. \end{aligned}$$

Calculator Solution Set the switch for radians. With an AOS, press

[7] [×] [π] [=] [÷] [6] [=] [sin] [1/x]

to get the result, −2. With RPN, press

[7] [ENTER] [π] [×] [6] [÷] [sin] [1/x]

to get −2.

Exercise 4.1

Express in degrees. (The given angle is understood to be in radians.) Do not use a calculator.

1. $\frac{\pi}{2}$
2. $\frac{\pi}{3}$
3. $\frac{\pi}{4}$
4. $\frac{\pi}{6}$

5. $\frac{\pi}{5}$ 6. $-\frac{7\pi}{9}$ 7. $\frac{21\pi}{10}$ 8. $\frac{19\pi}{18}$
9. $\frac{10\pi}{9}$ 10. $\frac{9\pi}{2}$ 11. -5π 12. $\frac{11\pi}{10}$
13. $\frac{15\pi}{8}$ 14. $\frac{17\pi}{20}$ 15. $\frac{13\pi}{18}$ 16. $\frac{10\pi}{3}$

Convert from radians to degrees; write the result to the nearest hundredth of a degree.

17. 5.7 18. 3.0 19. 0.866 20. -9.19

Express in radians, leaving the result in terms of π. Do not use a calculator.

21. 240° 22. 135° 23. 30° 24. 300°
25. $-600°$ 26. 350° 27. 315° 28. 198°
29. 585° 30. 3600° 31. 800° 32. 405°

Convert to radian measure, obtaining the result to four decimal places.

33. 269.77° 34. 20.20° 35. 348.95° 36. 83.06°

Find the exact values of the following. (The instructor may wish to permit the use of a calculator.)

37. $\cos \frac{\pi}{6}$ 38. $\sin\left(-\frac{\pi}{4}\right)$ 39. $\sin \frac{4\pi}{3}$ 40. $\cos\left(-\frac{5\pi}{3}\right)$
41. $\sin \frac{5\pi}{4}$ 42. $\cos \frac{2\pi}{3}$ 43. $\cos \frac{5\pi}{6}$ 44. $\sin \frac{7\pi}{6}$
45. $\tan \frac{11\pi}{3}$ 46. $\cot \frac{11\pi}{6}$ 47. $\tan\left(-\frac{7\pi}{4}\right)$ 48. $\sec \frac{3\pi}{4}$
49. $\sec(-5\pi)$ 50. $\tan \frac{3\pi}{2}$ 51. $\csc \frac{5\pi}{2}$ 52. $\cot \frac{\pi}{2}$

Evaluate the following to four decimal places. (The angle is understood to be in radians.)

53. cos 1.0001 54. sin 1.4242
55. tan 0.3700 56. cos 0.5533

57. Through how many radians does the minute hand of a clock rotate in 50 min? In 7 h?
58. Express in radians the smaller angle made by the hands of a clock at 4:00. At 2:30. At 10:12.
59. One angle of a triangle is $3\pi/10$. Another angle is 24°. Express the third angle in radians.
60. Through how many radians does the hour hand of a clock rotate in 8 h? In 1 week?

4.2 LENGTH OF A CIRCULAR ARC

In a circle of radius r, let an arc of length s be intercepted by a central angle θ. Then as we see in Figure 4.3,

$$s = r\theta, \qquad \textbf{where } \theta \textbf{ is in radians.}$$

To prove this, recall that in any circle the arc intercepted by a central angle is proportional to this angle. Hence if 1 radian intercepts an arc equal to the radius, then θ radians intercept an arc equal to θ times the radius. Remember that *the central angle must be measured in radians.*

In the equation $s = r\theta$, if two of the quantities s, r, θ are known, the third can be found.

Example A circle has a radius of 100 cm. (a) How long is the arc intercepted by a central angle of 72°? (b) How large is the central angle that intercepts an arc of 30 cm?

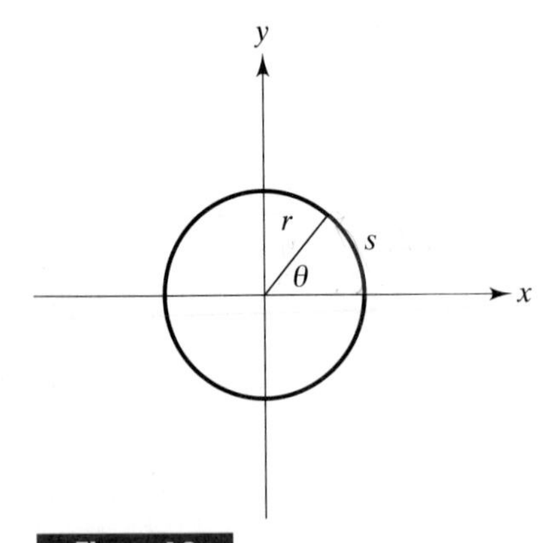

Figure 4.3

Solution* (a) Since

$$180° = \pi \text{ radians},$$

$$1° = \frac{\pi}{180} \text{ radian},$$

and

$$72° = 72 \cdot \frac{\pi}{180} \text{ radians} = \frac{2\pi}{5} \text{ radians}.$$

Using

$$s = r\theta, \qquad \text{where } \theta \text{ is in radians},$$

$$s = 100 \cdot \frac{2\pi}{5} = \mathbf{40\pi\ cm} \qquad (\text{or } \mathit{126\ cm}).$$

(b) Using

$$s = r\theta,$$

$$30 = 100\theta,$$

$$\theta = \frac{30}{100} = \frac{\mathbf{3}}{\mathbf{10}} \textbf{ radian.}$$

The color results are exact; the italicized answer is merely a three-figure approximation.

4.3 THE WRAPPING FUNCTION

Up to this point, most of our discussions have dealt with the trigonometric functions of *angles*. For example, the sine function was defined as sin = $\{(\theta, \sin\theta)\}$, where θ is a certain number of degrees. We shall now show that the six trigonometric functions may be defined as functions of real numbers—with no reference whatsoever to angles or triangles. To this end we introduce the *wrapping* function.

*In all discussions and problems in this chapter, all figures are to be considered as exact. They are not approximations. For the sake of uniformity, write all approximate results with three-figure accuracy. It may be convenient to use $1/\pi = 0.3183$.

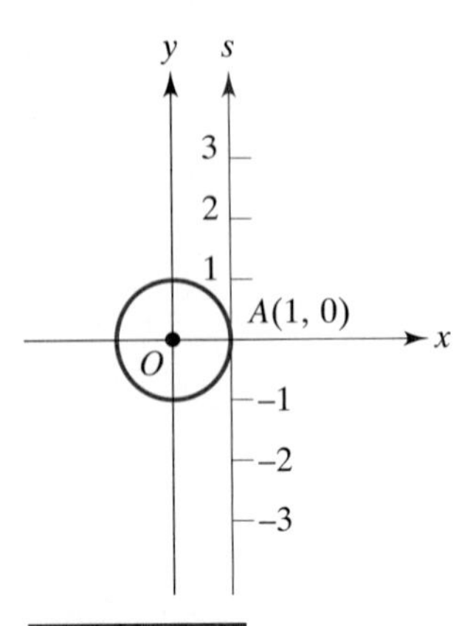

Figure 4.4

Consider the graph of a unit circle (radius 1) with center at the origin (Figure 4.4). The equation of this circle is $x^2 + y^2 = 1$. We intend to show that every real number can be associated with a unique point on the circumference of this circle. We begin by drawing a real-number line (the s axis) tangent to the unit circle at $A(1, 0)$, using OA as the unit and A as the zero point, and making the positive direction upward (Figure 4.4). Now imagine that this real-number line is completely flexible (like a tape measure). Think of wrapping the *positive* part of the s axis in a *counterclockwise* direction about the unit circle, and wrapping the negative part about the circle in a clockwise direction (Figure 4.5).

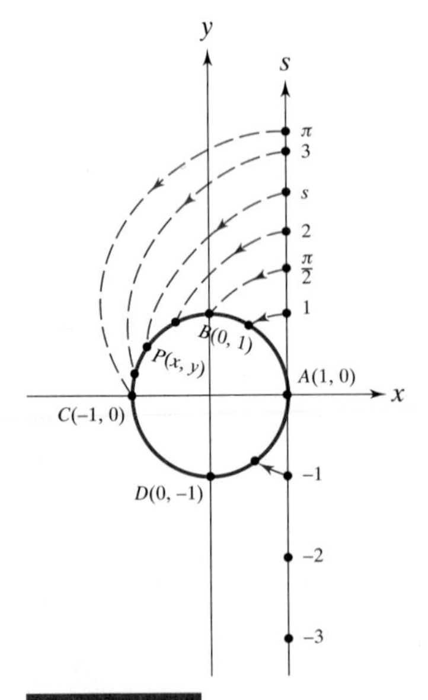

Figure 4.5

Obviously the number 0 on the s axis is associated with the point A on the unit circle. Using $c = 2\pi r$, we find the circumference of the circle to be 2π. Hence the real number $s = \pi/2$ (which is $\frac{1}{4}$ of the circle's circumference) is associated with the point $B(0, 1)$, the real number π is associated with the point $C(-1, 0)$, and the number $3\pi/2$ is associated with the point $D(0, -1)$. Because the circle's circumference is 2π, the portion of the s axis from 0 to 2π will wrap the circle one time, the portion of the s axis from 2π to 4π will wrap the circle a second time, etc. Each of the numbers 0, 2π, 4π, 6π, etc., is associated with the point $A(1, 0)$. In general, if the real number s is associated with the point P, then all the real numbers $s + 2\pi k$, where k is an integer (such as $s + 2\pi, s + 4\pi, \ldots, s - 2\pi, s - 4\pi, \ldots$), are associated with P. However, each positive real number is associated with one and only one point on the unit circle.

In similar fashion, wrapping the negative part of the s axis in a clockwise direction about the unit circle enables us to associate each negative number with a unique point on the circle.

We define the wrapping function as

$$W = \{[s, (x, y)]\},$$

where s is any real number and x and y are the coordinates of the associated point on the unit circle.

4.4 THE CIRCULAR FUNCTIONS

If s is a real number that is associated with the point $P(x, y)$ on the unit circle, the *circular* functions are defined as follows:

$$\begin{aligned}
\sin s &= y, \\
\cos s &= x, \\
\tan s &= \frac{y}{x}, \\
\cot s &= \frac{x}{y}, \\
\sec s &= \frac{1}{x}, \\
\csc s &= \frac{1}{y}.
\end{aligned}$$

Thus the circular functions are defined as functions of real numbers.

Because the s axis is wrapped about the unit circle, we know that if the real number s is associated with the point $P(x, y)$ on the unit circle (Figure 4.6), then the length of the arc from A to P is s. In order to establish a relationship between the circular functions and the trigonometric functions of an angle expressed in radians, we shall use the equation $s = r\theta$, θ in radians, to observe that the radian measure of angle AOP is $s/r = s/1 = s$. Using the definitions of the trigonometric functions of a general angle (Section 1.5), we find that

$$\text{The sine of angle } AOP = \frac{y}{r} = \frac{y}{1} = y = \sin s.$$

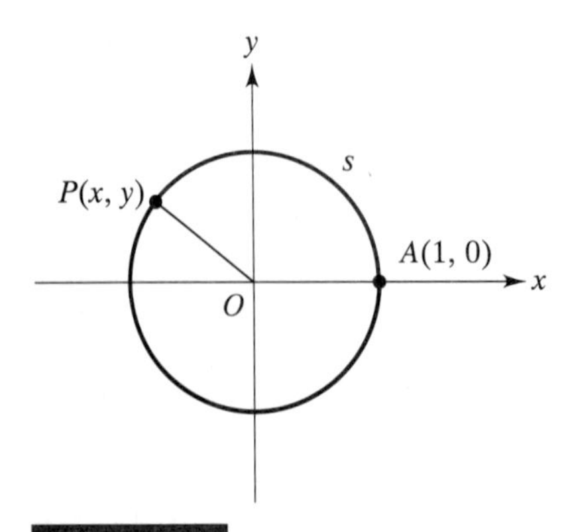

Figure 4.6

Hence $\sin s$ (the sine of the real number s) $= \sin s$ (radians). Similar statements can be made for the other five trigonometric functions. In other words, if θ is an angle with radian measure s, then any trigonometric function of θ is equal to the corresponding circular function of the real number s. Consequently, every identity in this text is valid for real numbers as well as for angles. Thus $\sin^2 s + \cos^2 s = 1$ holds true for every real number s.

Example Use the definition of the circular functions to find the values of $\sin s$ and $\cos s$ when (a) s is the real number $\pi/4$, (b) s is the real number $-3\pi/4$.

Solution (a) In Figure 4.7, the real number 0 corresponds to $A(1, 0)$ and the real number $\pi/2$ corresponds to $B(0, 1)$, since $\pi/4$ is halfway from 0 to $\pi/2$, it corresponds to a point C on the circular arc halfway from A to B. By symmetry the coordinates of C are equal and positive. This means that $x = \cos \pi/4 = \sin \pi/4 = y$ and that $\sin^2 \pi/4 + \cos^2 \pi/4 = x^2 + x^2 = 2x^2 = 1$, so $x^2 = \frac{1}{2}$ and $x = \sqrt{2}/2$. Thus $\sin \pi/4 = \cos \pi/4 = \sqrt{2}/2$.

(b) By similar reasoning, the real number $-3\pi/4$ corresponds to $D(y, y)$ where $y < 0$. It follows that $y = -\sqrt{2}/2$ and hence if s is the real number $-3\pi/4$, then $\sin s = \cos s = -\sqrt{2}/2$.

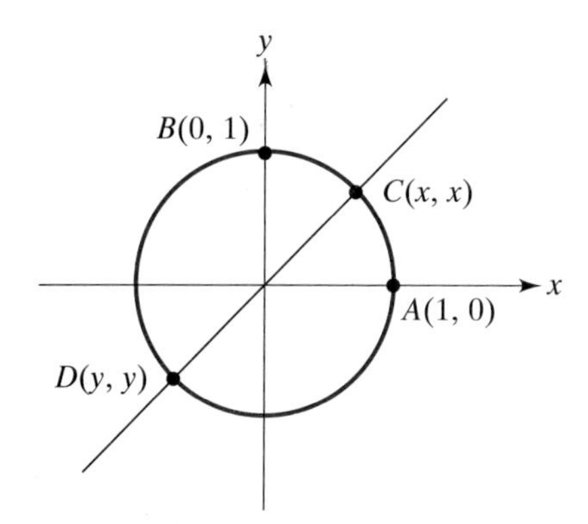

Figure 4.7

Exercise 4.2

Unless directed to the contrary, round off approximate results to three-figure accuracy.

1. Find the radius of a circle on which (a) a central angle of 2.3 radians intercepts an arc of 6.0 cm, (b) a central angle of 57° intercepts an arc of 20.0 in.
2. On a circle of diameter 4.80 m, how long is the arc intercepted by a central angle of (a) $\frac{3}{4}$ radian, (b) 10°, (c) 1.35π radians.
3. Find the number of radians in the central angle that intercepts an arc of 56.2 cm on a circle of diameter 41.4 cm. Express the angle in degrees.
4. A locomotive wheel turns through 310° as it moves a distance of 4.13 m. Find the diameter of the wheel.
5. A pendulum of length 60.0 cm swings through an arc of 18.0 cm. Through what angle, in degrees, does it swing?
6. Use the definitions to find the values of the circular functions sin, cos and tan at the real numbers -2π, $-\pi$, 0, and $3\pi/4$.

In Exercises 7 through 12, assume that the earth is a sphere of radius 4000 mi. Write results with two-figure accuracy.

7. Find the distance from the equator to Syracuse, New York (latitude 43° N). How far is Syracuse from the North Pole? (See Figure 4.8.)

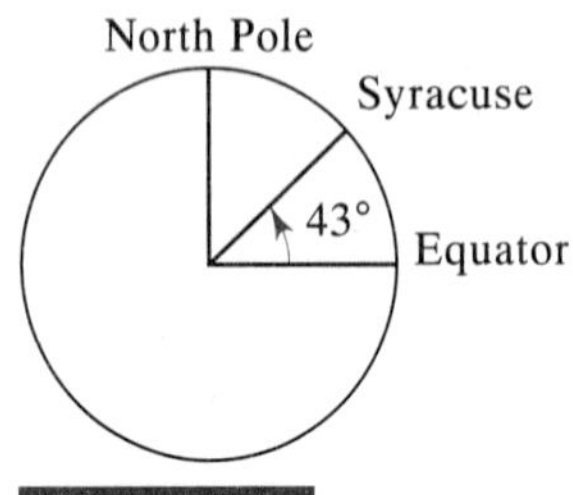

Figure 4.8

8. Alta Loma, California, is 2400 mi from the equator. Find the latitude of Alta Loma.
9. Billings, Montana (latitude 46° N) is due north of Easter Island (latitude 27° S). How far is Billings from Easter Island?
10. Kearney, Nebraska (latitude 41° N) is 1540 mi north of Mexico City. Find the latitude of Mexico City.
11. The North Pole is 140 mi closer to Barcelona, Spain, than to St. Louis, Missouri. If the latitude of Barcelona is 41° N, find the latitude of St. Louis.
12. The latitude of the Cape of Good Hope, at the southern tip of Africa, is 34° S. The latitude of Cape Horn, at the southern tip of South America, is 56° S. How much closer to the South Pole is Cape Horn?

In Exercises 13 and 14, assume that the earth is a sphere of radius 6400 km. Write results with two-figure accuracy.

13. Giamame, Somalia, and Pontiana, Borneo, are on the equator. Pontiana is 7400 km east of Giamame (longitude 43° E). Find the longitude of Pontiana. 43° E means 43° east of the prime meridian.
14. Quito, Equador (longitude 78.5° W), and Thomson's Falls, Kenya (longitude 36° E), are located near the equator. Find the distance between the cities in kilometers. 78.5° W means 78.5° west of the prime meridian.

In Exercises 15 through 18, since the angle is small, we can assume, with little error, that the chord is equal to its subtended arc.

15. A railroad boxcar, known to be 4 m high, subtends an angle of 2° at the eye of an observer on level ground. How far is the boxcar from the observer? (See Figure 4.9.)

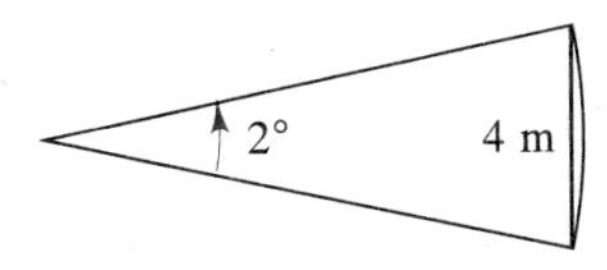

Figure 4.9

16. The moon as viewed from the earth subtends an angle of about 0.517°. If the moon's diameter is 2160 mi, find the distance from the earth to the moon.
17. The sun as seen from the earth subtends an angle of 0.532°. If the sun is 93,000,000 mi from the earth, find the sun's diameter.
18. With minor restrictions, a person with normal vision can, without strain, read print if it subtends an angle of at least 0.17° at his eye. Could such a person read $\frac{1}{2}$-in. print at a distance of 12 ft?
19. In Figure 4.10, if $AB = 9$, find the length of arc RST. Assume that all given distances are exact. Write your result with three-figure accuracy.

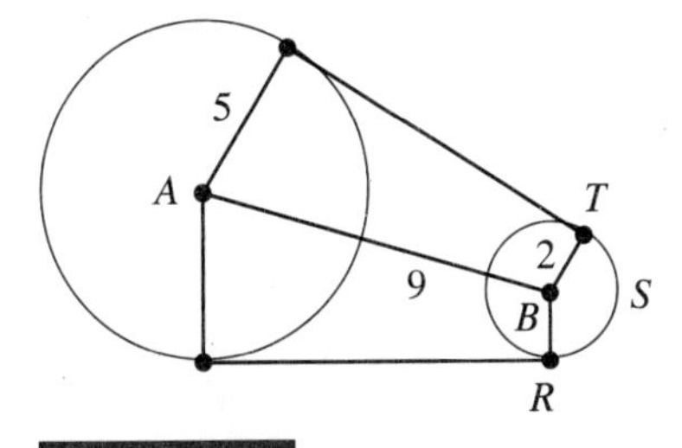

Figure 4.10

20. On a circle of radius 20.0 cm, the length of arc AB is 38.4 cm. If the tangent lines at A and B intersect at C, find the length of CB.
21. An equilateral triangle is inscribed in a circle. Find the length of a side of the triangle if each side subtends an arc of 28.0 cm.
22. In Figure 4.11, angle ABC is called an inscribed angle, whereas angle AOC is a central angle. By geometry, the inscribed angle that intercepts arc AC is equal to one-half of the central angle that intercepts arc AC. Find the radius of a circle on which an inscribed angle of 1.10 radians intercepts an arc of 72.6 cm.

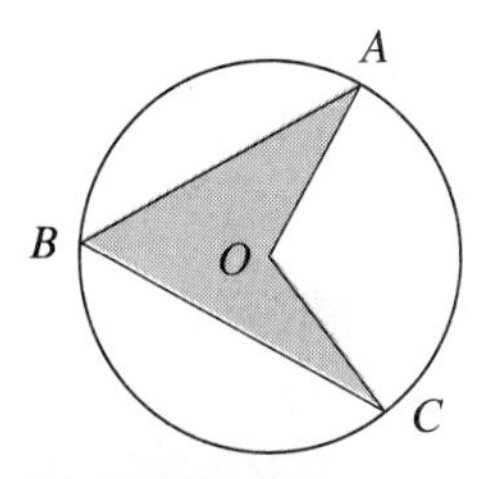

Figure 4.11

23. A sector of a circle is the part of the circle bounded by an arc and the radii drawn to its extremities (Figure 4.12). By geometry, the area of a sector is equal to one-half its arc times the radius of the circle. Show that

$$\textbf{Area of sector } OAB = \tfrac{1}{2}r^2\theta, \qquad \textbf{where } \theta \textbf{ is in radians.}$$

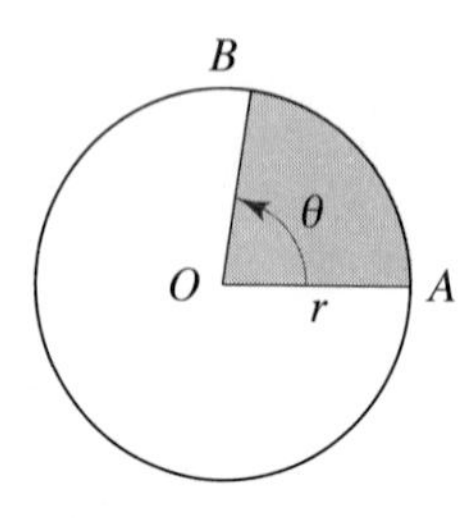

Figure 4.12

4.5 LINEAR AND ANGULAR VELOCITY

Consider a point P moving with constant speed on the circumference of a circle with radius r and center O. If P traverses a distance of s linear units (meters, feet, miles, etc.) in t time units (seconds, minutes, etc.), then $s/t = v$ is called the linear velocity of P. If the radius OP swings through θ angular units (degrees, radians, etc.), in t time units, then $\theta/t = \omega$ is called the angular velocity of P. If, further, θ is in radians and ω is in radians per unit of time, then we can divide $s = r\theta$ by t, and get

$$\frac{s}{t} = r \cdot \frac{\theta}{t},$$

or

$$v = r\omega,$$

provided that ω is in radians per time unit. This means that the linear velocity of a point on the circumference of a circle is equal to the radius times the angular velocity of the point, in radians per unit of time.

The angular velocity of a rotating body is quite often expressed in revolutions per minute (rpm). This can be readily converted into radians per minute, by remembering that one revolution represents 2π radians.

Example 1 A flywheel 6 ft in diameter makes 40 rpm. (a) Find its angular velocity in radians per second. (b) Find the speed of the belt that drives the flywheel.

Solution (a) Since 40 rpm represents $40(2\pi) = 80\pi$ radians per minute,

$$\omega = \frac{80\pi}{60} = \frac{4\pi}{3} \text{ radians per second.}$$

(b) The speed of the belt, if it does not slip, is equal to the linear velocity of a point on the rim of the flywheel. Using $v = r\omega$, we have

$$v = 3\left(\frac{4\pi}{3}\right) = 4\pi \text{ ft/s} \doteq 12.6 \text{ ft/s}.$$

Example 2 A VHS tape cartridge has 246 m of magnetic tape and records for 2 h on the SP speed. Each spool has a radius of 10 mm. Find the angular velocity in rpm of the empty takeup spool at the beginning of a recording.

Solution We know that $\omega = v(1/r)$, so that upon observing that $v =$ 123 m/h and multiplying by unity in various forms to convert to rpm we get

$$\omega = \frac{123 \text{ m}}{\text{hour}} \cdot \frac{1}{.01 \text{ m}} = \frac{12{,}300 \text{ radians}}{\text{hour}} \cdot \frac{1 \text{ rev}}{2\pi \text{ radians}}$$

$$= \frac{12{,}300 \text{ rev}}{2\pi \text{ h}} \cdot \frac{1 \text{ h}}{60 \text{ min}} = \frac{12{,}300 \text{ rev}}{376.99 \text{ min}}$$

$$= 325 \text{ rpm.}$$

Exercise 4.3

Unless directed to the contrary, round off approximate results to three-figure accuracy.

1. The wheel of a turbine makes 2100 rpm. What is its angular velocity in radians per second?
2. How many revolutions per minute are made by a wheel that has an agular velocity of 30,000 radians per hour?

3. The wheels of a truck are making 630 rpm as the truck travels at 60 mi/h. Find the diameter of the wheels in inches.
4. Find the linear speed, due to the rotation of the earth, of a point on the equator. Use 3960 mi as the radius at the equator. Express the result in miles per hour.
5. Find the linear speed, due to the rotation of the earth, of the State Tower in Syracuse (latitude 43° N). Use 3960 mi as the radius of the earth. (See Figure 4.8.) Express the result in miles per hour.
6. Two flywheels with diameters of 40 in and 60 in are driven by a moving belt. Find the speed of the belt in feet per second, and find the angular velocity of the larger wheel in revolutions per minute when the smaller wheel makes 48 rpm.
7. If the VHS cartridge in Example 2 is recording on EP (6 hours of recording) and if the full spool has a radius of 40 mm, including the tape, what is the angular velocity of the full spool at the beginning of a recording?
8. The diameters of the front and rear tires of a tractor are 36 in. and 64 in., respectively. Find the number of revolutions per minute of the front and rear tires when the tractor is traveling 9 mi/h.
9. A cogwheel is driven by a chain that travels at 61 cm/s. Find the diameter of the wheel if it makes 72 rpm.
10. A bicycle is driven by the pedals, which are attached to the sprocket wheel, which has a diameter of 10 in. A chain connects the sprocket to a smaller cogwheel, whose diameter is 3 in. This wheel is fastened to the bicycle's rear wheel, whose diameter is 22 in. Find the speed of the bicycle in miles per hour when the sprocket is making 1 rps (revolution per second).
11. On rewinding a VHS tape, the takeup spool rotates at 540 rpm. Find the speed of the tape in meters per minute and the angular velocity of the other spool when the takeup spool has a 20-mm thickness of tape on its 10-mm radius and the other spool has a 10-mm thickness of tape on it.
12. A rolling wheel travels at 15 km/h while turning at the rate of 3 revolutions per second (rps). Find the number of centimeters in the diameter of the wheel.
13. An earth satellite travels in a circular orbit of radius 4250 mi. Find the speed of the satellite if it makes one revolution around the earth every 1.5 h.
14. A satellite travels in a circular orbit that is concentric with the earth. If the satellite's speed is 13,200 mi/h and it orbits the earth (makes one revolution around the earth) in 2 h, how high is the satellite above the earth? Assume that the earth is a sphere of radius 4000 mi.

KEY TERMS

Radian, arc length, central angle, wrapping function, circular function, angular velocity, linear velocity

REVIEW EXERCISES

1. Convert 270° and −300° to radians.
2. Convert $11\pi/9$ and $7\pi/13$ to degrees.
3. Find exact values of (a) $\sin 2\pi/3$ (b) $\cos -\pi/4$ (c) $\tan \pi/6$.
4. Through how many radians does the hour hand of a clock rotate in a day? If the hand is 6 in long, how far does the tip travel in a week? What is the linear velocity in feet per second of a bug on the tip of the hand?
5. If the radius of the earth is 4000 mi, how far is Fort Worth, Texas (latitude 33° N), from the equator? How far is Fort Worth from the North Pole? Fort Worth is due south of Winnipeg, Manitoba (latitude 50° N). What is the distance between them?
6. A long-playing record has grooves from $2\frac{1}{2}$ to $5\frac{1}{2}$ in. from the center and rotates at $33\frac{1}{3}$ rpm on the turntable. How fast, in feet per second, is the groove passing the stylus at the beginning of a record? How fast is it passing at the end?

5

Trigonometric Identities

5.1 THE FUNDAMENTAL RELATIONS

In Section 1.7 we discussed the problem of determining all the trigonometric functions of an angle if one of them is given. This was a geometric process involving the construction of a triangle of reference for the angle. We shall now consider purely analytic relations among the functions themselves. These relations are of considerable importance in other branches of mathematics as well as in engineering and physics. Very often, the key to solving some important problem will be your ability to replace a given mathematical expression with an equivalent one that is more readily usable. Many such problems will involve the trigonometric functions. Consequently, the basic transformations that you will learn to use in this chapter are among the most important things that you should carry with you from this course into more advanced work in mathematics and science.

For any number θ, the following eight fundamental relations are true:

$$\csc \theta = \frac{1}{\sin \theta} \tag{1}$$

$$\sec \theta = \frac{1}{\cos \theta} \tag{2}$$

$$\cot \theta = \frac{1}{\tan \theta} \tag{3}$$

$$\tan \theta = \frac{\sin \theta}{\cos \theta} \tag{4}$$

$$\cot \theta = \frac{\cos \theta}{\sin \theta} \tag{5}$$

$$\sin^2 \theta + \cos^2 \theta = 1 \tag{6}$$

$$1 + \tan^2 \theta = \sec^2 \theta \tag{7}$$

$$1 + \cot^2 \theta = \csc^2 \theta \tag{8}$$

Relations (1) through (8) are true for all values of θ for which the functions actually exist. For example, (4) has no meaning if $\theta = \pi/2$. In that event, $\cos \pi/2 = 0$ and $\tan \pi/2$ is undefined.

The first three relations were discussed in Section 1.6. To prove (4), observe from Figure 5.1 that for any θ such that $\cos \theta \neq 0$,

$$\frac{\sin \theta}{\cos \theta} = \frac{y/r}{x/r} = \frac{y}{x} = \tan \theta .$$

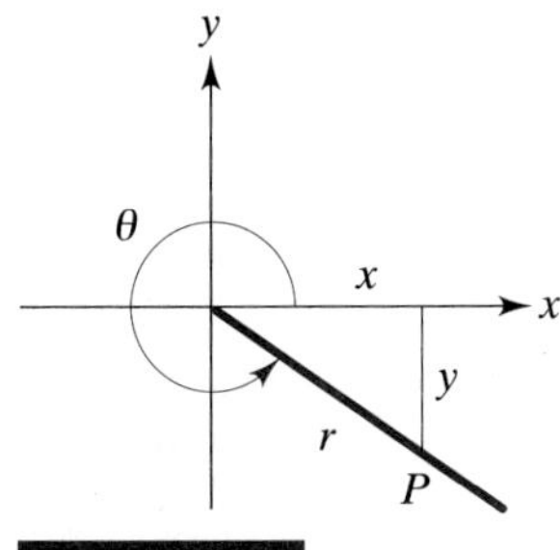

Figure 5.1

Relation (5) follows from (3) and (4).

We have already proved the very important relation (6) in Section 1.6. We may easily derive (7) from (6) by dividing both sides by $\sin^2 \theta$ to get

$$\frac{\sin^2 \theta}{\sin^2 \theta} + \frac{\cos^2 \theta}{\sin^2 \theta} = \frac{1}{\sin^2 \theta}$$

or
$$1 + \tan^2 \theta = \sec^2 \theta .$$

Similarly, division of (6) by $\cos^2 \theta$ yields (8).

The student should be able to recognize these eight fundamental relations in other forms. For example,

(1) may be reduced to $\sin \theta \csc \theta = 1$.
(2) may be reduced to $\cos \theta = 1/\sec \theta$.
(6) may be reduced to $\sin \theta = \pm\sqrt{1 - \cos^2 \theta}$.
(6) may be reduced to $\cos \theta = \pm\sqrt{1 - \sin^2 \theta}$.
(7) may be reduced to $\sec \theta = \pm\sqrt{1 + \tan^2 \theta}$.

The sign that should be chosen in the last three equations is determined by the quadrant in which θ lies. In Q I and Q IV, $\cos\theta = \sqrt{1 - \sin^2\theta}$; but $\cos\theta = -\sqrt{1 - \sin^2\theta}$ in Q II and Q III.

Again the student is warned against the careless habit of writing *sin* instead of *sin θ*. Equation (6) says that the square of the sine of any angle plus the square of the cosine of *that same angle* is equal to 1. It could just as well have been written

$$\sin^2 A + \cos^2 A = 1\,,$$

or

$$\sin^2 7B + \cos^2 7B = 1\,.$$

Example 1 Prove or disprove:

$$\sin^4 5A + 2\sin^2 5A\cos^2 5A + \cos^4 5A = 1\,.$$

Solution Since the given equation involves only sines and cosines, we shall attempt to derive it from (6), which says that for all values of θ,

$$\sin^2\theta + \cos^2\theta = 1\,.$$

Square both sides of the equation:

$$(\sin^2\theta + \cos^2\theta)^2 = 1^2,$$

$$\sin^4\theta + 2\sin^2\theta\cos^2\theta + \cos^4\theta = 1\,.$$

Now let $\theta = 5A$:

$$\sin^4 5A + 2\sin^2 5A\cos^2 5A + \cos^4 5A = 1\,.$$

This proves that the statement is true for all values of A.

Example 2 Prove or disprove:

$$\sin A + \cos A = 1\,.$$

Solution We can demonstrate that this equation is not generally true by setting A equal to some specific angle and then showing that the two sides of the equation are not numerically equal. Choosing $A = \pi/6$, we have

$$\sin\frac{\pi}{6} + \cos\frac{\pi}{6} = 1,$$

$$0.5 + 0.866 = 1,$$

$$1.366 = 1,$$

which is clearly false.

This proves conclusively that the given equation is not true for all values of A. It is, however, true for some values of A; for example, if $A = \pi/2$, we have

$$\sin\frac{\pi}{2} + \cos\frac{\pi}{2} = 1,$$

$$1 + 0 = 1,$$

which is true.

Example 3 Simplify $\sqrt{\csc^2 3B - 1}$.

Solution Since $1 + \cot^2 3B = \csc^2 3B$, it follows that $\cot 3B = \pm\sqrt{\csc^2 3B - 1}$. This implies that $\sqrt{\csc^2 3B - 1} = \pm\cot 3B$. If $3B$ is an angle in Q I or Q III, then $\cot 3B > 0$. In this case $\sqrt{\csc^2 3B - 1} = \cot 3B$. But if $3B$ is in Q II or Q IV, then $\cot 3B < 0$. By definition, $\sqrt{\csc^2 3B - 1}$ means the *principal* square root of $\csc^2 3B - 1$. It is positive or zero but never negative. In this case $\sqrt{\csc^2 3B - 1} = -\cot 3B$. Hence

$$\sqrt{\csc^2 3B - 1} = \cot 3B, \quad \text{if } 3B \text{ is in Q I or Q III},$$

and

$$\sqrt{\csc^2 3B - 1} = -\cot 3B, \quad \text{if } 3B \text{ is in Q II or Q IV}.$$

The student should note carefully that a general statement can be *disproved* by citing one instance in which it is not true. Such an instance is called a *counterexample.* But a general statement cannot be proved by merely showing that it is true for one or more special cases. It must be proved for all cases.

Exercise 5.1

Use the eight fundamental relations to write each of the following expressions as a single trigonometric function of some angle.

1. $\dfrac{\cos 2B}{\sin 2B}$
2. $\dfrac{\sin 2\pi/5}{\cos 2\pi/5}$
3. $\dfrac{1}{\sec 3C}$
4. $\dfrac{1}{\csc(A + B)}$
5. $\sqrt{1 - \sin^2 7\pi/10}$
6. $\sqrt{1 + \tan^2 A}$
7. $-\sqrt{\csc^2 6\pi/5 - 1}$
8. $\sqrt{1 - \cos^2 9\pi/4}$
9. $\cos 7\pi/5 \tan 7\pi/5$
10. $-\sqrt{1 + \cot^2 17\pi/10}$

Decide which of the following statements are valid consequences of the eight fundamental relations. If the statement is true, cite proof; if it is false, correct it.

11. $7 \cot^2 A = \dfrac{1}{7 \tan^2 A}$
12. $\tan B = \dfrac{1}{\cot A}$
13. $\cos A = \dfrac{1}{\csc A}$
14. $\sin \theta = \dfrac{1}{\cos \theta}$
15. $(3 \sin B \csc B)^2 = 9$
16. $\dfrac{\cos 7B}{\sin 7B} = \cot B$
17. $(2 \cos \theta)^3 = \dfrac{8}{\sec^3 \theta}$
18. $\dfrac{1}{\csc \theta} = \sin \theta$
19. $\sin B + \cos B = 1$
20. $\dfrac{\sin^5 \theta}{(\cos \theta)^5} = \tan^5 \theta$
21. $1 + 2 \tan^2 \theta + \tan^4 \theta = \sec^4 \theta$
22. $\cot^2 3\theta = \dfrac{\cos^2 3\theta}{\sin^2 3\theta}$
23. $(4 \sec A)^2 - (4 \tan A)^2 = 16$
24. $(\cot^2 \theta - \csc^2 \theta)^3 = -1$
25. $\csc \theta = -\sqrt{\cot^2 \theta + 1}$ holds for θ in Q III and Q IV
26. $\sin \theta = \sqrt{1 - \cos^2 \theta}$ holds for θ in Q I and Q IV
27. $\cos \theta = -\sqrt{1 + \sin^2 \theta}$ holds for θ in Q II and Q III
28. $\tan \theta = \sqrt{\sec^2 \theta - 1}$ holds for θ in Q I and Q IV
29. $\sqrt{\tan^2 100°} = \tan 100°$
30. $\sqrt{\tan^2 200°} = \tan 200°$

31. $\cot \frac{\pi}{3} = \frac{\cos \pi/3}{\sin \pi/3} = \frac{\frac{1}{2}}{\frac{1}{2}\sqrt{3}} = \frac{1}{\sqrt{3}} = \frac{1}{\sqrt{3}} \cdot \frac{\sqrt{3}}{\sqrt{3}} = \frac{\sqrt{3}}{3}$

32. $\sec \frac{\pi}{4} = \frac{1}{\cos \pi/4} = \frac{1}{\frac{1}{2}\sqrt{2}} = \frac{2}{\sqrt{2}} = \sqrt{2}$

5.2 ALGEBRAIC OPERATIONS WITH THE TRIGONOMETRIC FUNCTIONS

The expression $\sin \theta$ represents a real number, whether θ is viewed as a real number or as an angle. For this reason it can be treated in the same way that we deal with numbers and letters (representing numbers) in algebra. For example, $\sin^3 \theta + \cos^3 \theta$ may be factored as the sum of two cubes. Since

$$a^3 + b^3 = (a + b)(a^2 - ab + b^2),$$

$$\begin{aligned}\sin^3 \theta + \cos^3 \theta &= (\sin \theta + \cos \theta)(\sin^2 \theta - \sin \theta \cos \theta + \cos^2 \theta) \\ &= (\sin \theta + \cos \theta)(1 - \sin \theta \cos \theta).\end{aligned}$$

Also $(\sec \theta + \tan \theta)^2$ may be expanded as the square of a binomial to equal "the square of the first plus twice the product plus the square of the last":

$$(\sec \theta + \tan \theta)^2 = \sec^2 \theta + 2 \sec \theta \tan \theta + \tan^2 \theta.$$

A glance at the fundamental relations reveals that the functions that occur most often are $\sin \theta$ and $\cos \theta$. Equations (1), (2), (4), and (5) express each of the other functions directly in terms of $\sin \theta$ and $\cos \theta$. Also, $\sin \theta$ and $\cos \theta$ are the only trigonometric functions that are defined for all θ. For these reasons it is advantageous in many problems to reduce an expression to sines and cosines.

Example 1 Express

$$\frac{3 \csc \theta}{5 \csc \theta - 6 \cot^2 \theta}$$

in terms of $\sin \theta$ and $\cos \theta$.

Solution

$$\frac{3 \csc \theta}{5 \csc \theta - 6 \cot^2 \theta} = \frac{3\left(\frac{1}{\sin \theta}\right)}{5\left(\frac{1}{\sin \theta}\right) - 6\left(\frac{\cos^2 \theta}{\sin^2 \theta}\right)}$$

$$= \frac{\frac{3}{\sin \theta}}{\frac{5 \sin \theta - 6 \cos^2 \theta}{\sin^2 \theta}}$$

$$= \frac{3}{\sin \theta} \cdot \frac{\sin^2 \theta}{5 \sin \theta - 6 \cos^2 \theta}$$

$$= \frac{3 \sin \theta}{5 \sin \theta - 6 \cos^2 \theta}.$$

To express this quantity in terms of just $\sin \theta$, replace $\cos^2 \theta$ with $(1 - \sin^2 \theta)$ to obtain

$$\frac{3 \sin \theta}{5 \sin \theta - 6 + 6 \sin^2 \theta}.$$

Example 2 Express each of the other trigonometric functions of θ in terms of $\sin \theta$.

Solution 1

$$\cos \theta = \pm\sqrt{1 - \sin^2 \theta}, \qquad \text{using (6).}$$

$$\tan \theta = \frac{\sin \theta}{\cos \theta} = \frac{\sin \theta}{\pm\sqrt{1 - \sin^2 \theta}}, \qquad \text{using (4), (6).}$$

$$\cot \theta = \frac{\cos \theta}{\sin \theta} = \frac{\pm\sqrt{1 - \sin^2 \theta}}{\sin \theta}, \qquad \text{using (5), (6).}$$

$$\sec \theta = \frac{1}{\cos \theta} = \frac{1}{\pm\sqrt{1 - \sin^2 \theta}}, \qquad \text{using (2), (6).}$$

$$\csc \theta = \frac{1}{\sin \theta}, \qquad \text{using (1).}$$

Solution 2 Place θ in standard position, as in Figure 5.2. In order to make $y/r = \sin \theta$, let $y = \sin \theta$ and $r = 1$. The Pythagorean theorem gives $x = \pm\sqrt{1 - \sin^2 \theta}$. Then

$$\cos\theta = \frac{x}{r} = \pm\sqrt{1-\sin^2\theta},$$

$$\tan\theta = \frac{y}{x} = \frac{\sin\theta}{\pm\sqrt{1-\sin^2\theta}},$$

and so forth.

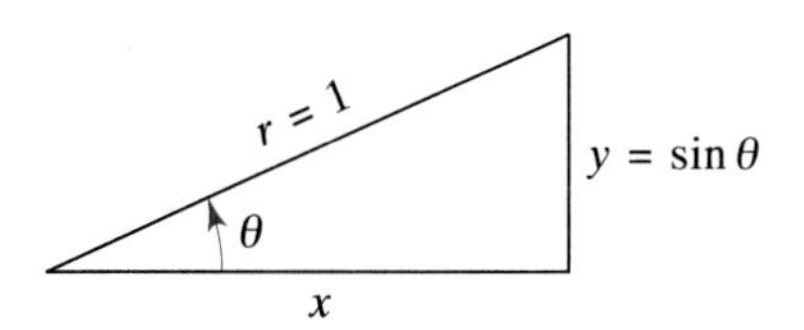

Figure 5.2

Exercise 5.2

Simplify each of the following.

1. $\dfrac{\sin^3\theta + \cos^3\theta}{\sin^2\theta - \sin\theta\cos\theta + \cos^2\theta}$
2. $\dfrac{\csc^4\theta - \cot^4\theta}{\csc^2\theta + \cot^2\theta}$
3. $\dfrac{\frac{1}{2} - (1/\sec^2\theta)}{1/(6\sec\theta) - 1/(3\sec^3\theta)}$
4. $\dfrac{1 - 2\tan\theta - 8\tan^2\theta}{1 - 5\tan\theta + 4\tan^2\theta}$

Reduce each of the following to an expression that involves no function except sin *θ* and cos *θ*. Simplify.

5. $\dfrac{\tan\theta + 5}{\sec\theta + 5\csc\theta}$
6. $\dfrac{\sqrt{\sec^2\theta - 1}}{\sec\theta}$
7. $\dfrac{8\sec^2\theta\csc^2\theta}{\sec^2\theta + \csc^2\theta}$
8. $\dfrac{\csc\theta - \sin\theta}{\cot\theta} - \dfrac{\cot\theta}{\csc\theta}$
9. Express $\sin \pi/17$ in terms of $\cos \pi/17$.
10. Express $\csc 6\pi/5$ in terms of $\cot 6\pi/5$.
11. Express $\sin 88°$ in terms of $\csc 88°$.
12. Express $\cos 111°$ in terms of $\sin 111°$.
13. Express $\tan 2B$ in terms of $\sec 2B$.
14. Express $\tan 3A$ in terms of $\cot 3A$.
15. Express each of the other trigonometric functions in terms of $\tan\theta$. Why is your expression for $\sin\theta$ not valid when $\theta = \pi/2$?
16. Express each of the other trigonometric functions in terms of $\cos\theta$.

5.3 IDENTITIES AND CONDITIONAL EQUATIONS

An identity is an equation that holds true for all permissible values of the letters involved.

The *permissible* values of the letters involved are all those values for which each side of the equation has meaning.

Illustration 1 $x^2 - 9 = (x + 3)(x - 3)$ holds true for all values of x.

Illustration 2 $x^2 + xy - 2y^2 = (x + 2y)(x - y)$ holds true for all values of x and y.

Illustration 3

$$x - \frac{x^2 - 7x}{x - 5} = \frac{2x}{x - 5}$$

holds true for all permissible values of x, that is, for all values of x except $x = 5$. When $x = 5$, each side of the equation involves a fraction whose denominator is zero. Such fractions have no meaning, and we say that their value does not exist.

Illustration 4 $\sin^2 \theta + \cos^2 \theta = 1$ holds true for all values of θ.

Illustration 5 $1 + \tan^2 \theta = \sec^2 \theta$ holds true for all permissible values of θ, that is, for all values of θ except $\pi/2$, $3\pi/2$, and numbers representing angles coterminal with them.

Illustration 6 The following "trick with numbers" illustrates a simple identity.

Choose any number except 0.
Multiply your number by 5.
To this number add the square of your original number.
Multiply your result by 2.
Divide the number you now have by the original number.
Subtract 10.
Divide by your original number.

If you have followed instructions, your result should be 2 regardless of your choice of original number. To prove this, let x be the original number. Then the numbers that follow are $5x$, $5x + x^2$, $2(5x + x^2)$ or $x(10 + 2x)$, $10 + 2x$, $2x$, and 2. The identity used is

$$\frac{\dfrac{2(5x + x^2)}{x} - 10}{x} = 2.$$

It holds for all values of x except 0. Try it for a fraction. For a negative number.

A conditional equation is an equation that *does not* hold true for all permissible values of the letters involved, but that does hold for some values of the unknowns.

Illustration 7 $2x - 7 = 3$ holds true for only one value of x, namely, $x = 5$.

Illustration 8 $x^2 - 8x + 15 = 0$ holds true for only two values of x, namely, $x = 3$ and $x = 5$.

Illustration 9 $x(x - 7)(x + 4) = 0$ holds true for only three values of x, namely, $x = 0$, $x = 7$, and $x = -4$.

Illustration 10 $\sin\theta = \cos\theta$ holds true for only two values of θ between 0 and 2π. They are $\theta = \pi/4$ and $\theta = 5\pi/4$.

An impossible equation is true for *no* values of the letters.

Illustration 11 $\sin\theta = 5 + \cos\theta$ holds true for no value of θ.

The difference between an identity and a conditional equation can easily be seen from the contrasting definitions:

$$\left\{\begin{array}{l}\textbf{An identity}\\ \textbf{A conditional equation}\end{array}\right\} \textbf{ is an equation that } \left\{\begin{array}{l}\textbf{holds true}\\ \textbf{does not hold true}\end{array}\right\}$$

for all permissible values of the letters involved.

An identity *says* that both sides of an equation are equal for all permissible values. The process by which we demonstrate that the two sides are identical is called "proving the identity." A conditional equation *asks,* "For what values of the unknowns is the left side of this equation equal to the right side?" The process by which these values are found is called "solving the equation."

5.4 TRIGONOMETRIC IDENTITIES

The eight fundamental relations are identities. By using them, we can prove other identities. For one who plans to go further in mathematics, or in subjects that involve mathematics, it is highly important to gain a certain amount of experience in proving identities. For this reason we place considerable emphasis on the following examples and problems.

It is most desirable to prove an identity by *transforming one side* of the equation *to the other side, which should be left unaltered.** *The side with which we work is usually the more complicated one.* There is no set rule for making these transformations. The following suggestions will, however, indicate the first step in most cases.

1. If one side involves only one function of the angle, express the other side in terms of this function.
2. If one side is factorable, factor it.
3. If one side has only one term in its denominator (and several terms in its numerator), break up the fraction.
4. If one side contains one or more indicated operations (such as squaring an expression, adding fractions, or multiplying two expressions), begin by performing these operations. This is especially helpful if this side involves only sines and cosines.
5. When working with one side, keep an eye on the other side to see which transformation will most easily reduce the first side to the other side. It is frequently helpful to multiply the numerator and denominator of a fraction by the same expression. If possible, avoid introducing radicals.
6. When in doubt, express the more complicated side in terms of sines and cosines and then simplify.

At each step, look for some combination that can be replaced by a simpler expression.

The following examples illustrate the suggestions.

*The instructor may wish to permit the student to reduce each side of the equation independently to a common third expression. This method is sometimes desirable when both sides are quite complicated.

Example 1 Prove the identity

$$3\cos^4\theta + 6\sin^2\theta = 3 + 3\sin^4\theta.$$

Proof We convert the left side to the right one.

$$\begin{aligned}3\cos^4\theta + 6\sin^2\theta &= 3(1 - \sin^2\theta)^2 + 6\sin^2\theta\\ &= 3 - 6\sin^2\theta + 3\sin^4\theta + 6\sin^2\theta\\ &= 3 + 3\sin^4\theta\end{aligned}$$

Example 2 Prove the identity

$$\sec^2\theta + \tan^2\theta = \sec^4\theta - \tan^4\theta.$$

Proof

$$\begin{aligned}\sec^4\theta - \tan^4\theta &= (\sec^2\theta + \tan^2\theta)(\sec^2\theta - \tan^2\theta)\\ &= (\sec^2\theta + \tan^2\theta)\cdot 1\\ &= \sec^2\theta + \tan^2\theta.\end{aligned}$$

Example 3 Prove the identity

$$\frac{\sin\theta + \cot\theta}{\cos\theta} = \tan\theta + \csc\theta.$$

Proof

$$\begin{aligned}\frac{\sin\theta + \cot\theta}{\cos\theta} &= \frac{\sin\theta}{\cos\theta} + \frac{\cot\theta}{\cos\theta}\\ &= \tan\theta + \frac{\dfrac{\cos\theta}{\sin\theta}}{\cos\theta}\\ &= \tan\theta + \frac{1}{\sin\theta}\\ &= \tan\theta + \csc\theta.\end{aligned}$$

Example 4 Prove the identity

$$\frac{\sin\theta}{1 + \cos\theta} + \frac{1 + \cos\theta}{\sin\theta} = 2\csc\theta.$$

Proof The left side indicates the addition of two fractions that involve only sines and cosines of θ. We begin by adding these fractions.

$$\frac{\sin\theta}{1+\cos\theta}+\frac{1+\cos\theta}{\sin\theta}=\frac{\sin^2\theta+1+2\cos\theta+\cos^2\theta}{(1+\cos\theta)\sin\theta}$$

$$=\frac{2+2\cos\theta}{(1+\cos\theta)\sin\theta}$$

$$=\frac{2(1+\cos\theta)}{(1+\cos\theta)\sin\theta}$$

$$=\frac{2}{\sin\theta}$$

$$=2\csc\theta.$$

Example 5 Prove the identity

$$\frac{\cot\theta}{\csc\theta-1}=\frac{\csc\theta+1}{\cot\theta}.$$

Proof Since the numerator of the right side is $\csc\theta+1$, let us multiply top and bottom of the left side by $\csc\theta+1$ to get

$$\frac{\cot\theta(\csc\theta+1)}{(\csc\theta-1)(\csc\theta+1)}=\frac{\cot\theta(\csc\theta+1)}{\csc^2\theta-1}$$

$$=\frac{\cot\theta(\csc\theta+1)}{\cot^2\theta}$$

$$=\frac{\csc\theta+1}{\cot\theta}.$$

Example 6 Prove the identity

$$\frac{\cot A+\csc B}{\tan B+\tan A\sec B}=\cot A\cot B.$$

Proof We convert to sines and cosines.

$$\frac{\cot A + \csc B}{\tan B + \tan A \sec B} = \frac{\dfrac{\cos A}{\sin A} + \dfrac{1}{\sin B}}{\dfrac{\sin B}{\cos B} + \dfrac{\sin A}{\cos A \cos B}}$$

$$= \frac{\dfrac{\sin B \cos A + \sin A}{\sin A \sin B}}{\dfrac{\sin B \cos A + \sin A}{\cos A \cos B}}$$

$$= \frac{\cos A \cos B}{\sin A \sin B}$$

$$= \cot A \cot B.$$

In trying to prove that an equation holds for all permissible values of the unknowns, we must *not* work with it as if it were an ordinary conditional equation. The key point, however, is that each step in the transformation from the expression on one side of the identity to the expression on the other side must be clearly justified as an *identity transformation.*

For example, to prove that $\cos\theta = \sin\theta/\tan\theta$, *it would not be proper to multiply both sides by* $\tan\theta$ to get $\cos\theta\tan\theta = \sin\theta$ and then to replace $\tan\theta$ with $\sin\theta/\cos\theta$ to get $(\cos\theta)\sin\theta/\cos\theta = \sin\theta$ or $\sin\theta = \sin\theta$. This does *not* prove that the given equation is an identity. It merely demonstrates that *if* $\cos\theta = \sin\theta/\tan\theta$, *then* (it is necessary that) $\sin\theta = \sin\theta$. We could use this same improper procedure to prove that $3 = 5$, because multiplying both sides by 0 gives us $3(0) = 5(0)$, which is a true statement. Another incorrect procedure is to square both sides of an equation and then claim that the given equation is an identity because the squares of its two sides are identically equal. With this sort of reasoning, we could claim that $-4 = 4$, because $(-4)^2 = 4^2$. If $a^2 = b^2$, it does not necessarily follow that $a = b$.

Exercise 5.3

Prove each of the following identities by reducing one side to the other.

1. $$\frac{3 - \dfrac{2x-6}{2x-2}}{1 - \dfrac{x-3}{x-1}} = x$$

2. $$\frac{\dfrac{y}{y+5} - \dfrac{3y-7}{y^2-25}}{3 - \dfrac{2y-8}{y-5}} = \frac{y-1}{y+5}$$

3. $\dfrac{\dfrac{5}{2} - \dfrac{x-8}{x+6}}{\dfrac{7}{4} - \dfrac{x-1}{x+6}} = 2$

4. $\dfrac{x + 9 - \dfrac{4x+12}{x+2}}{x + 8 - \dfrac{4x+18}{x+3}} = \dfrac{x+3}{x+2}$

5. $\dfrac{\sec\theta}{\cos\theta} - \dfrac{\tan\theta}{\cot\theta} = 1$

6. $\dfrac{\sec A}{\csc A} = \tan A$

7. $\cos^2 2B - \sec^2 7B + \sin^2 2B + \tan^2 7B = 0$

8. $\dfrac{1 + \tan C}{1 + \cot C} = \tan C$

9. $\sin^4\theta + \sin^2\theta + 3\cos^2\theta = \cos^4\theta + 2$

10. $3\sec^2\theta + \sec^4\theta = \tan^4\theta + 5\tan^2\theta + 4$

11. $\dfrac{\cot^2\theta + 8\csc\theta + 8}{\cot^2\theta + 9\csc\theta + 15} = \dfrac{\csc\theta + 1}{\csc\theta + 2}$

12. $7\cot^4\theta + 6\csc^2\theta = 7\csc^4\theta - 8\csc^2\theta + 7$

13. $\dfrac{3\sin\theta + 5\cos\theta\tan\theta}{\cos\theta} = 8\tan\theta$

14. $\dfrac{\cos\theta + \cot\theta\cos\theta + \sin\theta}{\csc\theta} = 1 + \sin\theta\cos\theta$

15. $\dfrac{4\cos^2\theta - 5\sin^2\theta}{6\sin\theta\cos\theta} = \dfrac{2}{3}\cot\theta - \dfrac{5}{6}\tan\theta$

16. $\dfrac{3 + 4\sin\theta + 5\tan\theta + 6\csc\theta}{\sin\theta} = 3\csc\theta + 4 + 5\sec\theta + 6\csc^2\theta$

17. $\csc^4 C - \csc^2 C = \cot^4 C + \cot^2 C$

18. $\dfrac{\cos^2\theta + 15}{20 - 9\sin\theta + \sin^2\theta} = \dfrac{4 + \sin\theta}{5 - \sin\theta}$

19. $\dfrac{\sec^3 A - \cos^3 A}{\sec A - \cos A} = 2 + \cos^2 A + \tan^2 A$

20. $\cos^6\theta + \sin^6\theta = 3\sin^4\theta - 3\sin^2\theta + 1$

21. $\dfrac{\sin A + \csc B}{\csc A + \sin B} = \dfrac{\sin A}{\sin B}$

22. $\dfrac{1 - \sin A}{\cos B} - \dfrac{\cos B}{1 + \sin A} = \dfrac{\sin^2 B - \sin^2 A}{\cos B(1 + \sin A)}$

23. $\dfrac{\tan A\tan B}{1 + \tan B} = \dfrac{\sin A\sin B}{\cos A(\cos B + \sin B)}$

24. $\dfrac{\tan A - \tan B}{1 + \tan A\tan B} = \dfrac{\sin A\cos B - \cos A\sin B}{\cos A\cos B + \sin A\sin B}$

25. $(\sin B + 2\cos B)^2 + (2\sin B - \cos B)^2 = 5$

26. $\dfrac{1}{\cot\theta + 2 + \tan\theta} = \dfrac{\tan\theta}{(1 + \tan\theta)^2}$

27. $\sec^2 \theta + \csc^2 \theta = \sec^2 \theta \csc^2 \theta$

28. $\dfrac{\sin \theta}{1 - \cos \theta} + \dfrac{1 - \cos \theta}{\sin \theta} = 2 \csc \theta$

29. $(1 + \sin^2 \theta)(1 - \sin^2 \theta) = 2 \cos^2 \theta - \cos^4 \theta$

30. $(a + b \cot \theta)(b + a \cot \theta) = (a^2 + b^2) \cot \theta + ab \csc^2 \theta$

31. $\dfrac{\cot \theta + 8 \tan \theta}{\cot \theta + 3 \tan \theta} = \dfrac{\csc^2 \theta + 7}{\csc^2 \theta + 2}$

32. $\cos \theta \csc \theta + \theta \tan^2 \theta - \sec \theta \cot \theta \cos \theta + \theta = \theta \sec^2 \theta$

In Exercises 33 through 40, either prove or disprove the statement. (Prove that the equation is, or is not, an identity.)

33. $\dfrac{1}{\tan \theta + \cot \theta} = 1$

34. $\dfrac{\sec C}{\tan C + \cot C} = \sin C$

35. $\sqrt{\sin^2 \theta} = \sin \theta$

36. $\sqrt{\tan^2 \theta + 1} = \sec \theta$

37. $\dfrac{\tan^3 \theta - \cot^3 \theta}{\tan \theta - \cot \theta} = \tan^2 \theta + \csc^2 \theta$

38. $(\tan \theta + \sec \theta)^2 + (\tan \theta - \sec \theta)^2 = 4 \sec^2 \theta + 2$

39. $\dfrac{\cot^2 \theta - 1}{\cot^2 \theta + 1} = \cos^2 \theta - \sin \theta$

40. $(\csc A - \cot A)^2 = \dfrac{1 - \cos A}{1 + \cos A}$

In Exercises 41 through 44, assume that $0 \le \theta \le \pi/2$. Prove each identity.

41. $\sqrt{\dfrac{\csc \theta - 1}{\csc \theta + 1}} = \dfrac{1 - \sin \theta}{\cos \theta}$

[*Hint:* Rationalize the left side by multiplying under the radical by $(\csc \theta - 1)/(\csc \theta - 1)$.]

42. $\sqrt{\dfrac{1 - \sin \theta}{1 + \sin \theta}} = \sec \theta - \tan \theta$

43. $\sqrt{\dfrac{\sec \theta + 1}{\sec \theta - 1}} = \dfrac{\tan \theta}{\sec \theta - 1}$

44. $\sqrt{\dfrac{\csc \theta + \cot \theta}{\csc \theta - \cot \theta}} = \dfrac{1}{\csc \theta - \cot \theta}$

45. State two nonpermissible values of θ for which the identity in Exercise 13 does not hold true.

46. State one nonpermissible value of A and two nonpermissible values of B for which the identity in Exercise 22 does not hold true.

47. State two nonpermissible values of A and four nonpermissible values of B for which the identity in Exercise 23 does not hold true.
48. State six nonpermissible values of C for which the identity in Exercise 8 does not hold true.

KEY TERMS

Identity, conditional equation, counterexample

REVIEW EXERCISES

Decide whether the equation is an identity:

1. $\sqrt{\sin^2 \theta} = \sin \theta$
2. $\sqrt{1 + \tan^2 \theta} = 1 + \tan \theta$
3. $\sin^4 \theta - \cos^4 \theta = \sin^2 \theta - \cos^2 \theta$
4. $\sqrt{1 - \cos^2 \theta} = \sin \theta$
5. $\sin 2\theta = 2 \sin \theta$
6. $\sin 2\theta = 2 \cos \theta$
7. Verify that $\sin 2\theta = 2 \sin \theta \cos \theta$ is satisfied for $\theta = \pi/6, \pi/4, \pi/3$, and $\pi/2$. Does this prove it is an identity?

Prove the identity:

8. $\tan A - \tan^2 A = \dfrac{\csc A - \sec A}{\csc A \cot A}$
9. $\sin^3 \theta(\cot^2 \theta + 2) - \dfrac{1}{\csc \theta} = \sin^3 \theta$
10. $\dfrac{1 + \sin C}{\cot^2 C} = \dfrac{\sin C}{\csc C - 1}$
11. $\dfrac{\cos A + \sec B}{\sec A + \cos B} + \dfrac{\cos A + \cos B}{\sec A + \sec B} = \dfrac{\sec B + \cos B}{\sec A}$
12. $\dfrac{\tan^2 A}{\cos^2 B} - \dfrac{\tan^2 B}{\cos^2 A} = \tan^2 A - \tan^2 B$
13. $(1 + \tan \theta + \sec \theta)^2 = 2(\sec \theta + 1)(\sec \theta + \tan \theta)$
14. $(\sin A - \cos A)(1 + \tan A + \cot A) = \sin A \tan A - \cos A \cot A$
15. $\dfrac{1 + (\cot \theta - \csc \theta)^2}{\sec \theta \csc \theta - \cot \theta \sec \theta} = 2 \cot \theta$
16. $\dfrac{\csc^2 \theta + \sin^2 \theta + 1}{\csc \theta + \sin \theta + 1} = \csc \theta + \sin \theta - 1$

6

Functions of Two Angles

6.1 FUNCTIONS OF THE SUM OF TWO ANGLES

The trigonometric identities in Chapter 5 are relations among the trigonometric functions of one angle. We shall now consider functions of an angle that is the sum of two given angles. It seems reasonable to say that if the functions of 30° and 45° are known, the functions of 75° can be obtained. For instance, is sin 30° + sin 45° = sin 75°? Obviously this is false, because

$$\frac{1}{2} + \frac{\sqrt{2}}{2} = 0.5 + 0.7 = 1.2,$$

which would make sin 75° greater than 1. This proves that, in general, $\sin(A + B) \neq \sin A + \sin B$. Likewise, $\sin 2A$ is not identically equal to $2 \sin A$, because $\sin 60° \neq 2 \sin 30°$. It is possible, however, to express $\sin(A + B)$ in terms of the functions of the separate angles A and B. And it is possible to express $\sin 2A$ in terms of the functions of A. These and other formulas will be developed in the following sections.

The purpose of this section is to prove the following identities:

$$\mathbf{\sin(A + B) = \sin A \cos B + \cos A \sin B,} \quad \mathbf{(1)}$$

$$\mathbf{\cos(A + B) = \cos A \cos B - \sin A \sin B.} \quad \mathbf{(2)}$$

We shall first derive a formula for $\cos(A + B)$. On a coordinate system (Figure 6.1), draw a unit circle with center at the origin. Place the angles A, $A + B$, and $-B$ in standard position. Let the terminal sides of these angles intersect the unit circle at points P_1, P_2, and P_3, respectively; let P_4 designate the point $(1, 0)$. If P_1 has coordinates (x_1, y_1), then $\cos A = x_1/1 = x_1$, and $\sin A = y_1/1 = y_1$. Hence the coordinates of P_1 are $(\cos A, \sin A)$. For the same reason, the coordinates of P_2 and P_3 are as indicated in Figure 6.1. Angle P_3OP_1 is $B + A$ because angle P_3OP_4 is B and angle P_4OP_1 is A. Therefore, triangles P_3OP_1 and P_4OP_2 are congruent. (Why?) Hence $P_3P_1 = P_4P_2$.

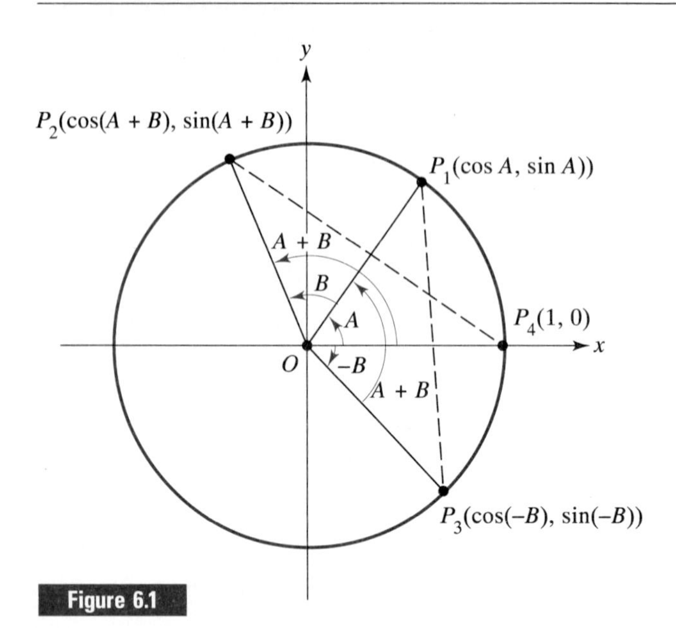

Figure 6.1

Applying the distance formula (Section 1.3), we get

$$P_3P_1 = \sqrt{[\cos A - \cos(-B)]^2 + [\sin A - \sin(-B)]^2}.$$

Squaring and using the identities $\cos(-B) = \cos B$ and $\sin(-B) = -\sin B$, we obtain

$$\begin{aligned}(P_3P_1)^2 &= (\cos A - \cos B)^2 + (\sin A + \sin B)^2 \\ &= \cos^2 A - 2 \cos A \cos B + \cos^2 B + \sin^2 A + 2 \sin A \sin B \\ &\quad + \sin^2 B \\ &= 2 - 2 \cos A \cos B + 2 \sin A \sin B,\end{aligned}$$

since $\cos^2 \theta + \sin^2 \theta = 1.$

Again using the distance formula, we find

$$\begin{aligned}(P_4P_2)^2 &= [\cos(A + B) - 1]^2 + [\sin(A + B) - 0]^2 \\ &= \cos^2(A + B) - 2 \cos(A + B) + 1 + \sin^2(A + B).\end{aligned}$$

Hence $(P_4P_2)^2 = 2 - 2 \cos(A + B),$

since $\cos^2(A + B) + \sin^2(A + B) = 1.$

Upon equating the expressions for $(P_4P_2)^2$ and $(P_3P_1)^2$, we get

$$2 - 2\cos(A + B) = 2 - 2\cos A \cos B + 2\sin A \sin B,$$

which reduces to

$$\mathbf{\cos(A + B) = \cos A \cos B - \sin A \sin B.} \tag{2}$$

Note carefully that this proof does not depend on the quadrants in which A, B, and $A + B$ happen to lie. The formula for $\cos(A + B)$ is valid for all values of A and B, positive or negative.

Before deriving a formula for $\sin(A + B)$, we shall need to obtain general expressions for $\cos(A - C)$, $\cos(90° - C)$, and $\sin(90° - C)$.

Since formula (2) for $\cos(A + B)$ holds for all A and B, we shall rewrite it with B replaced by $-C$. Hence

$$\begin{aligned}\cos(A - C) &= \cos[A + (-C)] \\ &= \cos A \cos(-C) - \sin A \sin(-C).\end{aligned}$$

Since $\cos(-C) = \cos C$ and $\sin(-C) = -\sin C$, we have

$$\cos(A - C) = \cos A \cos C + \sin A \sin C. \tag{6.1}$$

If we replace A with $90°$ in formula (6.1), we find that

$$\cos(90° - C) = \cos 90° \cos C + \sin 90° \sin C.$$

Since $\cos 90° = 0$ and $\sin 90° = 1$, we have

$$\cos(90° - C) = \sin C. \tag{6.2}$$

Moreover, since $C = 90° - (90° - C)$, we use formula (6.1) to get

$$\begin{aligned}\cos C &= \cos[90° - (90° - C)] \\ &= \cos 90° \cos(90° - C) + \sin 90° \sin(90° - C) \\ &= 0 \cdot \cos(90° - C) + 1 \cdot \sin(90° - C).\end{aligned}$$

Hence
$$\cos C = \sin(90° - C),$$

or
$$\sin\left(\frac{\pi}{2} - C\right) = \cos C. \tag{6.3}$$

Formulas (6.2) and (6.3) were established for any acute angle C in Section 1.9. The foregoing arguments prove the formulas for *all* values of C. For example, if $C = 250°$, formula (6.2) says that $\cos(90° - 250°) = \sin 250°$, which is true because $\cos(90° - 250°) = \cos(-160°) = \cos 160° = -\cos 20° = -\sin 70°$, whereas $\sin 250° = -\sin 70°$, by the related-angle theorem (Section 2.5).

To derive a formula for $\sin(A + B)$, we shall reverse (6.2) and replace C with $(A + B)$ to get

$$\begin{aligned}\sin(A + B) &= \cos[90° - (A + B)]\\ &= \cos[(90° - A) - B]\\ &= \cos(90° - A)\cos B + \sin(90° - A)\sin B.\end{aligned}$$

Using formulas (6.2) and (6.3), we may write

$$\mathbf{\sin(A + B) = \sin A \cos B + \cos A \sin B.} \qquad \textbf{(1)}$$

It is impossible to overemphasize the importance of these results. In addition to learning Equations (1) and (2), the student should be able to state them in words. These statements are

(1) The sine of the sum of two angles is equal to the sine of the first times the cosine of the second, plus the cosine of the first times the sine of the second.

(2) The cosine of the sum of two angles is equal to the cosine of the first times the cosine of the second, minus the sine of the first times the sine of the second.

It is equally important for the student to be able to use these formulas backward. For example, when

$$\cos 7\theta \cos 2\theta - \sin 7\theta \sin 2\theta$$

is encountered, it should be recognized as the expansion of $\cos(7\theta + 2\theta)$ or $\cos 9\theta$.

Example 1 Compute $\sin 75°$ and $\cos 75°$ from the functions of 30° and 45°.

Solution

$$\begin{aligned}
\sin 75^\circ &= \sin(30^\circ + 45^\circ) \\
&= \sin 30^\circ \cos 45^\circ + \cos 30^\circ \sin 45^\circ \\
&= \frac{1}{2} \cdot \frac{\sqrt{2}}{2} + \frac{\sqrt{3}}{2} \cdot \frac{\sqrt{2}}{2} \\
&= \frac{\sqrt{2}}{4} + \frac{\sqrt{6}}{4} = \frac{\sqrt{2} + \sqrt{6}}{4} \\
&\doteq \frac{1.414 + 2.449}{4} = \frac{3.863}{4} \doteq 0.966.
\end{aligned}$$

$$\begin{aligned}
\cos 75^\circ &= \cos(30^\circ + 45^\circ) \\
&= \cos 30^\circ \cos 45^\circ - \sin 30^\circ \sin 45^\circ \\
&= \frac{\sqrt{3}}{2} \cdot \frac{\sqrt{2}}{2} - \frac{1}{2} \cdot \frac{\sqrt{2}}{2} \\
&= \frac{\sqrt{6} - \sqrt{2}}{4} \doteq 0.259.
\end{aligned}$$

Notice that these decimal approximations agree with the "story" as presented in Sections 1.5 and 1.6: Angles near 90° have sines close to 1 and cosines close to 0.

Example 2 Given $\sin A = \frac{5}{13}$, with A in Q I; and $\cos B = -\frac{4}{5}$, with B in Q II. Find (*a*) $\sin(A + B)$, (*b*) $\cos(A + B)$, (*c*) the quadrant in which $A + B$ lies.

Solution First find $\cos A$ and $\sin B$ by drawing triangles of reference (Figure 6.2). Hence

$$\cos A = \tfrac{12}{13} \qquad \text{and} \qquad \sin B = \tfrac{3}{5}.$$

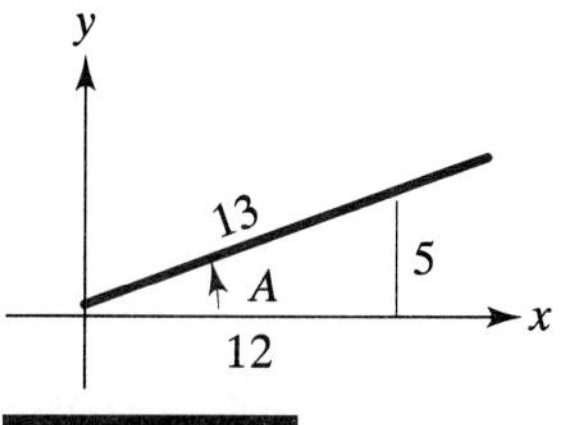

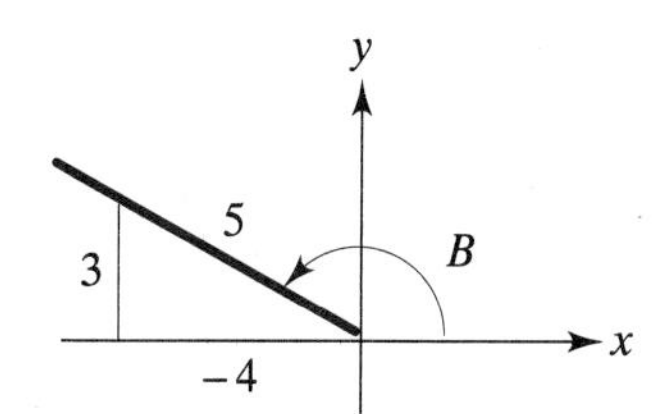

Figure 6.2

(a)
$$\sin(A + B) = \sin A \cos B + \cos A \sin B$$
$$= \left(\frac{5}{13}\right)\left(\frac{-4}{5}\right) + \left(\frac{12}{13}\right)\left(\frac{3}{5}\right)$$
$$= -\frac{20}{65} + \frac{36}{65} = \frac{16}{65}.$$

(b)
$$\cos(A + B) = \cos A \cos B - \sin A \sin B$$
$$= \left(\frac{12}{13}\right)\left(\frac{-4}{5}\right) - \left(\frac{5}{13}\right)\left(\frac{3}{5}\right)$$
$$= -\frac{48}{65} - \frac{15}{65} = -\frac{63}{65}.$$

(c) The angle $A + B$ lies in Q II because $\sin(A + B)$ is positive and $\cos(A + B)$ is negative.

Exercise 6.1

1. Compute cos 255° from the functions of 45° and 210°.
2. Compute cos 345° from the functions of 300° and 45°.
3. Compute sin 165° from the functions of 30° and 135°.
4. Compute sin 255° from the functions of 225° and 30°.
5. Use a calculator to show that

$$\sin 15° + \sin 20° \neq \sin 35°$$

$$\cos 10° + \cos 40° \neq \cos 50°$$

Simplify the following by reducing each one to a single term.

6. $\sin(80° + \theta)\cos(70° - \theta) + \cos(80° + \theta)\sin(70° - \theta)$
7. $\cos\frac{4\pi}{5}\cos\frac{\pi}{5} - \sin\frac{4\pi}{5}\sin\frac{\pi}{5}$
8. $\sin\frac{6\theta}{7}\cos\frac{8\theta}{7} + \sin\frac{8\theta}{7}\cos\frac{6\theta}{7}$
9. $\sin 3\theta \sin\theta - \cos 3\theta \cos\theta$

Prove the following identities.

10. $\cos(\pi + \theta) = -\cos\theta$
11. $\sin\left(\frac{\pi}{2} + \theta\right) = \cos\theta$

12. $\cos\left(\dfrac{5\pi}{2} + \theta\right) = -\sin\theta$
13. $\sin C \cos(D - C) + \cos C \sin(D - C) = \sin D$
14. $\sin(\pi/3 + \theta) + \cos(5\pi/6 + \theta) = 0$
15. $\cos(135° + \theta) + \sin(315° + \theta) = -\sqrt{2}\cos\theta$
16. $(\sin A \cos B + \cos A \sin B)^2 + (\sin A \sin B - \cos A \cos B)^2 = 1$
17. Express $\cos 100°$ in terms of functions of $80°$ and $20°$.
18. Express $\sin 38°$ in terms of functions of $11°$ and $27°$.
19. Prove that

$$\frac{\cos 3\theta}{\sin 6\theta} - \frac{\sin 3\theta}{\cos 6\theta} = \frac{\cos 9\theta}{\sin 6\theta \cos 6\theta}.$$

20. Prove that

$$\frac{\sin 4\theta}{\sin\theta} + \frac{\cos 4\theta}{\cos\theta} = \sin 5\theta \sec\theta \csc\theta.$$

21. Given $\sin A = \frac{1}{3}$ with A in Q I, and $\sin B = \frac{2}{3}$ with B in Q I. Find $\sin(A + B)$. Check your result by using a calculator or tables to approximate A, B, and $A + B$ as determined by their sines.
22. Given $\sin(A - B) = \frac{7}{25}$ with $A - B$ in Q I, and $\sin B = \frac{4}{5}$ with B in Q I. Find $\sin A$.
23. Given $\cos A = \frac{9}{41}$ with A in Q I, and $\sin B = -\frac{4}{5}$ with B in Q III. Find (*a*) $\sin(A + B)$, (*b*) $\cos(A + B)$, (*c*) the quadrant in which $A + B$ lies.
24. Given $\sin A = \frac{15}{17}$ with A in Q II, and $\tan B = \frac{5}{12}$ with B in Q III. Find (*a*) $\sin(A + B)$, (*b*) $\cos(A + B)$, (*c*) the quadrant in which $A + B$ lies.
25. Prove that

$$\csc(A + B) = \frac{\csc A \csc B}{\cot A + \cot B}.$$

26. Prove the identity

$$\sin(A + B + C) = \sin A \cos B \cos C + \cos A \sin B \cos C + \cos A \cos B \sin C - \sin A \sin B \sin C.$$

Hint: $\sin(A + B + C) = \sin[(A + B) + C]$.

27. Prove the identity

$$\cos(A + B + C) = \cos A \cos B \cos C - \sin A \sin B \cos C - \sin A \cos B \sin C - \cos A \sin B \sin C.$$

28. If A and B are complementary angles, prove that $\sin(5A + 8B) = \cos 3B$.
29. If A and B are complementary angles, prove that $\cos(3A + 2B) = -\cos A$.

In Exercises 30 through 32, either prove or disprove the statement. (Prove that the equation is, or is not, an identity.)

30. $\sin(\theta + n\pi) = (-1)^n \sin \theta, \quad (n \text{ an integer}).$
31. $\sin\left(\theta + \dfrac{n\pi}{2}\right) = (-1)^{(n-1)/2} \cos \theta, \quad (n \text{ an odd integer}).$
32. $\cos\left(\theta + \dfrac{n\pi}{2}\right) = (-1)^{(n+1)/2} \sin \theta, \quad (n \text{ an odd integer}).$

6.2 tan(*A* + *B*)

Since

$$\tan \theta = \frac{\sin \theta}{\cos \theta},$$

$$\tan(A + B) = \frac{\sin(A + B)}{\cos(A + B)} = \frac{\sin A \cos B + \cos A \sin B}{\cos A \cos B - \sin A \sin B}.$$

Dividing top and bottom of this fraction by $\cos A \cos B$, we get

$$\tan(A + B) = \frac{\dfrac{\sin A \cos B}{\cos A \cos B} + \dfrac{\cos A \sin B}{\cos A \cos B}}{\dfrac{\cos A \cos B}{\cos A \cos B} - \dfrac{\sin A \sin B}{\cos A \cos B}}$$

$$= \frac{\dfrac{\sin A}{\cos A} + \dfrac{\sin B}{\cos B}}{1 - \dfrac{\sin A}{\cos A} \cdot \dfrac{\sin B}{\cos B}},$$

or

$$\boldsymbol{\tan(A + B) = \frac{\tan A + \tan B}{1 - \tan A \tan B}}. \quad \textbf{(3)}$$

It is important to note that this relationship, and all others developed in this chapter, is valid for only those angles for which the functions involved are defined and for which the denominators are not 0.

Stated in words, we have

(3) The tangent of the sum of two angles is equal to the sum of their tangents, divided by 1 minus the product of their tangents.

If A and B are any two angles, then

$$\sin(A - B) = \sin A \cos B - \cos A \sin B, \tag{4}$$

$$\cos(A - B) = \cos A \cos B + \sin A \sin B, \tag{5}$$

and

$$\tan(A - B) = \frac{\tan A - \tan B}{1 + \tan A \tan B}. \tag{6}$$

Proof Recall that $\sin(-B) = -\sin B$, $\cos(-B) = \cos B$, and $\tan(-B) = -\tan B$. Then

$$\begin{aligned} \sin(A - B) &= \sin[A + (-B)] \\ &= \sin A \cos(-B) + \cos A \sin(-B) \\ &= \sin A \cos B + (\cos A)(-\sin B) \\ &= \sin A \cos B - \cos A \sin B. \end{aligned}$$

Equations (5) and (6) are proved by a similar method. The student should state Equations (4), (5), and (6) in words.

In comparing (1), (4), (2), and (5),

$$\begin{aligned} \sin(A + B) &= \sin A \cos B + \cos A \sin B, \\ \sin(A - B) &= \sin A \cos B - \cos A \sin B, \\ \cos(A + B) &= \cos A \cos B - \sin A \sin B, \\ \cos(A - B) &= \cos A \cos B + \sin A \sin B, \end{aligned}$$

we notice that "the sines have the same sign, but the cosines have different signs."

There is no need for formulas involving the cotangent, secant, and cosecant of $A + B$ and $A - B$ because they can be readily expressed in terms of their reciprocals.

6.3 REDUCTION OF $a \sin \theta + b \cos \theta$ TO $k \sin(\theta + H)$

Any two real nonzero numbers a and b are proportional to two other numbers that represent the cosine and sine, respectively, of some properly chosen angle H. These other two numbers are $a/\sqrt{a^2 + b^2}$ and $b/\sqrt{a^2 + b^2}$. We may write

$$\cos H = \frac{a}{\sqrt{a^2 + b^2}} \quad \text{and} \quad \sin H = \frac{b}{\sqrt{a^2 + b^2}}, \tag{6.4}$$

as indicated by Figure 6.3. Then

$$\begin{aligned} a \sin \theta + b \cos \theta &= \sqrt{a^2 + b^2}\left[(\sin \theta)\frac{a}{\sqrt{a^2 + b^2}} + (\cos \theta)\frac{b}{\sqrt{a^2 + b^2}}\right] \\ &= \sqrt{a^2 + b^2}\,(\sin \theta \cos H + \cos \theta \sin H). \end{aligned}$$

Hence $\quad \boldsymbol{a \sin \theta + b \cos \theta = \sqrt{a^2 + b^2} \sin(\theta + H)},$

where H is an angle that satisfies both equations in (6.4).

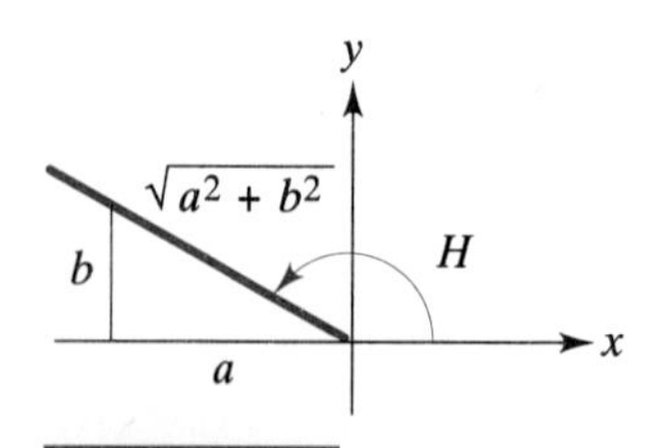

Figure 6.3

Formulas (6.4) imply that $\tan H = b/a$. In many cases it will be easiest to obtain H through its tangent, though then we must be careful to choose an angle in the correct quadrant as determined by the signs of a and b.

Example Express $4 \sin \theta + 3 \cos \theta$ in the form $k \sin(\theta + H)$, where k and H are constants and $k > 0$.

Solution Multiply and divide the given expression by $\sqrt{4^2 + 3^2} = 5$ to get

$$5(\tfrac{4}{5} \sin \theta + \tfrac{3}{5} \cos \theta) \qquad \text{or} \qquad 5[(\sin \theta)\tfrac{4}{5} + (\cos \theta)\tfrac{3}{5}].$$

Put $\frac{4}{5} = \cos H$; then $\frac{3}{5} = \sin H$. Our expression becomes

$$5(\sin \theta \cos H + \cos \theta \sin H) = 5 \sin(\theta + H).$$

Using a calculator, we find that if $\cos H = \frac{4}{5}$ and $\sin H > 0$, then $H = 36.87°$ or 0.6435 radian. Hence

$$\begin{aligned} 4 \sin \theta + 3 \cos \theta &= 5 \sin(\theta + 36.9°) \\ &= 5 \sin(\theta + 0.644), \qquad \text{approx.} \end{aligned}$$

Exercise 6.2

1. Compute tan 75° from the functions of 30° and 45°.
2. Compute tan 165° from the functions of 225° and 60°.
3. Compute sin 195° from the functions of 240° and 45°.
4. Compute cos 15° from the functions of 45° and 30°.

Simplify by reducing to a single term.

5. $\sin 288° \cos 18° - \cos 288° \sin 18°$
6. $\cos 110° \cos 20° + \sin 110° \sin 20°$
7. $\dfrac{\tan 100° + \tan 230°}{1 - \tan 100° \tan 230°}$
8. $\dfrac{\tan(246° + \theta) - \tan(126° + \theta)}{1 + \tan(246° + \theta) \tan(126° + \theta)}$

Prove the following identities.

9. $\tan\left(\theta - \dfrac{\pi}{3}\right) = \dfrac{\tan \theta - \sqrt{3}}{1 + \sqrt{3} \tan \theta}$
10. $\sin(s - \pi) = -\sin s$
11. $\cos\left(\dfrac{\pi}{2} - s\right) = \sin s$
12. $\tan\left(\theta + \dfrac{\pi}{4}\right) = \dfrac{\tan \theta + 1}{1 - \tan \theta}$
13. Given $\tan A = \frac{4}{3}$ with A in Q III, and $\sin B = \frac{24}{25}$ with B in Q I. Find (a) $\sin(A - B)$, (b) $\cos(A - B)$, (c) $\tan(A + B)$, (d) $\tan(A - B)$.
14. Given $\cos A = \frac{12}{13}$ with A in Q IV, and $\tan B = -\frac{8}{15}$ with B in Q II. Find (a) $\sin(A - B)$, (b) $\cos(A - B)$, (c) $\tan(A + B)$, (d) $\tan(A - B)$.

15. Given $\tan A = \frac{1}{2}$ with A in Q I. If $A + B = 3\pi/4$, find $\tan B$.
16. Given $\tan(C + D) = 10$ with $C + D$ in Q I, and $\tan C = 7$ with C in Q I. Find $\tan D$.
17. Express $\cos 50°$ in terms of functions of $70°$ and $20°$.
18. Express $\tan 127°$ in terms of $\tan 40°$ and $\tan 87°$.
19. Express $\tan 6°$ in terms of $\tan 8°$ and $\tan 2°$.
20. Express $\sin 194°$ in terms of functions of $244°$ and $50°$.

Prove the following identities.

21. $\dfrac{\sin \theta}{\csc 2\theta} - \dfrac{\cos 2\theta}{\sec \theta} = -\cos 3\theta$

22. $\sec(A - B) = \dfrac{\sec A \sec B}{1 + \tan A \tan B}$

23. $\tan(\theta + 135°)\tan(\theta - 315°) = -1$

24. $\sin\left(\dfrac{\pi}{6} - s\right) + \cos\left(\dfrac{\pi}{3} - s\right) = \cos s$

25. $\cot(A - B) = \dfrac{\cot A \cot B + 1}{\cot B - \cot A}$

26. $\cot(A + B) = \dfrac{\cot A \cot B - 1}{\cot A + \cot B}$

27. $\dfrac{\sin(A + B) + \sin(A - B)}{\cos(A + B) + \cos(A - B)} = \tan A$

28. $\dfrac{\cos(A - B)}{\cos(A + B)} = \dfrac{\cot A + \tan B}{\cot A - \tan B}$

29. $\tan(A + B + C) = \dfrac{\tan A + \tan B + \tan C - \tan A \tan B \tan C}{1 - \tan A \tan B - \tan A \tan C - \tan B \tan C}$

30. $\tan(A + B - C) = \dfrac{\tan A + \tan B - \tan C + \tan A \tan B \tan C}{1 - \tan A \tan B + \tan A \tan C + \tan B \tan C}$

31. Use a calculator to verify the identity in Problem 30 for the acute angles A, B, and C for which $\tan A = 5$, $\tan B = 4$, and $\tan C = 2$.
32. Prove that

$$\frac{\tan(A - B) + \tan C}{1 - \tan(A - B)\tan C} = \frac{\tan A - \tan(B - C)}{1 + \tan A \tan(B - C)}$$

33. If n is an integer, prove that $\cos(n\pi - \theta) = (-1)^n \cos \theta$.
34. If A and B are complementary angles, prove that $\tan(A - B) = \cot 2B$.
35. If A and B are complementary angles, prove that $\cos(A - B) = \sin 2A = \sin 2B$.

36. If A and B are supplementary angles and n is any real number, prove that

$$\tan(A + nB) = \tan(n - 1)B.$$

37. Express $15 \sin \theta + 8 \cos \theta$ in the form $k \sin(\theta + H)$.
38. Express $-9 \sin \theta - 40 \cos \theta$ in the form $k \sin(\theta + H)$.
39. Express $-\sin \theta + \cos \theta$ in the form $k \sin(\theta + H)$.
40. Express $6 \sin \theta - 8 \cos \theta$ in the form $k \sin(\theta + H)$.
41. Express $12 \sin\left(\theta + \frac{5\pi}{6}\right)$ in the form $a \sin \theta + b \cos \theta$.
42. Express $4 \sin\left(\theta + \frac{7\pi}{4}\right)$ in the form $a \sin \theta + b \cos \theta$.

6.4 DOUBLE-ANGLE FORMULAS

If A is any angle, then

$$\mathbf{\sin 2A = 2 \sin A \cos A,} \tag{7}$$

$$\mathbf{\cos 2A = \cos^2 A - \sin^2 A,} \tag{8a}$$

$$\mathbf{= 1 - 2 \sin^2 A,} \tag{8b}$$

$$\mathbf{= 2 \cos^2 A - 1.} \tag{8c}$$

Proof

$$\begin{aligned} \sin 2A &= \sin(A + A) \\ &= \sin A \cos A + \cos A \sin A \\ &= 2 \sin A \cos A. \end{aligned}$$

$$\begin{aligned} \cos 2A &= \cos(A + A) \\ &= \cos A \cos A - \sin A \sin A \\ &= \cos^2 A - \sin^2 A \\ &= (1 - \sin^2 A) - \sin^2 A \\ &= 1 - 2 \sin^2 A. \end{aligned}$$

Stated in words, Equation (7) says:

(7) **The sine of twice an angle is equal to twice the sine of the angle times the cosine of the angle.**

The student should state the three forms of Equation (8) in words.

Equations (7) and (8) are called *double-angle formulas* because the angle on the left side is double the angle on the right side. Equation (7) could have been written

$$\sin B = 2 \sin \frac{B}{2} \cos \frac{B}{2},$$

because B is the double of $B/2$. In other words, Equations (7) and (8) are used to express the sine and cosine of an angle in terms of the functions of an angle that is half as large. To illustrate,

$$\sin \frac{\pi}{3} = 2 \sin \frac{\pi}{6} \cos \frac{\pi}{6} = 2\left(\frac{1}{2}\right)\left(\frac{\sqrt{3}}{2}\right) = \frac{\sqrt{3}}{2},$$

$$\cos 14\theta = 2 \cos^2 7\theta - 1,$$

$$\cos \frac{8\pi}{9} = 1 - 2 \sin^2 \frac{4\pi}{9}.$$

Furthermore, Equation (7) implies that $\sin A \cos A = \frac{1}{2} \sin 2A$.

6.5 HALF-ANGLE FORMULAS

If θ is any angle, then

$$\sin \frac{\theta}{2} = \pm \sqrt{\frac{1 - \cos \theta}{2}}, \tag{9}$$

$$\cos \frac{\theta}{2} = \pm \sqrt{\frac{1 + \cos \theta}{2}}. \tag{10}$$

The choice of the sign in front of the radical is determined by the quadrant in which $\theta/2$ lies.

Proof Equation (8*b*) says

$$\cos 2A = 1 - 2 \sin^2 A.$$

Let $A = \theta/2$; then $2A = \theta$, and

$$\cos \theta = 1 - 2 \sin^2 \frac{\theta}{2}.$$

Transposing, we obtain

$$2 \sin^2 \frac{\theta}{2} = 1 - \cos \theta$$

$$\sin^2 \frac{\theta}{2} = \frac{1 - \cos \theta}{2}$$

$$\sin \frac{\theta}{2} = \pm \sqrt{\frac{1 - \cos \theta}{2}}.$$

By a similar method, Equation (10) can be derived from Equation (8*c*).

Equations (9) and (10) are called *half-angle formulas* because the angle on the left side is half the angle on the right side. To illustrate,

$$\sin \frac{\pi}{6} = \sqrt{\frac{1 - \cos \frac{\pi}{3}}{2}} = \sqrt{\frac{1 - \frac{1}{2}}{2}} = \sqrt{\frac{\frac{1}{2}}{2}} = \sqrt{\frac{1}{4}} = \frac{1}{2},$$

$$\cos 170° = -\sqrt{\frac{1 + \cos 340°}{2}}.$$

Here the minus sign is chosen because 170° is in Q II and has a negative cosine.

$$\sin C = \pm \sqrt{\frac{1 - \cos 2C}{2}}.$$

Notice that in the double-angle formulas the large angle is on the left side, whereas in the half-angle formulas the small angle is on the left side. In fact, the half-angle formulas are merely the double-angle formulas used backward (reading from right to left).

It is desirable to read the left side of Equation (9) as "the sine of half of θ," rather than to say "the sine of θ over 2," which might be construed as $(\sin \theta)/2$.

Example 1 Prove the identity

$$\cos 3\theta = 4 \cos^3 \theta - 3 \cos \theta.$$

Note that the right side involves $\cos \theta$ only.

Proof

$$\begin{aligned}
\cos 3\theta &= \cos(2\theta + \theta) \\
&= \cos 2\theta \cos \theta - \sin 2\theta \sin \theta \\
&= (2 \cos^2 \theta - 1) \cos \theta - (2 \sin \theta \cos \theta) \cdot \sin \theta \\
&= 2 \cos^3 \theta - \cos \theta - 2 \sin^2 \theta \cos \theta \\
&= 2 \cos^3 \theta - \cos \theta - 2(1 - \cos^2 \theta) \cos \theta \\
&= 2 \cos^3 \theta - \cos \theta - 2 \cos \theta + 2 \cos^3 \theta \\
&= 4 \cos^3 \theta - 3 \cos \theta .
\end{aligned}$$

This identity is frequently used in proving that it is impossible to trisect a general angle with ruler and compass.

Example 2 Express $\cos 20\theta$ in terms of $\sin 5\theta$.

Solution Since 20θ is the double of 10θ, and 10θ is the double of 5θ, we shall employ double-angle formulas:

$$\begin{aligned}
\cos 20\theta &= 1 - 2 \sin^2 10\theta \\
&= 1 - 2(2 \sin 5\theta \cos 5\theta)^2 \\
&= 1 - 8 \sin^2 5\theta \cos^2 5\theta \\
&= 1 - 8 \sin^2 5\theta(1 - \sin^2 5\theta) \\
&= 1 - 8 \sin^2 5\theta + 8 \sin^4 5\theta .
\end{aligned}$$

Example 3 By using half-angle formulas, reduce $\sin^4 A$ to an expression involving no even exponents.

Solution Upon squaring both sides of Equation (9), we obtain

$$\sin^2 \frac{\theta}{2} = \frac{1 - \cos \theta}{2} = \frac{1}{2}(1 - \cos \theta) .$$

Replacing $\theta/2$ with A gives $\sin^2 A = \frac{1}{2}(1 - \cos 2A)$. This equation enables us to change the exponent from 2 to 1, with a doubling of the angle. Hence

$$\begin{aligned}
\sin^4 A = (\sin^2 A)^2 &= [\tfrac{1}{2}(1 - \cos 2A)]^2 \\
&= \tfrac{1}{4}(1 - 2 \cos 2A + \cos^2 2A) .
\end{aligned}$$

If $\cos^2 2A$ is replaced by $\frac{1}{2}(1 + \cos 4A)$, we get

$$\sin^4 A = \tfrac{3}{8} - \tfrac{1}{2}\cos 2A + \tfrac{1}{8}\cos 4A.$$

Such transformations are helpful in the calculus.

Exercise 6.3

1. Use a calculator to show that

$$2 \sin 33° \neq \sin 66°,$$

and that
$$\cos 80° \neq 2 \cos 40°.$$

2. Use double-angle formulas to compute $\sin 2\pi/3$ and $\cos 2\pi/3$ from the functions of $\pi/3$.
3. Compute $\sin 157\frac{1}{2}°$ and $\cos 157\frac{1}{2}°$ from the functions of 315°.
4. Compute $\sin \pi/12$ and $\cos \pi/12$ from the functions of $\pi/6$.

Simplify by reducing to a single term involving only one function of an angle.

5. $\sqrt{\dfrac{1 - \cos 14B}{2}}$
6. $-\sqrt{\dfrac{1 + \cos 222°}{2}}$
7. $5\cos^2 \dfrac{\theta}{4} - 5\sin^2 \dfrac{\theta}{4}$
8. $2\sin^2 B - 1$
9. $18\cos^2 C - 9$
10. $\cos^2 3\theta + \sin^2 3\theta$
11. $\sin 18° \cos 18°$
12. $10 \sin 6\theta \cos 6\theta$
13. Given $\sin A = -\frac{15}{17}$ with $\pi < A < 3\pi/2$, find the exact values of $\sin A/2$ and $\sin 2A$.
14. Given $\cos(C/2) = -\frac{2}{3}$ with $\pi/2 < C/2 < \pi$, find the exact values of $\cos(C/4)$ and $\cos C$.
15. Given $\cos B = \frac{5}{8}$ with $3\pi/2 < B < 2\pi$, find the exact values of $\cos(B/2)$ and $\cos 2B$.
16. Given $\sin 2B = \frac{3}{5}$ with $0 < 2B < \pi/2$, find the exact values of $\sin B$ and $\sin 4B$.
17. Express $\cos 10D$ in terms of $\sin 5D$.
18. Express $\sin B$ in terms of functions of $B/2$.
19. Express $\sin 3C$ in terms of a function of $6C$.
20. Express $\cos 4A$ in terms of a function of $8A$.
21. Express $\cos 100°$ in terms of a function of 200°.

22. Express $\sin 66°$ in terms of a function of $132°$.
23. Express $\sin 320°$ in terms of $\sin 160°$.
24. Express $\cos 250°$ in terms of $\cos 125°$.

Identify as true or false and give reasons. Do not use a calculator.

25. $16 \sin^4 \theta \cos^4 \theta = \sin^4 2\theta$
26. $8 - 16 \sin^2 3B = 8 \cos 6B$
27. $\left(\dfrac{1 + \cos 80°}{2}\right)^{3/2} = \cos^3 40°$
28. $\sqrt{2(1 - \cos 7A)} = \pm 2 \sin \dfrac{7A}{2}$
29. $\cos^2 10° = \cos 20° + \sin^2 10°$
30. $\sqrt{\dfrac{1 + \cos 2B}{18}} = \pm \dfrac{1}{3} \cos B$
31. $\cos^2 \dfrac{C}{3} - \sin^2 \dfrac{C}{3} = \dfrac{1}{3} \cos 2C$
32. $\dfrac{3}{4} \sin \dfrac{3A}{2} = \dfrac{3}{2} \sin \dfrac{3A}{4} \cos \dfrac{3A}{4}$
33. $\sqrt{\dfrac{1 - \cos 400°}{2}} = \sin 200°$
34. $\sin^2 B - \cos^2 B = \cos(-2B)$
35. $\frac{1}{2} \cos 18B + \sin^2 9B = \frac{1}{2}$
36. $\cos^2 2\theta = 1 - 4 \cos^2 \theta + 4 \cos^4 \theta$
37. Reduce $\sin^4 5B$ to $\frac{3}{8} - \frac{1}{2} \cos 10B + \frac{1}{8} \cos 20B$.
38. Reduce $\cos^4 A$ to $\frac{3}{8} + \frac{1}{2} \cos 2A + \frac{1}{8} \cos 4A$.
39. Reduce $\sin^2 \theta \cos^2 \theta$ to $\frac{1}{8}(1 - \cos 4\theta)$.

Prove each of the following identities by transforming one side to the other.

40. $\sec 10B = \dfrac{1}{1 - 2 \sin^2 5B}$
41. $2 \csc 2C = \sec C \csc C$
42. $\tan C + \cot C = 2 \csc 2C$
43. $\sin 3\theta = 3 \sin \theta - 4 \sin^3 \theta$
44. $\cos 4\theta = 1 - 8 \sin^2 \theta \cos^2 \theta$
45. $\dfrac{\sin^3 \theta - \cos^3 \theta}{\sin^2 \theta - \cos^2 \theta} = \dfrac{2 + \sin 2\theta}{2(\sin \theta + \cos \theta)}$

46. $\dfrac{\cos 3\theta}{\sin \theta} + \dfrac{\sin 3\theta}{\cos \theta} = 2 \cot 2\theta$

47. $\sin^2 2A - \cos^2 3A = \sin^2 3A - \cos^2 2A$

48. $\dfrac{\cot C + \tan C}{\cot C - \tan C} = \sec 2C$

49. $\dfrac{\sin A + \sin 2A}{1 + \cos A + \cos 2A} = \tan A$

50. $\sec^2 \dfrac{B}{2} = \dfrac{2 \tan B}{\tan B + \sin B}$

51. $\sec C = \dfrac{\sec^2(C/2)}{2 - \sec^2(C/2)}$

52. $\sin 2B + \cos 2B = \dfrac{(1 + \cot B)^2 - 2}{1 + \cot^2 B}$

53. $\cos 4\theta + \cos 8\theta = (\cos 4\theta + 1)(2 \cos 4\theta - 1)$

54. $\sin 5C = 5 \sin C - 20 \sin^3 C + 16 \sin^5 C$

55. $\boldsymbol{\tan 2A = \dfrac{2 \tan A}{1 - \tan^2 A}}$

56. $\boldsymbol{\tan \dfrac{\theta}{2} = \dfrac{1 - \cos \theta}{\sin \theta} = \dfrac{\sin \theta}{1 + \cos \theta}}$

57. $\dfrac{\cot 6\theta + \cot 2\theta}{\cot 6\theta - \cot 2\theta} = -2 \cos 4\theta$

58. $4(\sin^6 A + \cos^6 A) = 4 - 3 \sin^2 2A$

59. $\dfrac{a \cos 2\theta + b \sin 2\theta - a}{b \cos 2\theta - a \sin 2\theta + b} = \tan \theta$

60. $\dfrac{\cos \dfrac{A}{2} - \sin \dfrac{A}{2}}{\cos \dfrac{A}{2} + \sin \dfrac{A}{2}} = \dfrac{1 - \sin A}{\cos A}$

61. In right triangle ABC, where $C = 90°$, prove that

$$\cos \frac{A}{2} = \sqrt{\frac{b + c}{2c}}.$$

62. Prove that the area of right triangle ABC, where $C = 90°$, is $\frac{1}{4}c^2 \sin 2A$.

63. Verify the identity in Exercise 43 for $\theta = 0, \pi/6, \pi/4, \pi/3$.

64. Verify the identity in Exercise 52 for $B = \pi/6, \pi/4, \pi/3$.

65. Verify the identity in Exercise 49 for $C = 0, \pi/6, \pi/4, \pi/3$.

66. Verify the identity in Exercise 42 for $C = \pi/6, \pi/4, \pi/3$.

67. Use a calculator to verify the identity in Exercise 59 for $a = 7$, $b = 9$, and $\theta = 1$ radian.
68. In proving the identity in Exercise 56, we can use the half-angle formulas (9) and (10) with a double sign preceding each radical. Prove that the sign of $\tan \theta/2$ is always the same as that of $(1 - \cos \theta)/\sin \theta$ for all permissible values of θ. [*Hint:* $1 - \cos \theta$ is never negative. Show that the sign of $\tan \theta/2$ is the same as the sign of $\sin \theta$, for all θ. Use $\sin \theta = 2 \sin \theta/2 \cos \theta/2 = (\tan \theta/2) \cdot 2 \cos^2 \theta/2$. Finish the argument.]

6.6 PRODUCT-TO-SUM FORMULAS; SUM-TO-PRODUCT FORMULAS*

It is sometimes necessary to convert a product of two trigonometric functions into a sum of two functions, and vice versa. For this reason we develop the following formulas. They are not nearly so important as the preceding 10 formulas.

When (1) and (4) are added, we get

$$\sin(A + B) + \sin(A - B) = 2 \sin A \cos B,$$

or

$$\mathbf{\sin A \cos B = \tfrac{1}{2}[\sin(A + B) + \sin(A - B)].} \qquad \mathbf{(11)}$$

Upon subtracting (1) and (4), we obtain

$$\mathbf{\cos A \sin B = \tfrac{1}{2}[\sin(A + B) - \sin(A - B)].} \qquad \mathbf{(12)}$$

Similarly, by adding and subtracting (2) and (5), we get

$$\mathbf{\cos A \cos B = \tfrac{1}{2}[\cos(A + B) + \cos(A - B)],} \qquad \mathbf{(13)}$$

and

$$\mathbf{\sin A \sin B = \tfrac{1}{2}[\cos(A - B) - \cos(A + B)].} \qquad \mathbf{(14)}$$

Equations (11), (12), (13), and (14) are used to convert a product of sines and cosines into a sum or difference of sines and cosines. They are used in certain problems in integral calculus.

When Equation (11) is used backward (from right to left), it converts a sum into a product. Thus

$$\sin(A + B) + \sin(A - B) = 2 \sin A \cos B. \qquad (6.5)$$

*This section may be omitted.

For convenience we shall change notation by making the substitutions

$$A + B = C \qquad \text{and} \qquad A - B = D.$$

Adding these two equations and dividing by 2, we then obtain $A = \frac{1}{2}(C + D)$. Subtracting and dividing by 2, we get $B = \frac{1}{2}(C - D)$. Substituting in (6.5), we obtain

$$\sin C + \sin D = 2 \sin \tfrac{1}{2}(C + D) \cos \tfrac{1}{2}(C - D). \tag{15}$$

Similarly,

$$\sin C - \sin D = 2 \cos \tfrac{1}{2}(C + D) \sin \tfrac{1}{2}(C - D), \tag{16}$$

$$\cos C + \cos D = 2 \cos \tfrac{1}{2}(C + D) \cos \tfrac{1}{2}(C - D), \tag{17}$$

$$\cos C - \cos D = -2 \sin \tfrac{1}{2}(C + D) \sin \tfrac{1}{2}(C - D). \tag{18}$$

Equations (15), (16), (17), and (18) are used to convert sums and differences into products. Equation (16) is sometimes employed in the derivation of the formula for the derivative of the sine in differential calculus.

Equation (1) should not be confused with Equation (15). The former deals with the sine of the sum of two angles; the latter deals with the sum of the sines of two angles.

Example 1 Reduce $\cos 5\theta \cos 3\theta$ to a sum.

Solution Using Equation (13), we get

$$\begin{aligned} \cos 5\theta \cos 3\theta &= \tfrac{1}{2}[\cos(5\theta + 3\theta) + \cos(5\theta - 3\theta)] \\ &= \tfrac{1}{2} \cos 8\theta + \tfrac{1}{2} \cos 2\theta. \end{aligned}$$

Example 2 Prove the identity

$$\frac{\sin 7\theta - \sin 3\theta}{\cos 7\theta + \cos 3\theta} = \tan 2\theta.$$

Proof

$$\begin{aligned} \frac{\sin 7\theta - \sin 3\theta}{\cos 7\theta + \cos 3\theta} &= \frac{2 \cos \frac{1}{2}(7\theta + 3\theta) \sin \frac{1}{2}(7\theta - 3\theta)}{2 \cos \frac{1}{2}(7\theta + 3\theta) \cos \frac{1}{2}(7\theta - 3\theta)} \\ &= \frac{\sin 2\theta}{\cos 2\theta} \\ &= \tan 2\theta. \end{aligned}$$

Exercise 6.4

Express each of the following as an algebraic sum of sines and cosines.

1. $6 \sin 5\theta \sin 9\theta$
2. $12 \cos 6\theta \cos \theta$
3. $2 \sin \dfrac{3\theta}{2} \cos \dfrac{5\theta}{2}$
4. $\sin 22° \sin 77°$
5. $100 \cos 75° \cos 25°$
6. $16 \cos 11° \sin 55°$

Express each of the following as a product.

7. $\sin 4\theta + \sin 5\theta$
8. $5 \cos B - 5 \cos(B + 60°)$
9. $\sin 13A - \sin 17A$
10. $\cos 70° + \sin 80°$

Prove the following identities. Do not use a calculator.

11. $\cos 50° + \sin 20° = \cos 10°$ [*Hint:* $\sin B = \cos(90° - B)$.]
12. $\dfrac{\cos 5\theta - \cos 7\theta}{\sin 8\theta + \sin 4\theta} = \dfrac{\sin \theta}{\cos 2\theta}$
13. $\dfrac{\sin 250° - \sin 110°}{\cos 100° - \cos 40°} = 2$
14. $\sin A + \cos B = 2 \sin\left(\dfrac{A - B}{2} + 45°\right) \cos\left(\dfrac{A + B}{2} - 45°\right)$
15. $\cos \theta + \cos 2\theta + \cos 3\theta = (2 \cos \theta + 1) \cos 2\theta$
16. $\dfrac{\cos 6\theta + \cos 12\theta}{\sin 14\theta - \sin 4\theta} + \dfrac{\sin 7\theta - \sin \theta}{\cos 9\theta + \cos \theta} = \dfrac{2 \cos 2\theta}{\sin 10\theta}$
17. $\cos^2 C - \cos^2 D = \sin(D + C) \sin(D - C)$
18. $1 - \tan 5B \tan 2B = \dfrac{2 \cos 7B}{\cos 7B + \cos 3B}$
19. $\sin 2A - \sin 4A + \sin 6A = 4 \sin A \cos 2A \cos 3A$
20. If $A + B + C = 90°$, prove that

$$\sin 2A + \sin 2B + \sin 2C = 4 \cos A \cos B \cos C.$$

[*Hint:* $\sin 2C = 2 \sin C \cos C$, $\sin(A + B) = \cos C$.]

21. Verify Equation (14) for $B = A$.
22. Verify Equation (17) for $D = 0$.

KEY TERMS

Sum and difference formulas, double-angle formulas, half-angle formulas, product-to-sum formulas

REVIEW EXERCISES

Prove or disprove Exercises 1 through 5.

1. $\sin(\pi + \theta) = -\sin\theta$
2. $\sin(\pi - \theta) = \sin\theta$
3. $\tan\theta + \tan\pi = \tan(\theta + \pi)$
4. $\cos(A - B) + \cos B = \cos A$
5. $\sin^2(A - B) = 1 - \cos^2(B - A)$
6. Express $5\sin\theta - 12\cos\theta$ in the form $k\sin(\theta + H)$.
7. Express $\tan\theta/6$ in terms of $\tan\theta/2$ and $\tan\theta/3$. Verify your expression when $\theta = \pi/2$.
8. Express $\sin 15\theta$ in terms of $\sin 5\theta$ and $\cos 5\theta$.
9. Express $\cos 15A - \cos 9A$ as a product.
10. Prove that in any triangle, *Mollweide's equations* hold:

$$\frac{a + b}{c} = \frac{\cos\frac{1}{2}(A - B)}{\sin\frac{1}{2}C},$$

and

$$\frac{a - b}{c} = \frac{\sin\frac{1}{2}(A - B)}{\cos\frac{1}{2}C}.$$

(*Hint:* Use the law of sines, Equation (15) of Section 6.6, and the double-angle formulas.)

7

Trigonometric Equations

7.1 CONDITIONAL TRIGONOMETRIC EQUATIONS

As we saw in Chapter 5, a conditional equation is an equation that is true for some, but *not all,* permissible values of the variables involved. Also in that chapter we learned to prove identities; that is, to demonstrate that the identity is true for all permissible values of the variable. In this chapter we shall discuss how to find all values of the variable for which a conditional equation is true. This is known as solving the equation.

A solution of a conditional trigonometric equation is a value of the variable that satisfies the equation. For example, $\theta = \pi$ is a solution of the equation $\sin\theta = 1 + \cos\theta$. Unless stated to the contrary, to solve a conditional trigonometric equation shall mean to find all solutions in the interval from 0 to 2π, that is, all θ such that $0 \leq \theta < 2\pi$.

7.2 SOLVING A TRIGONOMETRIC EQUATION ALGEBRAICALLY

The process of finding the solutions of a trigonometric equation involves algebraic as well as trigonometric methods. There is no general rule, but the following suggestions will take care of most cases.

(A) *If only one function of a single variable is involved, solve algebraically for the values of the function.* Then determine the corresponding variables.

Example 1 Solve for θ:

$$4\cos^2\theta = 3.$$

Solution

$$\cos^2\theta = \tfrac{3}{4}.$$

Hence

$$\cos\theta = \pm\frac{\sqrt{3}}{2}.$$

$\cos\theta = \dfrac{\sqrt{3}}{2}$	$\cos\theta = -\dfrac{\sqrt{3}}{2}$
$\theta = \dfrac{\pi}{6}, \dfrac{11\pi}{6}$	$\theta = \dfrac{5\pi}{6}, \dfrac{7\pi}{6}$

Therefore, arranging the solutions in order of size,

$$\theta = \frac{\pi}{6}, \frac{5\pi}{6}, \frac{7\pi}{6}, \frac{11\pi}{6}.$$

Example 2 Solve for θ:

$$4\sin^2\theta = 3\sin\theta.$$

Solution Transpose $3\sin\theta$ to make the right side 0; then factor:

$$\sin\theta(4\sin\theta - 3) = 0.$$

Set each factor equal to zero:

$\sin\theta = 0$	$4\sin\theta - 3 = 0$
$\theta = 0, \pi$	$\sin\theta = \frac{3}{4} = 0.75$
	$\theta = 0.8481, 2.2935$

Hence $\qquad \theta = 0, 0.8481, 2.2935, \pi.$

The student is warned to guard against dividing both sides of this equation by the variable factor $\sin\theta$. Had this been done, the solutions 0 and π would have been lost.

Example 3 Solve for θ:

$$\sin^2\theta - 5\sin\theta - 3 = 0.$$

Solution The left side is not factorable. The roots of the quadratic equation $ax^2 + bx + c = 0$ are

$$x = \frac{-b \pm \sqrt{b^2 - 4ac}}{2a}.$$

In this case $a = 1$, $b = -5$, $c = -3$, and $x = \sin \theta$. The formula gives us

$$\sin \theta = \frac{5 \pm \sqrt{25 + 12}}{2} = \frac{5 \pm 6.0828}{2}.$$

$\sin \theta = 5.5414$	$\sin \theta = -0.5414$
Impossible	(Related angle is 0.572 radian)
	θ is in Q III, IV
	$\theta = 3.714, 5.711$

Hence $$\theta = 3.714, 5.711.$$

(*B*) *If one side of the equation is zero and the other side is factorable, set each such factor equal to zero* and solve the resulting equations.

Example 4 Solve for θ:

$$\cos 2\theta \csc \theta - 2 \cos 2\theta = 0.$$

Solution Factor the left side:

$$\cos 2\theta(\csc \theta - 2) = 0.$$

Set each factor equal to zero:

$\cos 2\theta = 0$	$\csc \theta - 2 = 0$
$2\theta = \frac{\pi}{2}, \frac{3\pi}{2}, \frac{5\pi}{2}, \frac{7\pi}{2}$	$\csc \theta = 2$
$\theta = \frac{\pi}{4}, \frac{3\pi}{4}, \frac{5\pi}{4}, \frac{7\pi}{4}$	$\sin \theta = \frac{1}{2}$
	$\theta = \frac{\pi}{6}, \frac{5\pi}{6}$

Hence $$\theta = \frac{\pi}{6}, \frac{\pi}{4}, \frac{3\pi}{4}, \frac{5\pi}{6}, \frac{5\pi}{4}, \frac{7\pi}{4}.$$

In order to solve $\cos 2\theta = 0$ for all values of θ on the interval $0 \leq \theta < 2\pi$, it is necessary to find all values of 2θ on the interval $0 \leq 2\theta < 4\pi$.

(*C*) *If several functions of a single variable are involved,* use the fundamental relations to *express everything in terms of a single function.* Then proceed as in (*A*).

Example 5 Solve for θ:

$$\sin^2\theta - \cos^2\theta - \cos\theta = 1.$$

Solution Replace $\sin^2\theta$ with $1 - \cos^2\theta$; collect terms:

$$2\cos^2\theta + \cos\theta = 0$$

$$\cos\theta(2\cos\theta + 1) = 0.$$

$\cos\theta = 0$	$2\cos\theta + 1 = 0$
$\theta = \frac{\pi}{2}, \frac{3\pi}{2}$	$\cos\theta = -\frac{1}{2}$
	$\left(\text{Related angle is } \frac{\pi}{6}\right)$
	θ is in Q II, III
	$\theta = \frac{2\pi}{3}, \frac{4\pi}{3}$

Hence $$\theta = \frac{\pi}{2}, \frac{2\pi}{3}, \frac{4\pi}{3}, \frac{3\pi}{2}.$$

Example 6 Solve for θ:

$$\sec\theta = \tan\theta - 1.$$

Solution Replace $\sec\theta$ with $\pm\sqrt{1 + \tan^2\theta}$; square both sides:

$$1 + \tan^2\theta = \tan^2\theta - 2\tan\theta + 1$$

$$2\tan\theta = 0;\ \tan\theta = 0.$$

Hence $$\theta = 0, \pi$$

Inasmuch as we squared the equation, we may have introduced some extraneous roots. Consequently, we must check these values in the original equation.

Check for $\theta = 0$:

$$\sec 0 = \tan 0 - 1,$$

$$1 = -1, \text{ which is false.}$$

Check for $\theta = \pi$:

$$\sec \pi = \tan \pi - 1$$

$$-1 = -1, \text{ which is true.}$$

Therefore $\theta = 0$ is an *extraneous* root and $\theta = \pi$ is a true root. The only solution of the equation is $\theta = \pi$.

(*D*) *If several variables are involved,* use the fundamental identities to *express everything in terms of a single variable.* Then proceed as in (*C*).

Example 7 Solve for θ:

$$\cos 2\theta = 3 \sin \theta + 2.$$

Solution It is not convenient to replace $\sin \theta$ with a function of 2θ because this would introduce the radical

$$\pm\sqrt{\frac{1 - \cos 2\theta}{2}}.$$

It is better to replace $\cos 2\theta$ with one of its three forms. Since the right side involves only $\sin \theta$, we choose the form $\cos 2\theta = 1 - 2 \sin^2 \theta$ in order to reduce everything immediately to the same function of a single angle. This gives us

$$1 - 2 \sin^2 \theta = 3 \sin \theta + 2,$$

or

$$2 \sin^2 \theta + 3 \sin \theta + 1 = 0.$$

Factor:

$$(2 \sin \theta + 1)(\sin \theta + 1) = 0.$$

$\sin \theta = -\frac{1}{2}$	$\sin \theta = -1$
$\theta = \frac{7\pi}{6}, \frac{11\pi}{6}$	$\theta = \frac{3\pi}{2}$

Hence

$$\theta = \frac{7\pi}{6}, \frac{3\pi}{2}, \frac{11\pi}{6}.$$

Exercise 7.1

Solve the following equations for θ on the interval $0 \le \theta < 2\pi$.

1. $\cos\left(\theta + \frac{\pi}{36}\right) = -\frac{\sqrt{2}}{2}$
2. $\tan\left(\theta - \frac{\pi}{15}\right) = \frac{1}{\sqrt{3}}$
3. $4\cos^2\theta = 3$
4. $\csc^2\theta = 2$
5. $3\tan^2\theta = 1$
6. $16\sin^4\theta = 9$
7. $(1 + \sqrt{2}\cos\theta)(1 + \tan\theta) = 0$
8. $(1 - \tan\theta)(\sqrt{3} + 2\cos\theta) = 0$
9. $(1 + \sqrt{2}\sin\theta)(\sec\theta - 2) = 0$
10. $(\tan\theta + \sqrt{3})(\csc\theta + \sqrt{2}) = 0$
11. $\tan^2\theta = \sqrt{3}\tan\theta$
12. $\sec^3\theta = 2\sec\theta$

Solve the following equations for θ on the interval $0 \le \theta < 2\pi$.

13. $\sin\theta = 1 - 2\sin^2\theta$
14. $2\sin\theta\cos\theta - 2\sin\theta + \cos\theta - 1 = 0$
15. $\sec\theta(\sin^2\theta - 3) = 0$
16. $2\cos^2\theta - 5\cos\theta + 2 = 0$
17. $2\cos\theta\tan\theta + \sqrt{3}\tan\theta = 0$
18. $\sqrt{3}\cot\theta - 2\sin\theta\cot\theta = 0$
19. $2\tan^2\theta - 3\sec\theta = 0$
20. $\cos\theta + 10\sec\theta = 7$
21. $6\csc\theta\sec\theta = \sec\theta$
22. $4\sin^2\theta = 4\cos\theta + 1$
23. $\cos 2\theta + 6\sin^2\theta = 4$
24. $1 + \sqrt{3}\cos\theta + \cos 2\theta = 0$
25. $\sin 2\theta = \sqrt{2}\cos\theta$
26. $\cos 2\theta = 2 - 2\sin^2\theta$
27. $\sin\theta = \sqrt{2}\cos\frac{\theta}{2}$
28. $\sin 4\theta\csc 2\theta + 8\sin^2\theta = 5$

Solve the following equations for θ on the interval $0 \le \theta \le 2\pi$.

29. $\cos 6\theta = 5\cos 3\theta + 2$
30. $(\sin 2\theta - \cos 2\theta)^2 = 1$
31. $\sec^2\theta + \csc^2\theta = \sec^2\theta\csc^2\theta$
32. $\frac{\tan 4\theta - \tan\theta}{1 + \tan 4\theta\tan\theta} = \sqrt{3}$
33. $2\sec\theta + \tan\theta + \cot\theta = 0$
34. $\cos\theta + 2 = \sqrt{3}\sin\theta$
35. $1 + \sin\theta = \sqrt{3}\cos\theta$
36. $\cos\theta = \sin\theta - 1$
37. $\tan\theta - \sqrt{3} = \sqrt{3}\sec\theta$
38. $\sqrt{3}\cot\theta = \csc\theta + 1$
39. $3\tan\theta\sec\theta = -2$
40. $\sin\theta + \cos\theta = -\sqrt{2}$

Solve for θ on the interval $0 \leq \theta < 2\pi$. In some instances a calculator will be needed. Find θ to the nearest hundredth of a radian.

41. $2 \sec^2 \theta - 5 \tan \theta + 1 = 0$
42. $\csc \theta \cot \theta = 1000 \sec \theta \tan \theta$
43. $25 \sin^2 \theta = 144 \cos^2 \theta$
44. $5 \sin \theta + 4 \cos \theta = 0$
45. $2 \cos^2 \theta + 4 = 7 \cos \theta$
46. $\cot^2 \theta - 4 \cot \theta - 3 = 0$
47. $\tan^2 \theta + 2 \tan \theta = 4$
48. $\sin^2 \theta = 8 \sin \theta + 3$
49. $3 \sin \theta + 4 \cos \theta = -2.5$ [*Hint:* Express the left side in the form $k \sin(\theta + H)$.]
50. $8 \sin \theta - 6 \cos \theta = 5$
51. $-15 \sin \theta + 8 \cos \theta = 17$
52. $\sin \theta + \sqrt{3} \cos \theta = 1$
53. $\sin(\cos \theta) = 0.2$
54. $\cos 4\theta - \cos\left(2\theta - \dfrac{\pi}{3}\right) = 0$
55. $\cos 5\theta + \cos \theta = 2 \cos 3\theta$
56. $\sin 4\theta + \sin 2\theta = \cos \theta$
57. $2 \cos^2 \theta + 1 = 3 \sec 2\theta$
58. $1 + \sin 2\theta = \tan\left(\theta + \dfrac{\pi}{4}\right)$
59. $3\sqrt{2} \sin \dfrac{\theta}{2} - \sqrt{1 - \cos \theta} = 6$
60. $\cos \theta - \sin \theta = \dfrac{\sqrt{6}}{2}$
61. $\cot \theta \csc \theta = -\sqrt{2}$
62. $\sin^2 \theta - 2 \cos^2 \theta + \sin \theta \cos \theta = 0$

KEY TERMS

Conditional equation, solution of a conditional equation

REVIEW EXERCISES

Solve the following equations for θ in the interval $0 \leq \theta < 2\pi$:

1. $\csc^2 \theta = 4$
2. $\cos 2\theta - 2 \cos^2 \theta = 0$
3. $\sin \theta = \sin \dfrac{\theta}{2}$
4. $\cos 6\theta + \sin^2 3\theta = 0$
5. $\sin^2 \theta = 6 \cos \theta + 6$
6. $\tan^2 \theta + \tan \theta - 6 = 0$
7. $\sin 4\theta = \sin\left(2\theta - \dfrac{\pi}{6}\right)$
8. $\sqrt{2} \sin \dfrac{\theta}{2} + \sqrt{1 - \cos \theta} = \sqrt{2}$

8

Graphs of the Trigonometric Functions

8.1 PERIODIC FUNCTIONS

We saw in Sections 1.6 and 4.4 that given any trigonometric function, its value at any θ is equal to its value at any number representing an angle coterminal with θ. Thus for any number θ, $\sin\theta = \sin(\theta + 2\pi) = \sin(\theta - 2\pi) = \sin(\theta + 2n\pi)$, where n is any integer. Furthermore, 2π is the smallest positive number p with the property that

$$\sin\theta = \sin(\theta + p),$$

for *all* θ. We say that 2π is the period of the sine function according to the definition:

A function $f(\theta)$ is said to be periodic and of period p, provided that

$$f(\theta + p) = f(\theta) \qquad \textbf{for all } \theta,$$

where p is the smallest positive constant for which this is true.

The cosine function is periodic with a period of 2π, because

$$\cos(\theta + 2\pi) = \cos\theta, \qquad \text{for all } \theta,$$

and because this statement is not true if 2π is replaced by any smaller positive number, such as π.

8.2 GRAPHS OF THE TRIGONOMETRIC FUNCTIONS

We wish to investigate the behavior of $\sin\theta$ as θ increases from 0 to 2π. For this purpose, place θ in standard position and choose the point P on the terminal side of θ and exactly 1 unit from the origin. According to Section 4.4 and as seen in Figure 8.1, the coordinates of P are $(\cos\theta, \sin\theta)$.

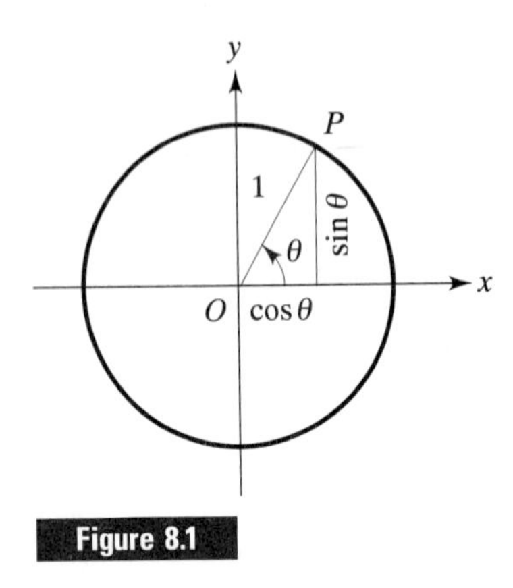

Figure 8.1

Visualize θ increasing from 0 to $\pi/2$. In that event, $\sin\theta$ increases from 0 to 1 and $\cos\theta$ decreases from 1 to 0. As θ increases from $\pi/2$ to π, $\sin\theta$ decreases from 1 to 0 while $\cos\theta$ decreases from 0 to -1. Continuing, as θ goes from π to $3\pi/2$, $\sin\theta$ decreases from 0 to -1 and $\cos\theta$ increases from -1 to 0. Finally, as θ moves from $3\pi/2$ to 2π, $\sin\theta$ increases from -1 to 0 while $\cos\theta$ increases from 0 to 1. We summarize this in the table:

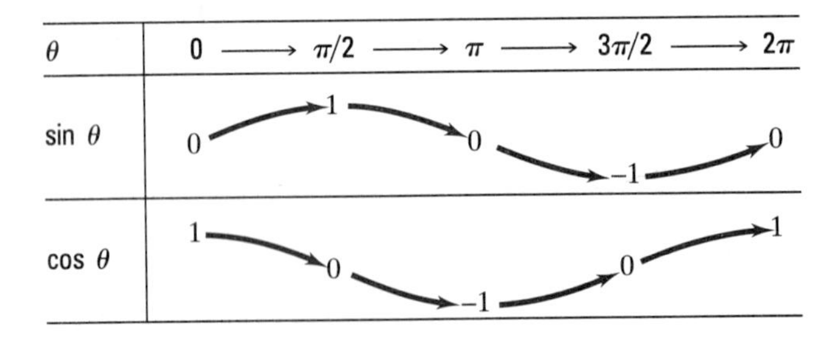

θ	0	$\longrightarrow$ $\pi/2$	$\longrightarrow$ π	$\longrightarrow$ $3\pi/2$	$\longrightarrow$ 2π
$\sin\theta$	0	1	0	-1	0
$\cos\theta$	1	0	-1	0	1

A complete "picture story" of the variation of the sine is presented by its graph. Let us draw a system of coordinate axes and label the horizontal axis as θ and the vertical axis as $\sin\theta$. Since the radian is the natural measure of angles, let the θ axis be laid off in radians. This means that the number 1 on the vertical scale and the number 1 (radian) on the horizontal scale are represented by the same distance. *Hence π on the θ axis should be π times as long as* 1 *unit on the* (*sin* θ) *axis.* Using trigonometric tables and the related-angle theorem, we form the following table:

θ in degrees	0°	30°	60°	90°	120°	150°	180°	210°	240°	270°	300°	330°	360°
θ in radians	0	$\frac{\pi}{6}$	$\frac{\pi}{3}$	$\frac{\pi}{2}$	$\frac{2\pi}{3}$	$\frac{5\pi}{6}$	π	$\frac{7\pi}{6}$	$\frac{4\pi}{3}$	$\frac{3\pi}{2}$	$\frac{5\pi}{3}$	$\frac{11\pi}{6}$	2π
$\sin\theta$	0	0.5	0.87	1	0.87	0.5	0	−0.5	−0.87	−1	−0.87	−0.5	0

After plotting these values on the coordinate axes, we obtain the curve in Figure 8.2.

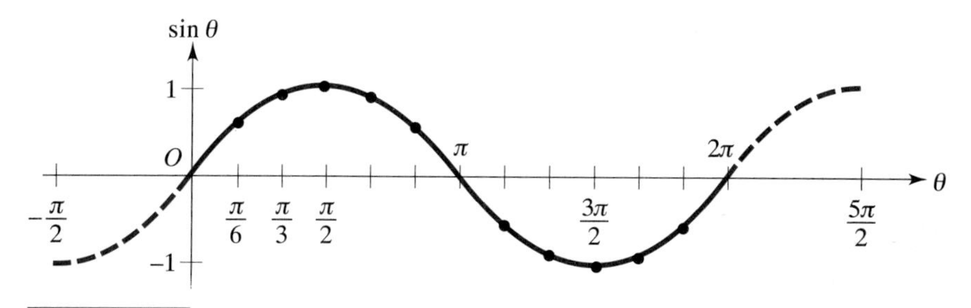

Figure 8.2

In a similar manner we may make a table of values for $\cos\theta$ and plot its graph as shown in Figure 8.3.

Students should practice drawing one period, $0 \leq \theta \leq 2\pi$, of each of these curves until they can make a sketch from memory.

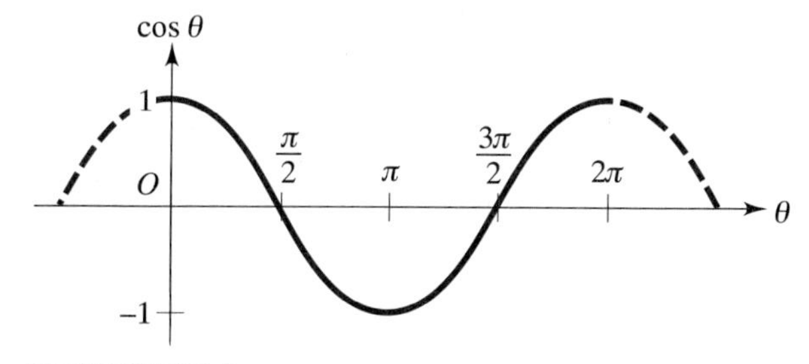

Figure 8.3

Furthermore, the sine curve recalls the related-angle theorem. For example, it is obvious from the curve (see Figure 8.4) that

$$\sin\frac{5\pi}{6} = \sin\frac{\pi}{6} = \frac{1}{2}$$

and

$$\sin\frac{5\pi}{4} = \sin\frac{7\pi}{4} = -\frac{\sqrt{2}}{2}.$$

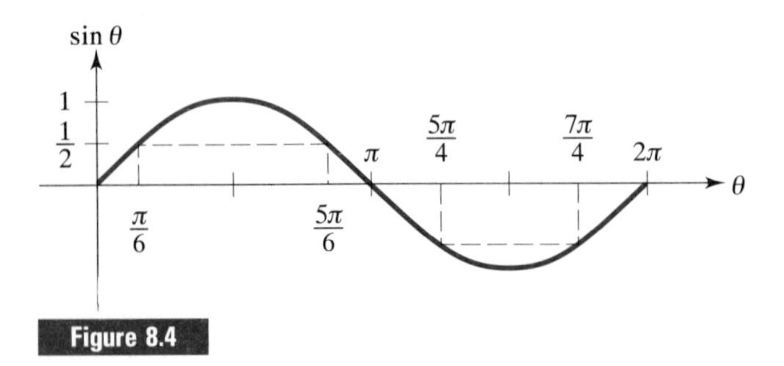

Figure 8.4

Because of its waveform, the sine curve is very important in the study of wave motion in electrical engineering and physics. The maximum distance of the curve from the θ axis is called the *amplitude* of the curve or wave. The period of the function representing the wave is called the period (or *wavelength*) of the curve. The sine curve has an amplitude of 1 and a period of 2π. The student can see from Figure 8.4 that $\sin\theta$ does not have a period that is less than 2π.

In discussing $\tan\theta$ we put θ in standard position and we draw the vertical line $x = 1$. For any θ for which $0 \leq \theta < \pi/2$, let P be the point of intersection of the terminal side of θ and the vertical line $x = 1$ as shown in Figure 8.5. Then $PA = \tan\theta$. We see that as θ increases from 0 to $\pi/2$, $\tan\theta$ increases from 0 and becomes large without bound as P moves up the line $x = 1$. We can make $\tan\theta$ as large as we wish by taking θ closer to $\pi/2$. Of course, $\tan\pi/2$ is undefined.

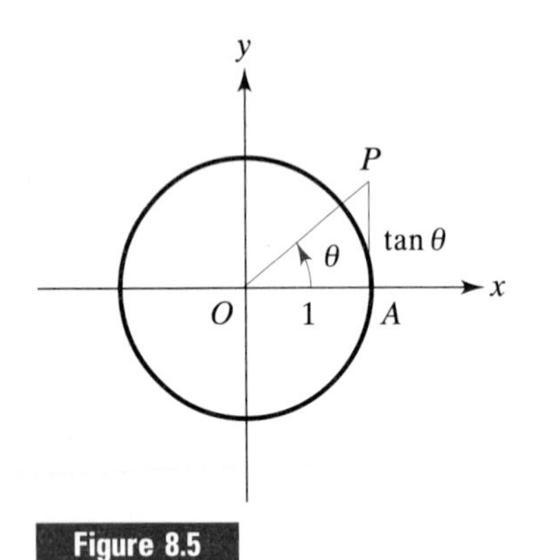

Figure 8.5

For $\pi/2 < \theta \leq \pi$, we make a similar construction, as shown in Figure 8.6, with the line $x = -1$ replacing the line $x = 1$. Since $PB/-1 = \tan\theta$, we see that $PB = -\tan\theta$. Thus for values of θ slightly larger than

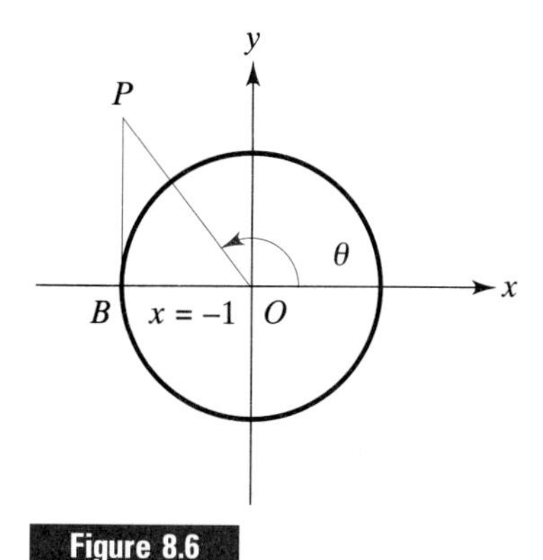

Figure 8.6

$\pi/2$, tan θ is negative and large in absolute value. As θ increases to π, tan θ *increases* to 0. We continue this analysis for $\pi \leq \theta \leq 2\pi$, $\theta \neq 3\pi/2$, and summarize the results in the chart:

θ	Q I $0 \to \pi/2$	Q II $\pi/2 \to \pi$	Q III $\pi \to 3\pi/2$	Q IV $3\pi/2 \to 2\pi$
sin θ	$0 \nearrow 1$	$1 \searrow 0$	$0 \searrow -1$	$-1 \nearrow 0$
cos θ	$1 \searrow 0$	$0 \searrow -1$	$-1 \nearrow 0$	$0 \nearrow 1$
tan θ	$0 \nearrow \infty$	$-\infty \nearrow 0$	$0 \nearrow \infty$	$-\infty \nearrow 0$
cot θ	$\infty \searrow 0$	$0 \searrow -\infty$	$\infty \searrow 0$	$0 \searrow -\infty$
sec θ	$1 \nearrow \infty$	$-\infty \nearrow -1$	$-1 \searrow -\infty$	$\infty \searrow 1$
csc θ	$\infty \searrow 1$	$1 \nearrow \infty$	$-\infty \nearrow -1$	$-1 \searrow -\infty$

With the aid of a table of values for tan θ, we are able to draw a graph for tan θ as seen in Figure 8.7. Note that the vertical lines $\theta = \pi/2 + n\pi$, for n an integer, are never touched by the curve, but that the graph gets arbitrarily close to those lines as θ gets very close to the values $\pi/2 + n\pi$. We say these vertical lines are **asymptotes** of the curve.

Similar analyses may be made for the other trigonometric functions. The graphs of cot θ, sec θ, and csc θ are shown in Figures 8.8, 8.9, and 8.10, respectively. Students should sketch each one of them by preparing a table of values, plotting the points, and then drawing a smooth curve through the points. It is important to emphasize that students should be able to make a quick sketch of sin θ, cos θ, and tan θ from memory.

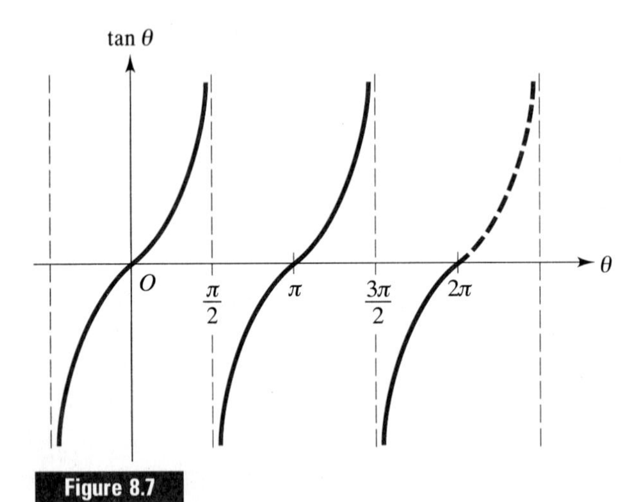

Figure 8.7

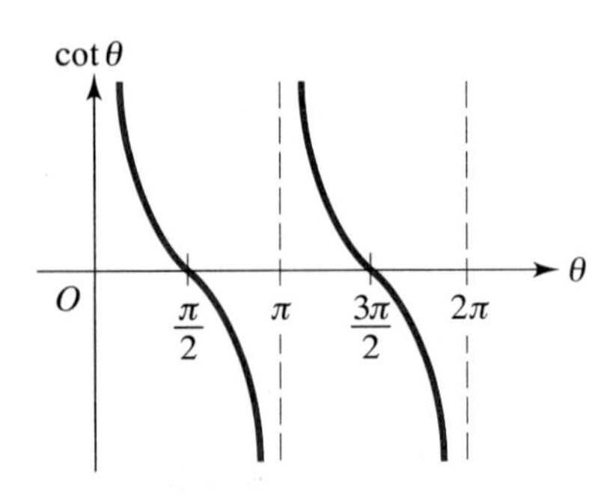

Figure 8.8

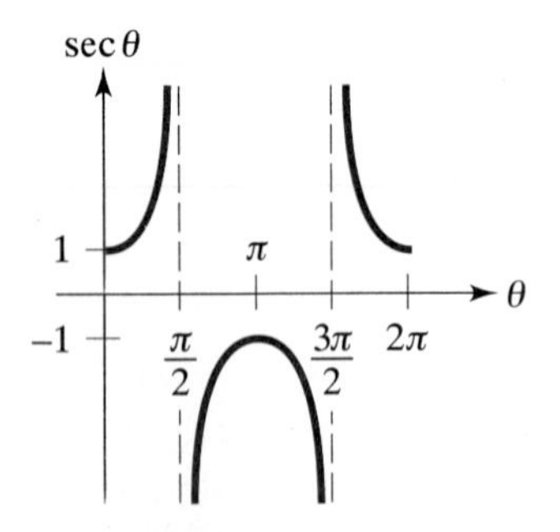

Figure 8.9

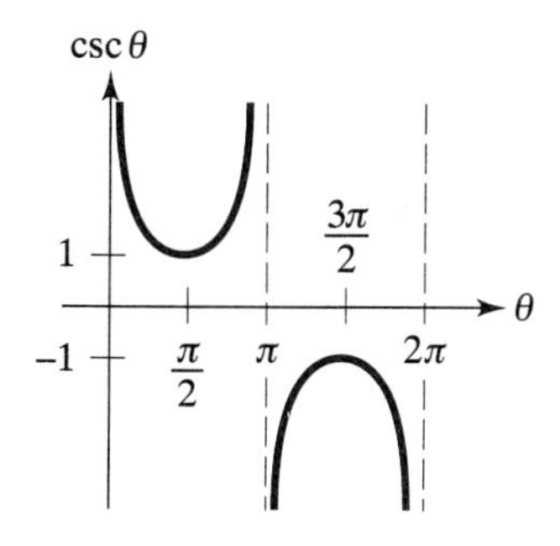

Figure 8.10

Example Solve the equation $\sin\theta = -\sqrt{3}/2$ for all values of θ on the interval $0 \le \theta < 2\pi$. (That is, solve for all values of θ that are equal to or greater than 0 and less than 2π.) Use the sine curve to identify the proper quadrants and check the related-angle theorem.

Solution The sine is negative (the sine curve lies below the θ axis) in Q III and Q IV. Remember that $\sin \pi/3 = \sqrt{3}/2$. Since $\sin\theta = -\sqrt{3}/2$, it follows that θ must be equal to those angles in Q III and Q IV that have $\pi/3$ for their related angle (see Figure 8.11).

In Q III: $$\theta = \pi + \frac{\pi}{3} = \frac{4\pi}{3}.$$

In Q IV: $$\theta = 2\pi - \frac{\pi}{3} = \frac{5\pi}{3}.$$

Hence if

$$\sin\theta = -\frac{\sqrt{3}}{2},$$

then

$$\theta = \frac{4\pi}{3}, \frac{5\pi}{3}.$$

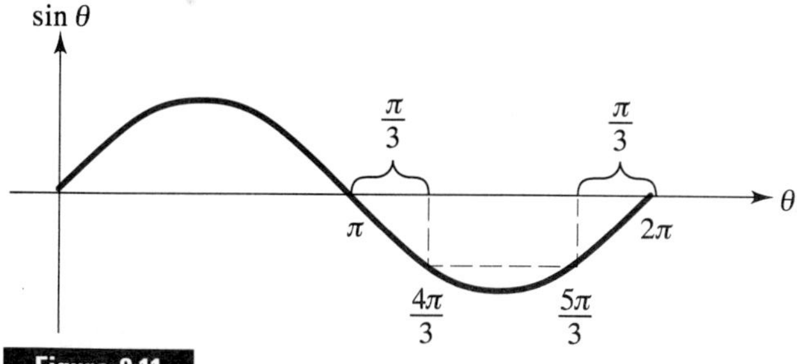

Figure 8.11

Exercise 8.1

1. Sketch the sine curve by drawing a smooth curve through the points obtained by assigning to θ the values $-\pi/2$, $-\pi/3$, $-\pi/4$, $-\pi/6$, 0, $\pi/6$, $\pi/4, \ldots, 2\pi$.
2. Sketch the cosine curve. Locate points in steps of $\pi/12$ from 0 to 2π.
3. Sketch the tangent curve. Locate points in steps of $\pi/12$ from $-\pi/2$ to 2π.
4. Sketch the cotangent curve.
5. Sketch the secant curve.
6. Sketch the cosecant curve.
7. Sketch $y = -\sin\theta$.
8. Sketch $y = -\cos\theta$.
9. What is the period of $\tan\theta$?
10. Discuss in detail the variation of the cosecant.

Solve the following equations for values of θ on the interval $0 \le \theta < 2\pi$. Use the curves to identify the proper quadrants and check the related-angle theorem. Do not use a calculator.

11. $\cos\theta = 0$
12. $\sin\theta = -\dfrac{\sqrt{3}}{2}$
13. $\sin\theta = 1$
14. $\cos\theta = \frac{1}{2}$
15. $\sin\theta = \frac{1}{2}$
16. $\cos\theta = \dfrac{\sqrt{2}}{2}$
17. $\cos\theta = -\frac{1}{2}$
18. $\sin\theta = -\dfrac{\sqrt{2}}{2}$
19. $\tan\theta = -\dfrac{1}{\sqrt{3}}$
20. $\csc\theta = -2$
21. $\sec\theta = \dfrac{2}{\sqrt{3}}$
22. $\cot\theta = \sqrt{3}$

Solve the following equations for values of s on the interval $0 \le s < 2\pi$. Do not use a calculator.

23. $\csc s = 0$
24. $\tan s = 0$
25. $\cot s = -1$
26. $\sec s = -1$
27. $\cos s = -\dfrac{\sqrt{3}}{2}$
28. $\sec s = 1$
29. $\sin s = \dfrac{\sqrt{3}}{2}$
30. $\tan s = 1$

Use a calculator to solve the following equations for values of θ on the interval $0 \le \theta < 2\pi$. Obtain results correct to the nearest hundredth of a radian.

31. $\sin \theta = \frac{5}{6}$
32. $\cos \theta = -\frac{3}{8}$
33. $\tan \theta = -15$
34. $\cot \theta = 30$
35. $\cot \theta = -0.12$
36. $\tan \theta = 0.77$
37. $\cos \theta = 0.7955$
38. $\sin \theta = -0.3015$
39. State the maximum and minimum values achieved by the following functions of θ:
    (a) $10 + 3 \cos \theta$
    (b) $-11 - 5 \sin^3 \theta$
    (c) $12 - \cos^2 \theta$
40. State the maximum and minimum values achieved by the following functions of θ, as in Exercise 39.
    (a) $3 - 4 \sin \theta$
    (b) $14 + 2 \cos^5 \theta$
    (c) $-8 + 9 \sin^4 \theta$

In Exercises 41 through 45, either prove or disprove the statement. (Prove that the equation is, or is not, an identity.)

41. $\sin n\pi = 0$ (n an integer)
42. $\cos n\pi = (-1)^n$ (n an integer)
43. $\cos \dfrac{n\pi}{2} = 0$ (n an odd integer)
44. $\sin \dfrac{n\pi}{2} = (-1)^{(n-1)/2}$ (n an odd integer)
45. $\tan \dfrac{n\pi}{4} = (-1)^{(n-1)/2}$ (n an odd integer)

8.3 THE GRAPH OF $y = a \sin(bx + c)$

The graph of $y = \sin x$ (Figure 8.2) is a wave that starts at 0, goes up to 1, drops back to 0, then to -1, and finally climbs back up to 0. We are discussing the changes that take place in y as x varies from 0 to 2π. The amplitude of the wave (i.e., the maximum distance from the x axis) is 1. The period, or wavelength, which is the same as the period of the function $\sin x$, is 2π. The graph of $y = 3 \sin x$ (Figure 8.12) is another wave, with amplitude 3 and period 2π. For any value of x, the magnitude of y in $y = 3 \sin x$ is 3 times the value of y in $y = \sin x$. The coefficient 3 merely *stretches* the *height* of the sine curve by the multiple 3 without affecting its period.

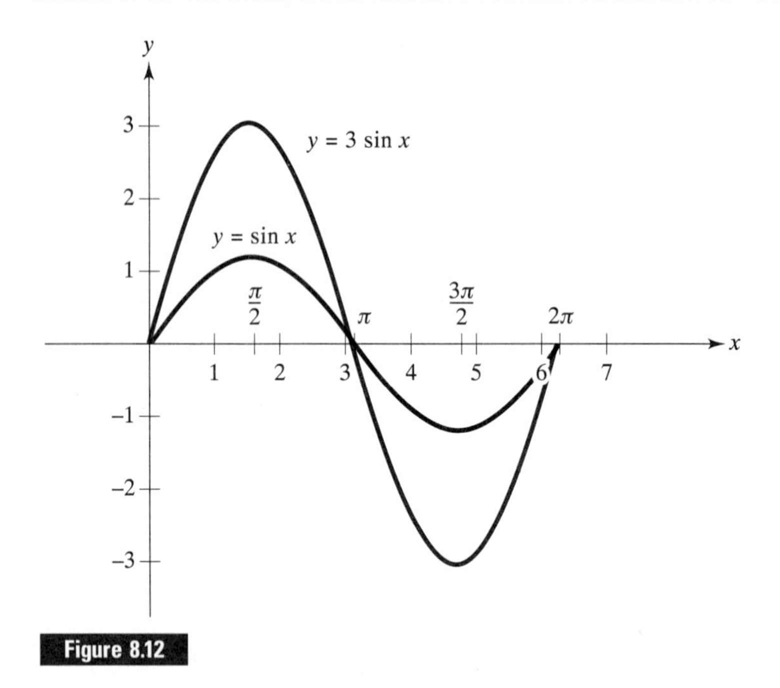

Figure 8.12

The graph of $y = \sin 2x$ is another sine wave (Figure 8.13) with amplitude 1 but with period π. As x varies from 0 to $\pi/4$, $2x$ changes from 0 to $\pi/2$, and $\sin 2x$ increases from 0 to 1. The coefficient 2 *compresses* the curve *horizontally* by the multiple $\frac{1}{2}$. The period of the function $\sin 2x$ is half the period of $\sin x$. (Things happen twice as fast.)

To generalize and combine our two previous observations, we may say that:

The graph of $y = a \sin bx$ is a sine wave with amplitude $|a|$ and period $2\pi/|b|$.

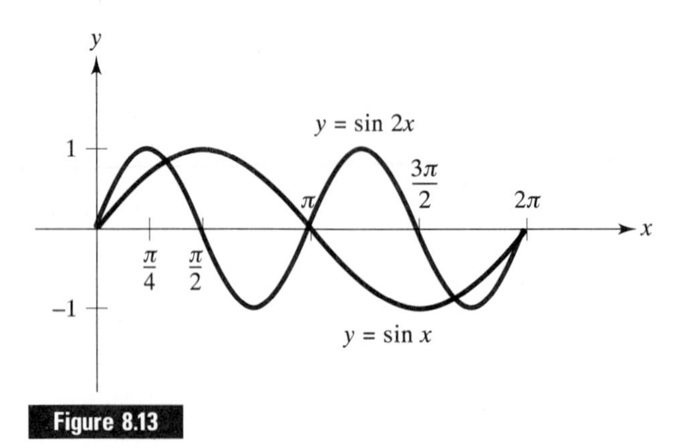

Figure 8.13

A similar statement can be made for $y = a \cos bx$. Figure 8.14 shows the graph of $y = 3 \sin 2x$.

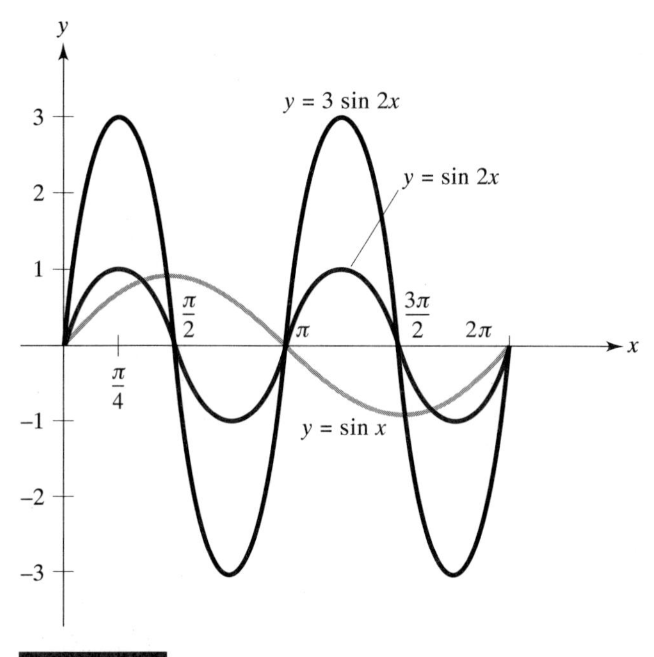

Figure 8.14

Consider the equation $y = \sin(x + \pi/6)$. When $x = 0$, $y = \sin \pi/6 = \frac{1}{2}$. If $x = \pi/3$, $y = \sin \pi/2 = 1$. As x varies from 0 to 2π, y takes on the same values assumed by y in the equation $y = \sin x$ *in the same order but starting at a different place.* The graph of $y = \sin(x + \pi/6)$ may be obtained (Figure 8.15) by shifting the graph of $y = \sin x$ to the *left,* $\pi/6$ units. If the reader does not follow the argument, he or she should plot as many points as necessary using $y = \sin(x + \pi/6)$. Suggested values to assign to x are 0, $\pi/3$, $5\pi/6$, $4\pi/3$, and others if needed.

The graph of $y = \sin(x - \pi/2)$ is the sine curve shifted $\pi/2$ units to the *right* (Figure 8.15). Another approach is to use the formula for $\sin(A - B)$ and find $\sin(x - \pi/2) = -\cos x$. The graph of $y = -\cos x$ is merely the reflection of the graph of $y = \cos x$ in the *x axis.* Place a two-sided mirror on the x axis. The graph of $y = -\cos x$ is the image of the graph of $y = \cos x$. For any given value of x, the y coordinate of $y = -\cos x$ is the negative of that of $y = \cos x$.

The graph of $y = \sin(x + k)$ may be obtained by shifting the graph of

$$y = \sin x \text{ a distance } |k| \text{ to the } \begin{Bmatrix} \textit{left} \text{ if } k \text{ is } \textit{positive} \\ \textit{right} \text{ if } k \text{ is } \textit{negative} \end{Bmatrix}.$$

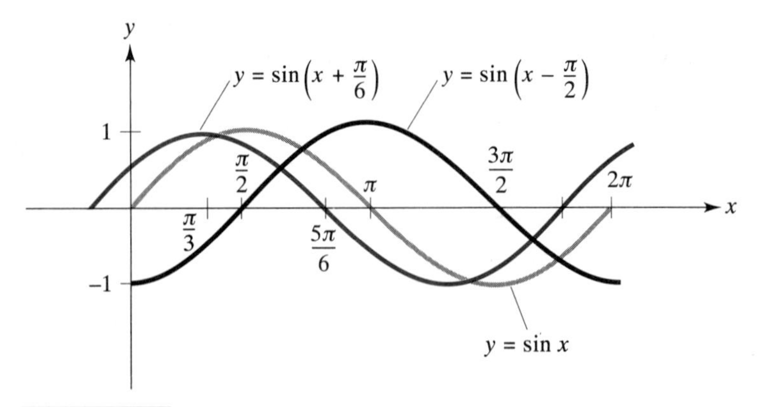

Figure 8.15

If you are in doubt, check with one point; set $x = 0$. The quantity $|k|$ is called the *phase displacement.*

Let us now consider the equation $y = 2.5 \sin(2x + \pi/3)$, which is equivalent to $y = 2.5 \sin 2(x + \pi/6)$. First, sketch the graph of $y = \sin x$ (Figure 8.16). Second, *compress* the curve *horizontally* by the multiple $\frac{1}{2}$ to get the graph of $y = \sin 2x$ (Figures 8.13 and 8.16). Third, *shift* this curve

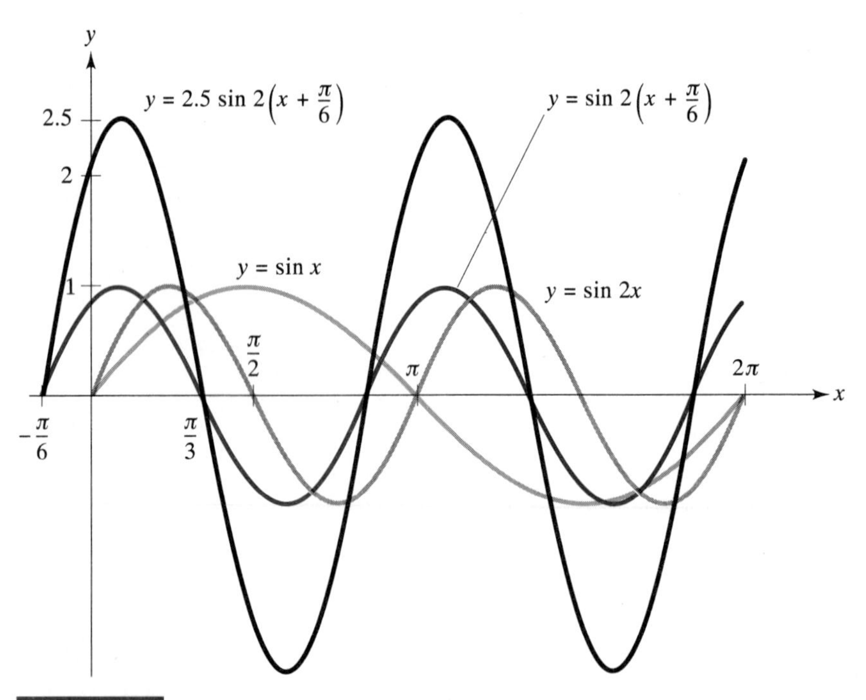

Figure 8.16

$\pi/6$ units to the left to obtain the graph of $y = \sin 2(x + \pi/6)$. Finally, *stretch* this curve vertically to 2.5 times its original height to get the graph of $y = 2.5 \sin 2(x + \pi/6)$ (Figure 8.16). Again the reader is encouraged to check the result by assigning a few values to x in the original equation, computing the corresponding values of y, and then plotting the points. Some of the strategic points are $(0, 2.2)$, $(\pi/12, 2.5)$, $(\pi/3, 0)$, $(7\pi/12, -2.5)$, $(5\pi/6, 0)$. For the function $2.5 \sin(2x + \pi/3)$, the amplitude is 2.5, the period is $2\pi/2 = \pi$, and the phase displacement is $\pi/3 \div 2 = \pi/6$.

The function $a \sin(bx + c)$ has amplitude $|a|$, period $2\pi/|b|$, and phase displacement $|c/b|$. The graph of $y = a \sin(bx + c)$ may be obtained from the graph of $y = a \sin bx$ by shifting it horizontally $|c/b|$ units: to the *left* if c/b is *positive,* to the *right* if c/b is *negative.* A similar statement holds true for $a \cos(bx + c)$.

8.4 SKETCHING CURVES BY COMPOSITION OF y COORDINATES

If the graphs of $y = f(x)$ and $y = g(x)$ are drawn to the same scale on the same set of axes, the graph of

$$y = f(x) + g(x)$$

can be sketched by the process of *composition of y coordinates.* For any value of x, we can determine y of the equation $y = f(x) + g(x)$ by finding graphically the *algebraic* sum of the y coordinates of the two equations $y = f(x)$ and $y = g(x)$. After a suitable number of points have been located by "adding the heights of the given curves," we connect them with a smooth curve to get the required graph.

Example 1 Graph the equation $y = x + \sin x$.

Solution First draw the graphs of $y = x$ and $y = \sin x$ on the same axes (x being measured in radians). Place a straightedge parallel to the y axis at M_1. Use a compass to add the segments M_1S_1 and M_1R_1. The sum is M_1P_1, thus locating P_1. To get point P_2, add the negative segment M_2S_2 to the positive segment M_2R_2; their *algebraic* sum is M_2P_2 (see Figure 8.17).

Example 2 Graph the equation $y = \sin x + \sin 2x$.

Solution After graphing the equations $y = \sin x$ and $y = \sin 2x$, we use the process of composition of y coordinates (Figure 8.18).

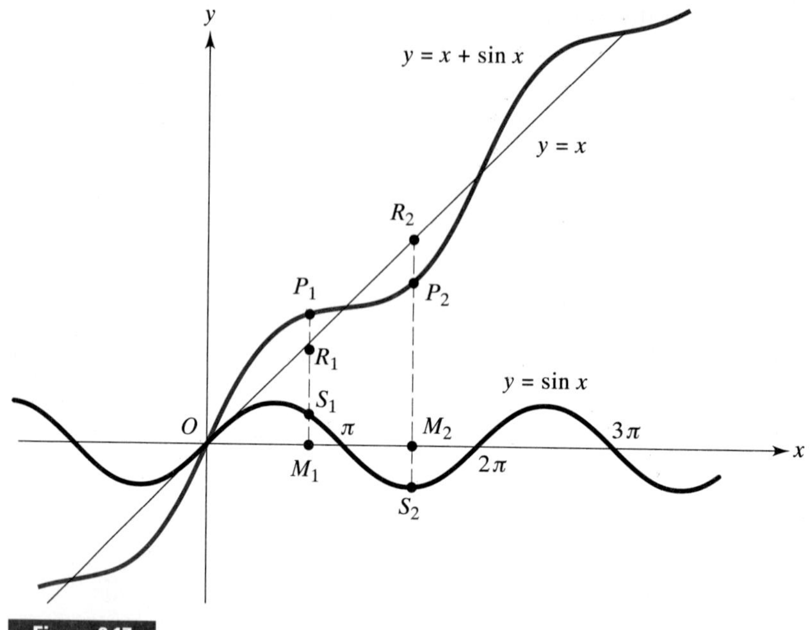

Figure 8.17

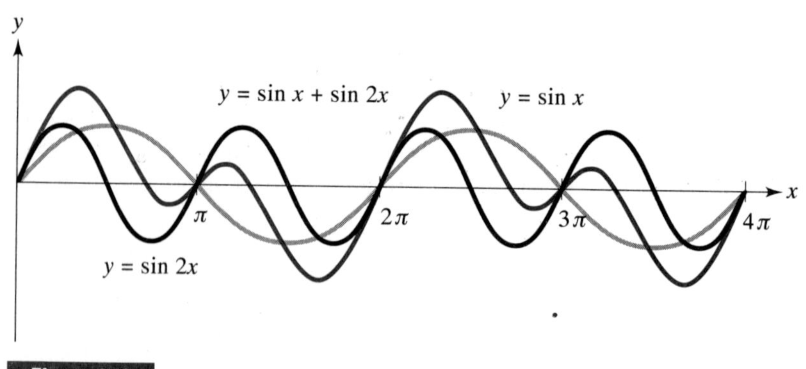

Figure 8.18

8.5 THE GRAPH OF $y = a \sin x + b \cos x$

One plan of attack in drawing the graph of $y = a \sin x + b \cos x$ is first to sketch (on the same coordinate system) the graphs of $y = a \sin x$ and $y = b \cos x$ and then to use the method of composition of y coordinates. A much shorter way, however, is to reduce $y = a \sin x + b \cos x$ to the form $y = k \sin(x + H)$, the graph of which is a sine wave with amplitude k, period 2π, and phase displacement $|H|$.

Example Sketch the graph of $y = 4 \sin x + 3 \cos x$.

Solution From the example of Section 6.3, we have $4 \sin x + 3 \cos x = 5 \sin(x + 0.64)$. Hence the graph of $y = 4 \sin x + 3 \cos x$ is the same as the graph (Figure 8.19) of $y = 5 \sin(x + 0.64)$.

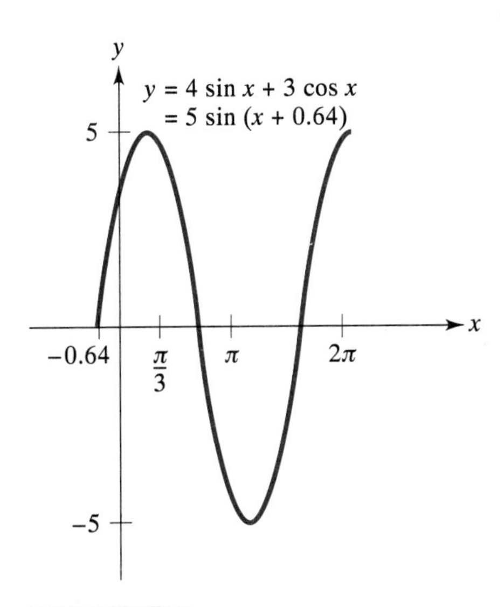

Figure 8.19

Exercise 8.2

Find the period and amplitude of the following functions of x.

1. $\sin 6x$
2. $4 \sin 3x$
3. $\frac{1}{3} \sin \frac{\pi x}{6}$
4. $8 \sin \frac{\pi x}{7}$
5. $9 \tan \frac{2\pi x}{3}$
6. $5 \cos \frac{\pi x}{6}$
7. $4 \cos \frac{7x}{8}$
8. $\tan 4x$

Find the period and amplitude of the function defined by the first equation in each of the following problems. Give the position of its graph (number of units right or left) relative to the graph of the second equation.

9. $y = 5 \sin(3x - 6)$ $\qquad y = 5 \sin 3x$
10. $y = 7.5 \sin\left(\frac{4x}{5} + \pi\right)$ $\qquad y = 7.5 \sin \frac{4x}{5}$

11. $y = 11 \sin\left(\frac{\pi x}{2} + \frac{\pi}{3}\right)$ $\qquad y = 11 \sin \frac{\pi x}{2}$

12. $y = \sin\left(4x - \frac{\pi}{2}\right)$ $\qquad y = \sin 4x$

Sketch graphs of the following equations on the specified intervals.

13. $y = 5 \sin 3x$ $\qquad x = 0$ to $x = \pi$
14. $y = 1.5 \cos 2x$ $\qquad x = 0$ to $x = \pi$
15. $y = \cos \frac{\pi x}{4}$ $\qquad x = 0$ to $x = 8$
16. $y = 4 \sin \frac{x}{2}$ $\qquad x = 0$ to $x = 4\pi$
17. $y = 3 \cos\left(x + \frac{\pi}{4}\right)$ $\qquad x = 0$ to $x = 2\pi$
18. $y = \sin(x + \pi)$ $\qquad x = 0$ to $x = 2\pi$
19. $y = 5 \sin\left(x - \frac{\pi}{6}\right)$ $\qquad x = 0$ to $x = 2\pi$
20. $y = 2 \cos\left(x - \frac{\pi}{3}\right)$ $\qquad x = 0$ to $x = 2\pi$
21. $y = 6 \sin\left(\frac{x}{3} + \frac{\pi}{12}\right)$ $\qquad x = 0$ to $x = 6\pi$
22. $y = 3 \sin\left(\frac{x}{2} - \frac{\pi}{6}\right)$ $\qquad x = 0$ to $x = 4\pi$
23. $y = 4 \sin\left(3x - \frac{\pi}{3}\right)$ $\qquad x = 0$ to $x = \pi$
24. $y = 2 \tan\left(x + \frac{\pi}{4}\right)$ $\qquad x = 0$ to $x = 2\pi$
25. $y = 5 \sin\left(2x + \frac{\pi}{3}\right)$ $\qquad x = 0$ to $x = \pi$
26. $y = -2 \sin x$ $\qquad x = 0$ to $x = 2\pi$
27. $y = -3 \cos 2x$ $\qquad x = 0$ to $x = 2\pi$
28. $y = \tan 3x$ $\qquad x = 0$ to $x = \pi$

Use composition of y coordinates in sketching graphs of the following equations.

29. $y = 1 + \cos x$
30. $y = -x + \sin x$
31. $y = \frac{x}{\pi} + \cos x$
32. $y = x^2 + \sin x$
33. $y = \cos x + \cos 2x$
34. $y = \sin 2x + 2 \sin x$
35. $y = \frac{1}{2} \cos 2x + \frac{1}{2}$

Graph the following equations after changing them to the form $y = k \sin(x + H)$.

36. $y = 4 \sin x - 3 \cos x$
37. $y = \sin x + \sqrt{3} \cos x$
38. $y = \sin x - 2 \cos x$
39. $y = \sin x + \cos x$
40. (a) Graph $y = \sin^2 x$.
    (b) Graph $y = \frac{1}{2} - \frac{1}{2} \cos 2x$.
    (c) Why are the graphs identical?

8.6 USING GRAPHING CALCULATORS*

In recent years we have seen the appearance of hand calculators with graphing capability. Since each model has its own peculiar keying requirements, we will not attempt to provide keying instructions for the reader. Nonetheless, each graphing calculator will require the user to specify the function(s) to be graphed and a viewing window, i.e., a *range* of x values and of y values to be displayed on the screen. Most models will allow the user to *zoom* into a subwindow to be enlarged to full screen size for better viewing. Finally, most models will allow the user to move the cursor and to identify the coordinates of its location. With these capabilities, we can use the graphing calculator to graph trigonometric functions and to solve trigonometric equations graphically.

Example 1 Graph $y = 3 \sin 2x + \cos x$, for $0 \leq x \leq 2\pi$.

Solution Enter the function $3 \sin 2\pi + \cos x$ and the values $0 \leq x \leq 2\pi$, $-5 \leq y \leq 5$ for the viewing window. The graph is seen in Figure 8.20. The reader will have to discern the coordinates of various points of interest on the graph by using the cursor.

*This section may be omitted.

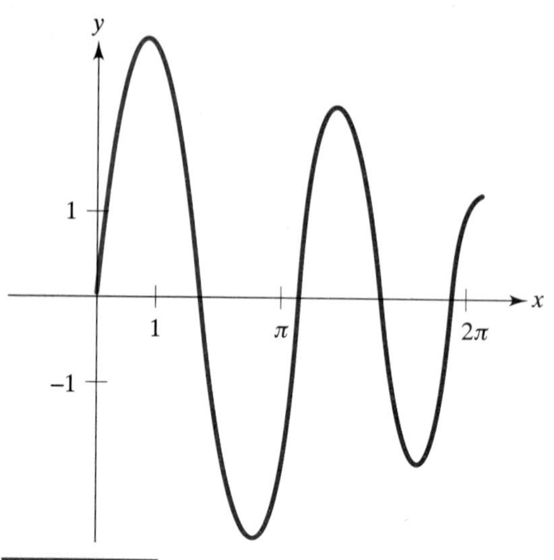

Figure 8.20

Example 2 Solve the equation $\sin x = 1 + \cos x$, for $0 \leq x \leq 2\pi$.

Solution Graph $y_1 = \sin x$ and $y_2 = 1 + \cos x$ for $0 \leq x \leq 2\pi$ and $-3 \leq y \leq 3$. The points of intersection of the two curves are the solutions, as seen in Figure 8.21. From what we know about the sine curve, we can read the points of intersection to be at $(\pi/2, 1)$ and $(\pi, 0)$.

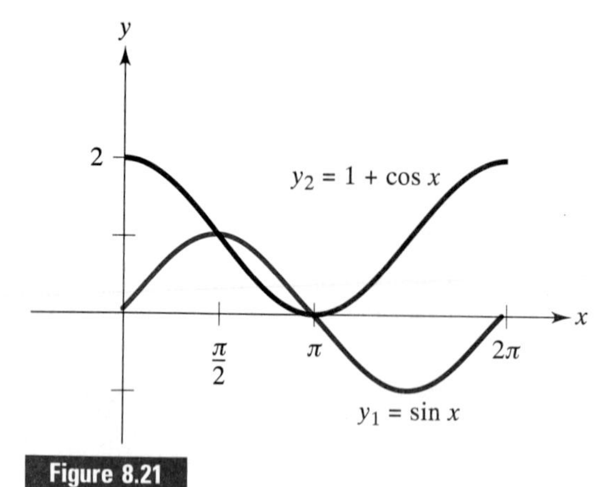

Figure 8.21

Exercise 8.3

Use a graphing calculator to sketch the graph of the following for $0 \le x \le 2\pi$.

1. $y = -3.7 \sin(x - 2.9) + \cos 3x$
2. $y = x^2/2 + 6 \sin 3x$
3. $y = \sin^3 x$
4. $y = \sin^4 x$
5. $y = -x + \sin 2x - \sin 4x$
6. $y = \sin^3 x + \cos^3 x - \sin 3x$

Solve the following trigonometric equations for $0 \le x < 2\pi$, specifying the solution to 4 decimal places.

7. $\sin x = 1 - 2 \sin^2 x$
8. $\sin x = \sqrt{2} + \cos x$
9. $\cos 6x = 5 \cos 3x + 2$
10. $\cos x = \dfrac{\sqrt{6}}{2} + \sin x$
11. $1 + \cos x = x + \sin x$
12. $x^2 - x = \sin x$
13. These are situations in which a graphing calculator is a great advantage and situations in which it is not as quick and effective as an analytical argument. Each of these is illustrated by trying to solve, for $0 \le x \le 2\pi$,
    (a) $x^3 - 2x = 2 \cos x$
    (b) $x = \sin x$.

KEY TERMS

Periodic function, amplitude, period, phase shift, composition of ordinates, viewing window

REVIEW EXERCISES

1. Sketch the cotangent curve.

Find the period, amplitude and phase shift of the following and sketch a graph for $0 \leq \theta \leq 2\pi$.

2. $y = -2 \sin\left(3\theta - \frac{\pi}{2}\right)$
3. $y = 3 \sin 2\theta$
4. $y = \tan\left(\theta + \frac{\pi}{4}\right)$
5. $y = -\cos\left(2\theta - \frac{\pi}{4}\right)$

Sketch a graph of the following exercises for $0 \leq x \leq 2\pi$.

6. $y = -x + \cos x$
7. $y = 3 \sin x + 4 \cos x$
8. $y = -\cos(x + \pi)$
9. $y = 1 - 2 \sin^2 x$

Use a graphing calculator to find all solutions to the following.

10. $x^3 = \sin x$
11. $x^2 = \cos x$
12. $x^4 = 2 + \cos x$

9

Inverse Trigonometric Functions

9.1 INVERSE FUNCTIONS

In Section 2.3 we began a discussion of inverse functions. Recall that two functions f and g are inverses of each other if for every u in the domain of f, $g(f(u)) = u$, and if for every v in the domain of g, $f(g(v)) = v$. For this to happen, both f and g must be one-to-one functions; that is to say, for each y in the range of f, the equation $y = f(u)$ must have *exactly one* solution u. Likewise, for each z in the range of g the equation $z = g(v)$ must have *exactly one* solution v. When f and g are inverse functions, the ordered pair (u, v) is the function f if and only if the ordered pair (v, u) is in the function g.

A graphical consequence of this last statement is that two inverse functions have graphs which are reflections of each other through the line $y = x$ bisecting the first and third quadrants.

Given a function f, in order to find a formula for f^{-1} we solve the equation $v = f(u)$ for u in terms of v. The result is a formula for the function $g(v) = f^{-1}(v)$ in terms of v.

Example If $f(u) = u^3 + 1$, for all real numbers u, find the inverse function $g(v) = f^{-1}(v)$.

Solution We set $v = f(u) = u^3 + 1$. Solving for u, we get

$$v - 1 = u^3$$

$$(v - 1)^{1/3} = u.$$

We have $g(v) = f^{-1}(v) = (v - 1)^{1/3}$. Note that

$$\begin{aligned} g(f(u)) &= g(u^3 + 1) = (u^3 + 1 - 1)^{1/3} \\ &= (u^3)^{1/3} \\ &= u, \end{aligned}$$

and that

$$\begin{aligned} f(g(v)) &= f((v-1)^{1/3}) \\ &= [(v-1)^{1/3}]^3 + 1 \\ &= v - 1 + 1 \\ &= v\,. \end{aligned}$$

Thus g and f are inverses.

Exercise 9.1

In the following problems, g designates the inverse of the function f.

1. Given

$$f(u) = \frac{7u+8}{9},$$

   (a) Find $g(v) = f^{-1}(v)$.
   (b) Show that $g[f(u)] = u$.

2. Given

$$f(u) = \frac{2u+3}{u-7}, \qquad u \neq 7,$$

   (a) Find $g(v) = f^{-1}(v)$.
   (b) Show that $g[f(u)] = u$, provided that $u \neq 7$.
   (c) Show that $f[g(v)] = v$, provided that $v \neq 2$.

3. Given

$$f(u) = 2 - \sqrt{u-3}, \qquad u \geq 3,$$

   (a) Find $g(v) = f^{-1}(v)$.
   (b) Show that $g[f(u)] = u$, provided that $u \geq 3$.
   (c) Show that $f[g(v)] = v$, provided that $v \leq 2$.

4. Given $f(u) = 1 + \sqrt[3]{u}$.
   (a) Find $g(v) = f^{-1}(v)$.
   (b) Show that $g[f(u)] = u$.

5. Use your calculator to find $\sin 3\pi/4$, then find $\sin^{-1}(\sin 3\pi/4)$. Do you get $3\pi/4$? Do the same for $\cos 3\pi/4$. What is happening?

INVERSE TRIGONOMETRIC FUNCTIONS

The equation

$$u = \sin \theta$$

says that u is a number representing the sine of the number θ. Another interpretation is

$$\theta \text{ is a number whose sine is } u.$$

This statement is usually written in the form

$$\theta = \arcsin u.$$

With this understanding, we can say

$$\arcsin \frac{1}{2} = \frac{\pi}{6}, \frac{5\pi}{6}, \frac{13\pi}{6}, -\frac{7\pi}{6}, \text{ etc.,}$$

because the sine of each of these numbers is $\frac{1}{2}$. Values of arcsin $\frac{1}{2}$ may be obtained by finding all the points on the sine curve that are $\frac{1}{2}$ unit above the θ axis (Figure 9.1).

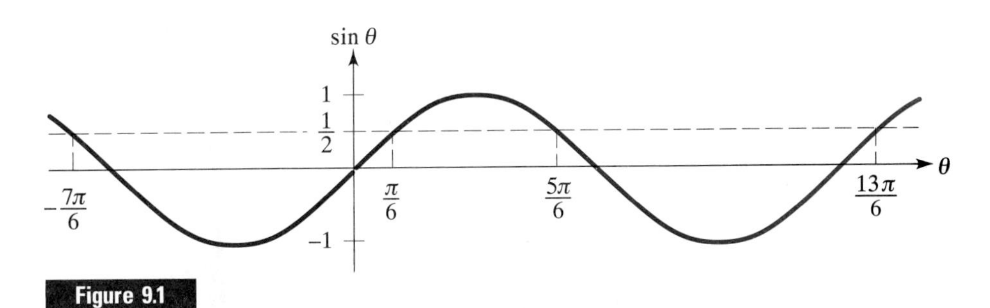

Figure 9.1

Similarly, arccos u denotes a number whose cosine is u, arctan u denotes a number whose tangent is u, and so on. The six inverse trigonometric relations are arcsin u, arccos u, arctan u, arccot u, arcsec u, arccsc u. Thus arccos $0 = \pi/2, 3\pi/2$, and so on, and arctan $1 = \pi/4, 5\pi/4$, and so on. Recall that a *relation* is a set of ordered pairs of numbers, whereas a *function* is a relation $\{(x, y)\}$ such that to each x there corresponds one and only one y.

The sine function is defined as the infinite set of ordered pairs $\{(\theta, \sin\theta)$, where θ is a real number$\}$. This trigonometric relation is a function because to each real number θ there corresponds one and only one number $\sin\theta$. Let us now consider the arcsin relation $\{(u, \arcsin u) | \quad -1 \le u \le 1\}$. As demonstrated above, to each value of u there corresponds more than one value of arcsin u. Therefore, the arcsin relation is not a function.

It will be quite useful to modify the arcsin relation in such a way as to make it a function. A careful inspection of the sine curve (Figure 9.2) shows that as θ varies from $-\pi/2$ to $\pi/2$, then $\sin\theta$ moves from -1 to $+1$, taking on all intervening values once and only once, that is, *exactly* once. Thus, if arcsin u is restricted to the closed interval $[-\pi/2, \pi/2]$, then the relation $\{(u, \arcsin u) | \quad -\pi/2 \le \arcsin u \le \pi/2\}$ is a function. The notation Arcsin u or $\mathrm{Sin}^{-1} u$ (read "inverse sine u") means *the* number (between $-\pi/2$ and $\pi/2$, inclusive) whose sine is u; that is,

$$-\frac{\pi}{2} \le \mathrm{Sin}^{-1} u \le \frac{\pi}{2}.$$

Thus,

$$\mathrm{Sin}^{-1}\frac{1}{2} = \frac{\pi}{6}.$$

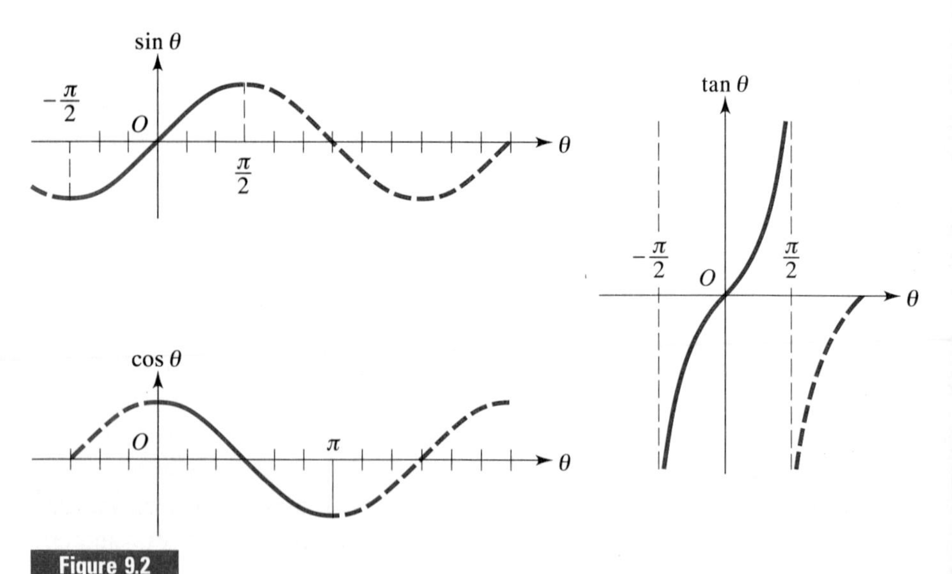

Figure 9.2

It should be carefully noted that the superscript -1 is *not* to be interpreted as an exponent; that is, $\text{Sin}^{-1} u$ is *not* $(\text{Sin } u)^{-1}$ or $1/\sin u$.

The symbols arccos u and arctan u may be modified to refer to functions rather than relations by defining Arccos $u = \text{Cos}^{-1} u$ and Arctan $u = \text{Tan}^{-1} u$ as follows:

$$0 \le \text{Cos}^{-1} u \le \pi; \qquad -\frac{\pi}{2} < \text{Tan}^{-1} u < \frac{\pi}{2}.$$

The following summary shows the domain, range, and defining equation for each of the previously discussed inverse trigonometric functions.

Function	Inverse Function	Defining Equation	Domain	Range
sin	Sin^{-1}	$y = \text{Sin}^{-1} x$	$-1 \le x \le 1$	$-\frac{\pi}{2} \le \text{Sin}^{-1} x \le \frac{\pi}{2}$
cos	Cos^{-1}	$y = \text{Cos}^{-1} x$	$-1 \le x \le 1$	$0 \le \text{Cos}^{-1} x \le \pi$
tan	Tan^{-1}	$y = \text{Tan}^{-1} x$	$-\infty < x < \infty$	$-\frac{\pi}{2} < \text{Tan}^{-1} x < \frac{\pi}{2}$

Figure 9.2 restates these definitions in terms of the sine, cosine, and tangent curves; the unbroken lines indicate the portions of the curves that are to be used.

Hence, by definition, $\text{Sin}^{-1} u$, $\text{Cos}^{-1} u$, and $\text{Tan}^{-1} u$ refer to functions rather than relations. Thus $\text{Cos}^{-1} u$ means *the* number between 0 and π whose cosine is u.

Illustrations

$$\text{Sin}^{-1}\left(-\frac{\sqrt{3}}{2}\right) = -\frac{\pi}{3}.$$

$$\text{Cos}^{-1}\left(-\frac{\sqrt{2}}{2}\right) = \frac{3\pi}{4}.$$

$$\text{Tan}^{-1}(-1) = -\frac{\pi}{4}.$$

The functions $\text{Cot}^{-1} u$, $\text{Sec}^{-1} u$, and $\text{Csc}^{-1} u$ are of less importance. The following table completes the definition of the inverse trigonometric functions.

Function	Inverse Function	Defining Equation	Domain	Range
cot	Cot^{-1}	$y = \text{Cot}^{-1} x$	$-\infty < x < \infty$	$0 < \text{Cot}^{-1} x < \pi$
sec	Sec^{-1}	$y = \text{Sec}^{-1} x$	$x \geq 1$	$0 \leq \text{Sec}^{-1} x < \dfrac{\pi}{2}$
			or $x \leq -1$	or $\pi \leq \text{Sec}^{-1} x < \dfrac{3\pi}{2}$
csc	Csc^{-1}	$y = \text{Csc}^{-1} x$	$x \geq 1$	$0 < \text{Csc}^{-1} x \leq \dfrac{\pi}{2}$
			or $x \leq -1$	or $\pi < \text{Csc}^{-1} x \leq \dfrac{3\pi}{2}$

For example, $\text{Sec}^{-1}\sqrt{2}$ is *the* number, between 0 and $\pi/2$, whose secant is $\sqrt{2}$; hence $\text{Sec}^{-1}\sqrt{2} = \text{Cos}^{-1}(1/\sqrt{2}) = \pi/4$. Also, $\text{Sec}^{-1}(-2)$ is *the* number, between π and $3\pi/2$, whose secant is -2. Hence, $\text{Sec}^{-1}(-2) = \pi + \text{Cos}^{-1}(\frac{1}{2}) = \pi + \pi/3 = 4\pi/3$.

Example 1 Evaluate to four decimal places: (a) $\text{Sin}^{-1}(-0.8763)$, (b) $\text{Cos}^{-1}(-0.4321)$, (c) $\text{Cot}^{-1} 3.271$.

Calculator Solution Most scientific (slide rule) calculators have a Change Sign key [+/−] that changes the sign of the displayed number. Switch the calculator to radian mode.

(a) Press

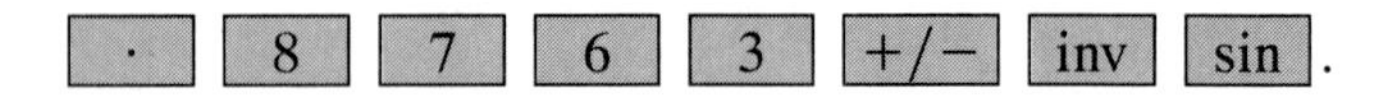

The displayed result is -1.068127636. Hence

$$\text{Sin}^{-1}(-0.8763) = -1.0681\,.$$

This is the radian measure of a negative fourth-quadrant angle $(-61.2°)$.

(b) Press

[.4321] [+/−] [inv] [cos].

The displayed result is 2.017616418. Therefore

$$\mathrm{Cos}^{-1}(-0.4321) = 2.0176\,.$$

This is the radian measure of a positive second-quadrant angle (115.6°).

(c) Since

$$\mathrm{Cot}^{-1}\, 3.271 = \mathrm{Tan}^{-1}\, \frac{1}{3.271},$$

press

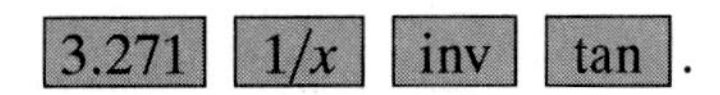

The displayed result is 0.296693375. Hence

$$\mathrm{Cot}^{-1}\, 3.271 = 0.2967\,.$$

Example 2 Sketch, on the same axes, $y = \sin x$, $y = \arcsin x$, and $y = \mathrm{Sin}^{-1}\, x$.

Solution We recall that the graph of $y = \mathrm{Sin}^{-1}\, x$ is the reflection of the graph of $y = \sin x$, for $-\pi/2 \leq x \leq \pi/2$, through the line $y = x$. The sketch is shown in Figure 9.3.

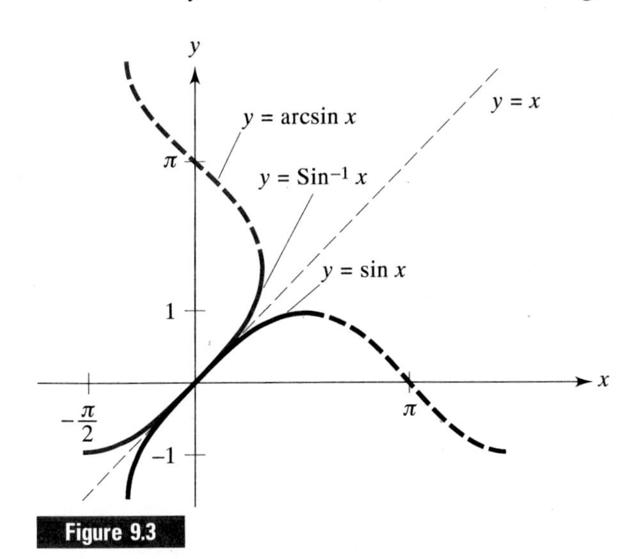

Figure 9.3

Exercise 9.2

Find the value of each of the following. Leave the results in terms of π. Do not use a calculator.

1. $\text{Sin}^{-1} \dfrac{\sqrt{3}}{2}$
2. $\text{Sin}^{-1} \left(-\dfrac{1}{2}\right)$
3. $\text{Cos}^{-1} \dfrac{\sqrt{2}}{2}$
4. $\text{Cos}^{-1} 0$
5. $\text{Cos}^{-1}(-1)$
6. $\text{Cos}^{-1} \dfrac{\sqrt{3}}{2}$
7. $\text{Sin}^{-1}\left(-\dfrac{\sqrt{2}}{2}\right)$
8. $\text{Sin}^{-1}\left(-\dfrac{\sqrt{3}}{2}\right)$
9. $\text{Sin}^{-1} \dfrac{\sqrt{2}}{2}$
10. $\text{Sin}^{-1}(-1)$
11. $\text{Cos}^{-1}(-\frac{1}{2})$
12. $\text{Cos}^{-1} \frac{1}{2}$
13. $\text{Cos}^{-1}\left(-\dfrac{\sqrt{3}}{2}\right)$
14. $\text{Sec}^{-1}(-\sqrt{2})$
15. $\text{Sin}^{-1} \frac{1}{2}$
16. $\text{Sin}^{-1} 1$
17. $\text{Tan}^{-1} 1$
18. $\text{Tan}^{-1} \sqrt{3}$
19. $\text{Tan}^{-1} 0$
20. $\text{Tan}^{-1}\left(-\dfrac{1}{\sqrt{3}}\right)$

Use a calculator to find the value of each of the following to four decimal places.

21. $\text{Cos}^{-1}(-0.9092)$
22. $\text{Sin}^{-1}(-0.4399)$
23. $\text{Tan}^{-1}(-14.30)$
24. $\text{Cos}^{-1}(-0.3355)$
25. $\text{Sin}^{-1} 0.8763$
26. $\text{Cos}^{-1} 0.1925$
27. $\text{Cos}^{-1}(-0.6691)$
28. $\text{Tan}^{-1} 0.5452$
29. Explain why $\text{Cos}^{-1} u$ could not be taken on the interval $-\pi/2$ to $\pi/2$.
30. Graph $y = \cos x$, $y = \arccos x$, and $y = \text{Cos}^{-1} x$ on the same axes.
31. Graph $y = \tan x$, $y = \text{Tan}^{-1} x$ on the same coordinate axes.

9.3 OPERATIONS INVOLVING INVERSE TRIGONOMETRIC FUNCTIONS

Since every element in the range of an inverse trigonometric function may be thought of as the radian measure of an angle, it is frequently convenient to place this angle in standard position and label its triangle of reference in accordance with the inverse function (Example 1). Sometimes it is advisable to replace the inverse functions with angle symbols, such as θ, A, B, and then try to express the problem in terms of ordinary functions.

Example 1 Evaluate $\cos(\text{Tan}^{-1} \frac{2}{3})$.

Solution We are asked to find the cosine of the angle whose tangent is $\frac{2}{3}$. Draw a right triangle with legs 2 and 3. The acute angle opposite the side 2 has a tangent of $\frac{2}{3}$. It can be labeled $\text{Tan}^{-1} \frac{2}{3}$. After finding that the hypotenuse is $\sqrt{13}$, we see (Figure 9.4) that

$$\cos\left(\text{Tan}^{-1} \frac{2}{3}\right) = \frac{3}{\sqrt{13}}.$$

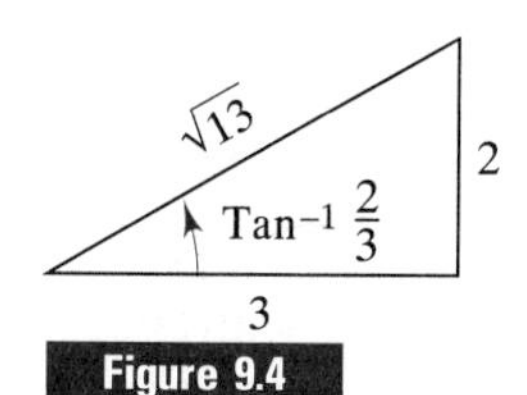

Figure 9.4

Example 2 Find the value of $\sin(\text{Sin}^{-1} u + \text{Cos}^{-1} v)$.

Solution Since $\text{Sin}^{-1} u$ and $\text{Cos}^{-1} v$ are angles, we have the sine of the sum of two angles. Let $A = \text{Sin}^{-1} u$ and $B = \text{Cos}^{-1} v$. Using

$$\sin(A + B) = \sin A \cos B + \cos A \sin B,$$

we have

$$\begin{aligned}\sin(\text{Sin}^{-1} u + \text{Cos}^{-1} v) &= \sin(\text{Sin}^{-1} u) \cos(\text{Cos}^{-1} v) \\ &\quad + \cos(\text{Sin}^{-1} u) \sin(\text{Cos}^{-1} v) \\ &= uv + \sqrt{1 - u^2}\sqrt{1 - v^2}.\end{aligned}$$

The value of $\cos(\text{Sin}^{-1} u)$ is found by use of the triangle in Figure 9.5. Notice that if u is negative, $\text{Sin}^{-1} u$ is in Q IV (its value in radians lies between $-\pi/2$ and 0) but $\cos(\text{Sin}^{-1} u)$ is still positive. Explain why $\sin(\text{Cos}^{-1} v)$ is always the positive radical $\sqrt{1 - v^2}$.

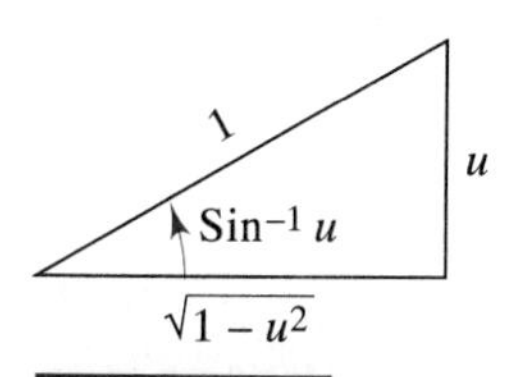

Figure 9.5

Example 3 Prove that

$$\text{Tan}^{-1}\frac{1}{2} + \text{Tan}^{-1}\frac{1}{3} = \frac{\pi}{4}.$$

Solution The left side is the sum of two acute angles, each less than $\pi/4$. Why? To prove that their sum is $\pi/4$, let us take the tangent of each side of the equation:

$$\tan(\text{Tan}^{-1}\tfrac{1}{2} + \text{Tan}^{-1}\tfrac{1}{3}) = \tan\frac{\pi}{4}$$

$$\frac{\tan(\text{Tan}^{-1}\tfrac{1}{2}) + \tan(\text{Tan}^{-1}\tfrac{1}{3})}{1 - \tan(\text{Tan}^{-1}\tfrac{1}{2})\tan(\text{Tan}^{-1}\tfrac{1}{3})} = 1$$

$$\frac{\tfrac{1}{2} + \tfrac{1}{3}}{1 - \tfrac{1}{2}\cdot\tfrac{1}{3}} = 1.$$

The proof is complete if we recall that two *acute* angles having the same tangent are equal. Notice that the formula for $\tan(A + B)$ was used in evaluating the left side.

Example 4 Solve for x:

$$\text{Sin}^{-1} 2x + \text{Cos}^{-1} x = \frac{\pi}{6}.$$

Solution Subtract $\text{Cos}^{-1} x$ from both sides of the equation. Then take the sine of both sides of the equation:

$$\sin(\text{Sin}^{-1} 2x) = \sin\left(\frac{\pi}{6} - \text{Cos}^{-1} x\right).$$

Let $A = \pi/6$ and $B = \text{Cos}^{-1} x$. Using

$$\sin(A - B) = \sin A \cos B - \cos A \sin B,$$

we obtain

$$2x = \sin\frac{\pi}{6}\cos(\text{Cos}^{-1} x) - \cos\frac{\pi}{6}\sin(\text{Cos}^{-1} x)$$

$$2x = \frac{1}{2}\cdot x - \frac{\sqrt{3}}{2}\cdot\sqrt{1 - x^2}$$

$$3x = -\sqrt{3}\cdot\sqrt{1 - x^2}.$$

Squaring both sides, we obtain

$$9x^2 = 3 - 3x^2, \qquad 4x^2 = 1, \qquad x = \pm\tfrac{1}{2}.$$

Inasmuch as we squared the equation, we must check all values to see if any are extraneous.

Check for $x = \frac{1}{2}$:

$$\operatorname{Sin}^{-1} 1 + \operatorname{Cos}^{-1} \frac{1}{2} = \frac{\pi}{6}$$

$$\frac{\pi}{2} + \frac{\pi}{3} = \frac{\pi}{6}, \text{ which is false.}$$

Check for $x = -\frac{1}{2}$:

$$\operatorname{Sin}^{-1}(-1) + \operatorname{Cos}^{-1}\left(-\frac{1}{2}\right) = \frac{\pi}{6}$$

$$-\frac{\pi}{2} + \frac{2\pi}{3} = \frac{\pi}{6}, \text{ which is true.}$$

Hence the only solution of the given equation is $x = -\frac{1}{2}$.

Exercise 9.3

Find the exact value of each of the following. Do not use a calculator.

1. $\cot(\operatorname{Tan}^{-1} u)$
2. $\cos(\operatorname{Sin}^{-1} u)$
3. $\tan(\operatorname{Cos}^{-1} u)$
4. $\sec(\operatorname{Sec}^{-1} u)$
5. $\cos(\operatorname{Csc}^{-1} 4)$
6. $\tan[\operatorname{Sin}^{-1}(-\frac{12}{13})]$
7. $\sin[\operatorname{Tan}^{-1}(-\frac{3}{2})]$
8. $\cos[\operatorname{Tan}^{-1}(-\frac{8}{15})]$
9. $\cos(\operatorname{Sin}^{-1} u + \operatorname{Cos}^{-1} v)$
10. $\sin(\operatorname{Cos}^{-1} u + \operatorname{Sin}^{-1} v)$
11. $\sin(\operatorname{Sin}^{-1} u - \operatorname{Cos}^{-1} v)$
12. $\cos(\operatorname{Sin}^{-1} u - \operatorname{Sin}^{-1} v$
13. $\sin\left(\frac{3\pi}{2} - \operatorname{Sin}^{-1} \frac{5}{6}\right)$
14. $\cos\left(\frac{\pi}{3} - \operatorname{Cos}^{-1} \frac{3}{7}\right)$
15. $\cos\left(\frac{3\pi}{4} + \operatorname{Sin}^{-1} \frac{1}{5}\right)$
16. $\sin\left(3\pi + \operatorname{Cos}^{-1} \frac{7}{8}\right)$

17. $\cos(2\ \mathrm{Cos}^{-1}\ u)$
(*Hint:* Let $A = \mathrm{Cos}^{-1}\ u$. We seek $\cos 2A$.)
18. $\sin(2\ \mathrm{Sin}^{-1}\ u)$
19. $\sin(2\ \mathrm{Cos}^{-1}\ u)$
20. $\cos(2\ \mathrm{Sin}^{-1}\ u)$
21. $\sin(\mathrm{Sin}^{-1}\ 2u)$
22. $\cos(2\ \mathrm{Cos}^{-1}\ 4u)$
23. $\mathrm{Cos}^{-1}\left(\cos\frac{7\pi}{6}\right)$
24. $\mathrm{Sin}^{-1}\left(\sin\frac{7\pi}{4}\right)$
25. $\mathrm{Sin}^{-1}\left[\sin\left(-\frac{\pi}{7}\right)\right]$
26. $\mathrm{Tan}^{-1}\left(\tan\frac{3\pi}{5}\right)$
27. $\mathrm{Tan}^{-1}\sqrt{2 - \sec\pi}$
28. $\mathrm{Cos}^{-1}(\tan\pi)$
29. $\mathrm{Cos}^{-1}\left[\cos\left(-\frac{\pi}{9}\right)\right]$
30. $\mathrm{Sin}^{-1}(-\cos 0)$

Assume that $u > 0$. Copy the following and fill in the blanks.

31. $\mathrm{Tan}^{-1}\dfrac{u}{\sqrt{1-u^2}} = \mathrm{Sin}^{-1}\_\_\_\_\_\_ = \mathrm{Cos}^{-1}\_\_\_\_\_\_,\ 0 < u < 1$

32. $\mathrm{Sin}^{-1}\dfrac{\sqrt{16-u^2}}{4} = \mathrm{Cos}^{-1}\_\_\_\_\_\_ = \mathrm{Tan}^{-1}\_\_\_\_\_\_,\ 0 < u < 4$

33. $\mathrm{Cos}^{-1}\dfrac{2}{\sqrt{u^2+2u+5}} = \mathrm{Sin}^{-1}\_\_\_\_\_\_ = \mathrm{Tan}^{-1}\_\_\_\_\_\_$

34. $\mathrm{Csc}^{-1}\dfrac{u}{\sqrt{u^2-36}} = \mathrm{Cot}^{-1}\_\_\_\_\_\_ = \mathrm{Sec}^{-1}\_\_\_\_\_\_,\ u > 6$

Without using a calculator, prove that each of the statements is true for all permissible values of any variables.

35. $\sin\left(\frac{1}{2}\mathrm{Cos}^{-1}\ u\right) = \sqrt{\frac{1-u}{2}}$

36. $\cos\left(\frac{1}{2}\mathrm{Cos}^{-1}\ 6a\right) = \sqrt{\frac{1+6a}{2}}$

37. $\cos\left(\frac{1}{2}\mathrm{Sin}^{-1}\ a\right) = \sqrt{\frac{1+\sqrt{1-a^2}}{2}}$

38. $\sin\left(\frac{1}{2}\mathrm{Sin}^{-1}\ 3u\right) = \pm\sqrt{\frac{1-\sqrt{1-9u^2}}{2}}$

39. $\mathrm{Tan}^{-1}\dfrac{1}{10} + \mathrm{Tan}^{-1}\dfrac{9}{11} = \dfrac{\pi}{4}$
40. $\mathrm{Tan}^{-1} 7 - \mathrm{Tan}^{-1} 2 = \mathrm{Tan}^{-1}\frac{1}{3}$
41. $\mathrm{Sin}^{-1}\frac{3}{5} + \mathrm{Sin}^{-1}\frac{9}{41} = \mathrm{Sin}^{-1}\frac{156}{205}$
42. $\mathrm{Sin}^{-1}\frac{1}{4} + \mathrm{Cos}^{-1}\frac{7}{8} = \mathrm{Sin}^{-1}\frac{11}{16}$
43. $2\ \mathrm{Cos}^{-1}\frac{1}{4} + \mathrm{Cos}^{-1}\frac{7}{8} = \pi$
44. $\frac{1}{2}\ \mathrm{Cos}^{-1}\frac{119}{169} - \mathrm{Sin}^{-1}\frac{3}{5} = \mathrm{Sin}^{-1}(-\frac{16}{65})$

Identify each statement as true or false and give reasons. (Consider all permissible values—positive, negative, and zero—of the letters involved.)

45. $\mathrm{Tan}^{-1}(-u) = -\mathrm{Tan}^{-1} u$
46. $\mathrm{Sin}^{-1}(-u) = -\mathrm{Sin}^{-1} u$
47. $\sin(2\ \mathrm{Sin}^{-1} u) = 2u$
48. $\sec(\mathrm{Sin}^{-1} u) = \dfrac{1}{u}$
49. $\mathrm{Sin}^{-1} u + \mathrm{Cos}^{-1} u = \dfrac{\pi}{2}$
50. $\mathrm{Cos}^{-1} u + \mathrm{Cos}^{-1}(-u) = \pi$
51. $\cos(2\ \mathrm{Cos}^{-1} u) = 2u^2 - 1$
52. $\mathrm{Tan}^{-1}(-1) = -\dfrac{\pi}{4}$
53. $\mathrm{Cos}^{-1} u = \mathrm{Sin}^{-1}\sqrt{1 - u^2}$
54. $\mathrm{Tan}^{-1} a + \mathrm{Tan}^{-1}\dfrac{1}{a} = \dfrac{\pi}{2}$
55. $\cot(\mathrm{Cos}^{-1} u) = \dfrac{u}{\sqrt{1 - u^2}}$
56. $\sin(\mathrm{Cos}^{-1} u) = \cos(\mathrm{Sin}^{-1} u)$

Solve for x.

57. $\mathrm{Sin}^{-1} 2x = \mathrm{Cos}^{-1} 3x$
58. $\mathrm{Cos}^{-1} x + \mathrm{Cos}^{-1}(1 - x) = \dfrac{\pi}{2}$
59. $\mathrm{Sin}^{-1}\left(\dfrac{x}{\sqrt{3}}\right) + \mathrm{Cos}^{-1}(-x) = \pi$
60. $\mathrm{Sin}^{-1}\sqrt{x + 1} = \mathrm{Cos}^{-1}\sqrt{x}$
61. $\mathrm{Sin}^{-1} x = 2\ \mathrm{Cos}^{-1} x$

62. $2\ \text{Sin}^{-1} x = \text{Cos}^{-1}(2 - 3x)$

63. $\tan(\text{Sin}^{-1} x) = \dfrac{x}{\sqrt{1 - x^2}}$

64. $\text{Sin}^{-1}(x\sqrt{2}) + \text{Sin}^{-1}(2x) = \dfrac{3\pi}{4}$

65. Given the equation

$$\text{Tan}^{-1} u = \frac{1}{2} \text{Cos}^{-1} \frac{1 - u^2}{1 + u^2},$$

(a) Prove that this equation is, or is not, an identity for $u \geq 0$.
(b) Prove that the equation does not hold for any value of $u < 0$.

KEY TERMS

Relation, function, inverse relation, inverse function

REVIEW EXERCISES

1. Given $f(u) = 3 - 2u$, find $g(u) = f^{-1}(u)$, and show $g(f(u)) = u$.
2. Find $\text{Tan}^{-1}(-0.6398)$ and $\text{Tan}^{-1} 1994$.
3. Find the value of $\text{Sin}^{-1}(-\sqrt{2}/2)$ and of $\text{Cos}^{-1}(-\sqrt{2}/2)$.
4. Sketch, on the same coordinate system, $y = \text{Cos}^{-1} x$ and $y = \cos x$.
5. Find $\text{Sin}^{-1}(\sin(-\pi/14))$ and $\cos(2 \cos^{-1} v)$
6. Find $\cos(\text{Sin}^{-1} u - \text{Cos}^{-1} v)$.
7. Is $\text{Sin}^{-1}(-u) = -\text{Sin}^{-1} u$?

10

Transcendental Functions

10.1 THE LOGARITHM FUNCTION

The trigonometric functions belong to a class of functions known as the *transcendental functions.* In this chapter we will examine two other very important transcendental functions, the **logarithm** and the **exponential function,** which are actually inverses of each other.

Long, drawn-out computations have always been sheer drudgery for people who deal with numbers. Some relief was obtained in the middle 1600s after a Scotsman, John Napier, and an Englishman, Henry Briggs, invented logarithms. By using logarithms they were able to replace the operations of multiplication and division with addition and subtraction. As a result, logarithms have been widely used in trigonometric computations for the past 300 years. The slide rule was a consequence of the invention of logarithms. In recent years the use of logarithms for computation has been replaced by the electronic calculator. Today the analytic aspects of logarithms and their applications in higher mathematics and science are far more important than their use in performing numerical computations.

A logarithm, as we shall see, is an exponent. Accordingly we recall the following laws of exponents, for any real numbers a, m, and n.

1. $a^m \cdot a^n = a^{m+n}$ — *Example:* $2^3 \cdot 2^4 = 2^7$
2. $(a^m)^n = a^{mn}$ — *Example:* $(2^3)^4 = 2^{12}$
3. $\dfrac{a^m}{a^n} = a^{m-n}$ — *Example:* $\dfrac{2^8}{2^2} = 2^6$
4. $a^{m/n} = (\sqrt[n]{a})^m = \sqrt[n]{a^m}$ — *Example:* $8^{4/3} = (\sqrt[3]{8})^4 = 2^4 = 16$
5. $a^0 = 1$, if $a \neq 0$ — *Example:* $(\frac{2}{3})^0 = 1$
6. $a^{-n} = \dfrac{1}{a^n}$ — *Example:* $3^{-2} = \dfrac{1}{3^2} = \dfrac{1}{9}$

Although these laws are true for all values of m and n and for positive and negative values of a, we shall have occasion to use them only for positive values of a.

We now state the definition of logarithm.

The logarithm of a number N to a given base b, written $\log_b N$, is the exponent that must be placed on the base to produce the number.

This exponent need not be rational. For example, $\log_{10} 2$ is an irrational number that, rounded off to five decimal places, becomes 0.30103. If the exponent 30,103/100,000 is placed on 10, the result is 2 *approximately.* Thus the logarithm of 9 to the base 3 (written $\log_3 9$) is 2, because $3^2 = 9$.

Illustrations

$\log_2 8 = 3$ because $2^3 = 8$.

$\log_7 1 = 0$ because $7^0 = 1$.

$\log_3 \frac{1}{3} = -1$ because $3^{-1} = \frac{1}{3}$.

$\log_2 \dfrac{1}{16} = -4$ because $2^{-4} = \dfrac{1}{2^4} = \dfrac{1}{16}$.

$\log_{25} 5 = \frac{1}{2}$ because $(25)^{1/2} = \sqrt{25} = 5$.

$\log_8 4 = \frac{2}{3}$ because $8^{2/3} = (\sqrt[3]{8})^2 = 2^2 = 4$.

$\log_6 6 = 1$ because $6^1 = 6$.

The definition of logarithm implies that

$$\textbf{if} \quad \log_b N = x,$$

$$\textbf{then} \quad b^x = N.$$

These two equations, the former logarithmic and the latter exponential, are equivalent. They say the same thing in two different ways. We shall assume in further discussions that N is a positive number and that b is a positive number different from 1. (Explain the necessity for such restrictions.) But x may be any real number: positive, negative, or zero.

Since, by the definition of logarithm, $\log_b N$ is the exponent that must be placed on b to produce N, it follows that if $\log_b N$ is applied as an exponent to the base b, then the result must be N:

$$b^{\log_b N} = N.$$

The two numbers 10 and $e = 2.718282\ldots$, an irrational number, are most often used as the bases for logarithms. When 10 is the base it is frequently omitted in writing so that log represents $\log_{10}$. log is called the **common logarithm**. e is the base of the **natural logarithm,** written ln.

As a consequence of the definition of logarithm, we have three properties or laws of logarithms, which are true for logarithms to any base.

Property 1 **The logarithm of a product is equal to the sum of the logarithms of the factors; that is,**

$$\log MN = \log M + \log N.$$

Proof Let $x = \log_b M$ and $y = \log_b N$.

Express in exponential form: $M = b^x$ and $N = b^y$.

Multiply the equations: $MN = b^x b^y = b^{x+y}$.

Change to logarithmic form: $\log_b MN = x + y$

$$\log_b MN = \log_b M + \log_b N.$$

The proof is similar for a product of more than two factors.

Illustrations $\log 35 = \log 5 \cdot 7 = \log 5 + \log 7$.

$\log 30 = \log 2 \cdot 3 \cdot 5 = \log 2 + \log 3 + \log 5$.

Property 2 **The logarithm of a fraction is equal to the logarithm of the numerator minus the logarithm of the denominator; that is,**

$$\log \frac{M}{N} = \log M - \log N.$$

Proof Let $x = \log_b M$ and $y = \log_b N$.

Express in exponential form: $M = b^x$ and $N = b^y$.

Divide the equations: $\frac{M}{N} = \frac{b^x}{b^y} = b^{x-y}$.

Change to logarithmic form: $\log_b \frac{M}{N} = x - y$

$$\log_b \frac{M}{N} = \log_b M - \log_b N.$$

Illustrations $\log \frac{2}{3} = \log 2 - \log 3.$

$$\log \frac{6}{35} = \log \frac{2 \cdot 3}{5 \cdot 7} = \log 2 + \log 3 - (\log 5 + \log 7).$$

Property 3 **The logarithm of the *k*th power of a number is equal to *k* times the logarithm of the number; that is,**

$$\log N^k = k \log N.$$

Proof Let $x = \log_b N.$

Express in exponential form: $N = b^x.$

Raise to the kth power: $N^k = (b^x)^k = b^{kx}.$

Change to logarithmic form: $\log_b N^k = kx$

$\log_b N^k = k \log_b N.$

Illustrations $\log 8 = \log 2^3 = 3 \log 2.$

$\log \sqrt{3} = \log 3^{1/2} = \frac{1}{2} \log 3.$

$$\log \frac{125}{49} = \log \frac{5^3}{7^2} = \log 5^3 - \log 7^2 = 3 \log 5 - 2 \log 7.$$

Note: Since $\sqrt[r]{M} = M^{1/r}$,

$$\log \sqrt[r]{M} = \frac{1}{r} \log M.$$

Exercise 10.1

Find the value of each of the following logarithms without using a calculator.

1. $\ln 1$
2. $\log_2 32$
3. $\log 10$
4. $\log \frac{1}{100}$
5. $\log_{81} \frac{1}{3}$
6. $\ln e^{-2}$
7. $\ln e^{1492}$
8. $\log_{3/10} \frac{100}{9}$

Find the unknown, N, b, or x in the following:

9. $\log N = 1$
10. $\ln N = 1$
11. $\log_b 4 = -1$
12. $\log_5 N = 0$
13. $x = e^{\ln 1776}$
14. $\log_b e^{11} = 11$
15. $x = \ln(\sin \pi/2)$
16. $\ln N = \cos \pi/2.$

Express in logarithmic form:

17. $e^q = r$
18. $10^{-3} = 0.001$
19. $2^{10} = 1024$
20. $e^3 = 20.0856$

Express in exponential form:

21. $\log_a \frac{1}{3} = -\frac{1}{2}$
22. $\ln N = x$
23. $\log 0.0001 = -4$
24. $\ln \pi = 1.1447$

Express as a single logarithm, assuming all logarithms are to the same base.

25. $\log(x + 7) + \log(x - 7)$
26. $3 \log a + 2 \log b - \frac{1}{2} \log c$
27. $\log 2 + \log \pi + \frac{1}{2}(\log n - \log m)$
28. $\frac{1}{2}[\log(s - a) + \log(s - b) + \log(s - c) - \log s]$

Identify as true or false and give reasons for your choice. Again assume all logarithms are to the same base.

29. $9^{8 \log_9 x + 7 \log_9 y} = x^8 y^7$
30. $7^{6 \log_7 x - 5 \log_7 y} = \dfrac{x^6}{y^5}$
31. $\log \dfrac{a^3 b}{c} = \dfrac{3 \log a + \log b}{\log c}$
32. $4 \log(\log x) = \log(\log x)^4$
33. $\log \sqrt{xy} = \sqrt{\log x + \log y}$
34. $\log N = \dfrac{1}{p} \log N^p$
35. $3^{(\log_3 x)/8} = \sqrt[8]{x}$
36. $10^{3 + \log_{10} 3} = 3000$
37. $\log_8 \dfrac{\sqrt[3]{4}}{2} = -\dfrac{1}{9}$
38. $\log_{81} 27^m = \dfrac{3m}{4}$

10.2 THE EXPONENTIAL FUNCTION

The exponential function is very useful in modeling various physical, biological, and abstract phenomena.

For a positive number $b \neq 1$, the *exponential function* with base b is defined by

$$y = f(x) = b^x,$$

for any real number x. The numbers 10 and e, the bases of the common and the natural logarithms, are most frequently used as bases for the exponential function.

We recall from Section 10.1 that for any $x > 0$, $e^{\ln x} = x$ and $\ln e^x = x$. Thus the exponential and the natural logarithm functions are inverse functions. Indeed, for many calculators the way to get the decimal expansion for e is to key

[1] [INV] [ln x].

Example 1 Which is larger, e^π or π^e?

Solution To calculate e^π we key [π] [INV] [ln x] to find that $e^\pi = 23.14069$.

Similarly, we calculate π^e by [π] [y^x] [1] [INV] [ln x] [=], yielding $\pi^e = 22.459$, showing that e^π is larger.

Example 2 Sketch on the same coordinate axes $y = e^x$ and $y = \ln x$.

Solution We make a table of values:

x	-3	-2	-1	0	1	2	3
e^x	0.05	0.14	0.37	1	2.72	7.39	20.9
$\ln x$					0	.69	1.10

Note that the negative x axis is an asymptote to the curve $y = e^x$, and that the negative y axis is an asymptote to $y = \ln x$. Recalling that the inverse functions have graphs which are reflections through the line $y = x$, we complete the sketch, as seen in Figure 10.1.

When the rate of growth or decay of a quantity is proportional to the amount of the quantity that is present, an exponential model (function) ex-

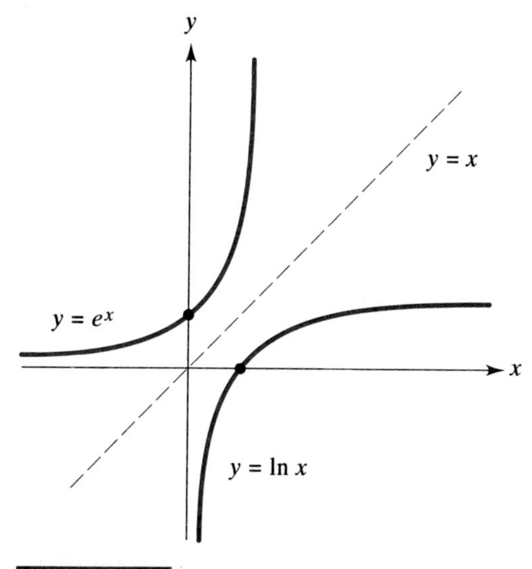

Figure 10.1

presses the total amount of the quantity at a given time. Such a function is given by

$$f(t) = ae^{bt},$$

where $a = f(0)$ and if $b > 0$ a growth is experienced. If $b < 0$, the amount is decreasing as t increases. Such is the case for radioactive decay. The growth of a bacteria in a culture or the unrestricted spread of a communicable disease in a society is also represented by an increasing exponential function.

Example 3 The AIDS virus was introduced in the United States in 1980, and in 1981 there were 380 diagnosed cases. By 1989, the total number of cases was 117,781. If the virus were left to grow unhindered in the population, a formula for the total number of cases in the United States t years after 1981 would be given by

$$f(t) = 380e^{0.7170t}.$$

If this equation models the situation how many cases are to be expected by 1996? How many new cases in 1996?

Solution In 1996, $t = 15$ so the total number of cases at that time would be

$$\begin{aligned} f(15) &= 380e^{(0.7170)(15)} \\ &= 380e^{10.755} \\ &= 380(46.864) = 17{,}808{,}230\,. \end{aligned}$$

The new cases in 1996 would number

$$\begin{aligned} f(15) - f(14) &= 17{,}808{,}230 - 8{,}694{,}240 \\ &= 9{,}113{,}990\,, \end{aligned}$$

which would be more than double the number of cases in 1995. This bleak fate will be avoided by factors which will inhibit the uncontrolled spread of the syndrome in the population, thereby rendering the exponential model inaccurate.

Exercise 10.2

Evaluate:

1. $e^{\sin \pi}$
2. $\sin e$
3. $e^{\cos \pi}$
4. $\cos e$
5. $\cos(\sin e)$
6. $e \cos 0$
7. $e^{\pi/2}$
8. $\pi^{e/2}$
9. $\cos^{-1}(\sin e)$
10. $\sin^{-1}(\cos e)$
11. $\sin^2 e^2$
12. $1 - \cos^2 e^2$
13. $e^{(\pi^0)}$
14. $(e^{\pi})^0$

Sketch a graph of the function in Exercises 15–20.

15. $y = e^{\sin x}$
16. $y = e^{\cos x}$
17. $y = \sin e^x$
18. $y = \cos e^x$
19. $y = -e^{-x}$
20. $y = -3e^{2x}$
21. The half-life of radium is about 1700 years. (That is, after 1700 years, only half of a present supply of radium will remain; the rest will have disintegrated.) If the initial amount of radium is 100 milligrams (mg) and N is the number of milligrams that are left after t years, then

$$N = 100(\tfrac{1}{2})^{t/1700}.$$

(a) How much radium will be present after 200 years?
(b) How many years must pass before only 90 mg remain?

22. When bacteria grow under ideal conditions, their rate of growth varies directly as the number of bacteria present. If the initial number of bacteria in a certain culture is 1000, and N is the number after t hours, then

$$N = 1000(2^t).$$

(a) How many bacteria will be present after 1.60 h?
(b) How many hours must pass before 9000 bacteria are present?

10.3 LOGARITHMIC AND EXPONENTIAL EQUATIONS

A logarithmic equation is an equation that contains the logarithm of some expression involving the unknown. Such an equation can usually be solved by rewriting it in one of the following two forms:

1. $\log M = \log N$, which implies that $M = N$;

or

2. $\log_b N = w$, which implies that $N = b^w$.

Example 1 Solve for x:

$$\log x + 3 \log a = \tfrac{1}{2} \log b.$$

Solution We shall rewrite the equation in the form $\log M = \log N$ and then state that $M = N$. (If two numbers have equal logarithms, the numbers must be equal.)

$$\begin{aligned}
\log x + 3 \log a &= \tfrac{1}{2} \log b \\
\log x + \log a^3 &= \log b^{1/2} \\
\log a^3 x &= \log \sqrt{b}, \\
a^3 x &= \sqrt{b}, \\
x &= \frac{\sqrt{b}}{a^3}.
\end{aligned}$$

Comment A common mistake is to say that if $\log A + \log B = \log C$, then $A + B = C$. This is incorrect because $\log A + \log B \neq \log(A + B)$. The proper way to handle this situation is to replace $\log A + \log B$ with $\log AB$. Then $\log AB = \log C$; $AB = C$.

Example 2 Solve for x:

$$\log_{10} x + \log_{10} a = 3 + 4 \log_{10} b.$$

Solution 1 We shall rewrite the equation in the form $\log_{10} N = w$ and then assert that $N = 10^w$ (definition of logarithm). Then

$$\begin{aligned}
\log_{10} x + \log_{10} a - \log_{10} b^4 &= 3 \\
\log_{10} \frac{xa}{b^4} &= 3, \\
\frac{xa}{b^4} &= 10^3, \\
x &= \frac{1000b^4}{a}.
\end{aligned}$$

This result may be checked by equating the logarithms of the two sides of the last equation and then showing that the resulting equation is equivalent to the given equation.

Solution 2 Replace 3 with $\log_{10} 1000$ and proceed as in Example 1. Then

$$\begin{aligned}
\log_{10} x + \log_{10} a &= \log_{10} 1000 + \log_{10} b^4 \\
\log_{10} xa &= \log_{10} 1000b^4 \\
xa &= 1000b^4 \\
x &= \frac{1000b^4}{a}.
\end{aligned}$$

An exponential equation is an equation in which a variable appears in an exponent. Such an equation can usually be solved by equating the logarithms of the two sides and then finding the roots of the resulting algebraic equation.

Example 3 Solve for x:

$$(9.55)^x = 0.0345\,.$$

Solution Take the logarithm of each side:

$$\begin{aligned}\log(9.55)^x &= \log 0.0345\\ x\log 9.55 &= \log 0.0345\,,\\ x &= \frac{\log 0.0345}{\log 9.55}\\ &= \frac{-1.4622}{0.98}\\ &= -1.49\,.\end{aligned}$$

Had we used natural logarithms, we would have found

$$\begin{aligned}x &= \frac{\ln 0.0345}{\ln 9.55}\\ &= \frac{-3.3668}{2.2565} = -1.49\,.\end{aligned}$$

Exercise 10.3

Solve the following equations for x.

1. $\log(3 - x) + \log(4 - x) = \log 20$
2. $\log(5 + x) + \log(7 - x) = \log 32$
3. $\log x + \log(x - 2) = \log(8 - 4x)$
4. $2\log(x - 1) - \log(3x - 7) = \log 2$
5. $\log_2(x^2 + 15) = 4 + \log_2(x - 3)$
6. $\log_{10}(x + 10) + \log_{10}(x + 3) - \log_{10}(x + 4) = 1$
7. $\log_3(x + 2) + \log_3(x - 22) = 4$
8. $\log_{10}(x + 11) + 2\log_{10} 5 = 2 + \log_{10} 7$
9. $a^{2x+3} = c^{4x-5}$
10. $y = \dfrac{1}{2}\log_b\left(\dfrac{1 + x}{1 - x}\right)$
11. $y = \log_{10}(\log_{10} x)$

12. $y = \ln(\ln x)$
13. $e^x = \pi$
14. $(5.37)^x = 0.482$

For Exercises 15 through 17, graph the two functions on the same set of coordinate axes.

15. $y = \ln x, y = e^x$
16. $y = \log_3 x, y = 3^x$
17. $y = e^{-x}, y = e^{x^2}$, for $-5 \le x \le 5$.
18. Use a graphing calculator to graph $y = e^x + \cos x$ and $y = e^x - \ln x$ on the same coordinate axes, for $-2 \le x \le 3$.

KEY TERMS

Logarithm, exponential

REVIEW EXERCISES

Find the unknown:

1. $\log_b 9 = -1$
2. $\ln e^{-4} = x$
3. $\log N = -4$

Solve for x:

4. $\ln(x - 2) + \ln(x + 2) = \ln(2x - 1)$
5. $e^x = 7.632$
6. $\ln x = 1.548$

Graph on the same coordinate axes:

7. $y = e^{-x}; y = -e^x$
8. $y = 2e^x, y = e^{2x}$

11

Polar Coordinates

11.1 PARAMETRIC EQUATIONS

The locus of a curve in the plane is sometimes described by expressing the x coordinate and the y coordinate of each point on the curve as functions of a third variable, which is called the parameter. For instance, if the parameter is θ and if $0 \le \theta \le 2\pi$, while $x(\theta) = \cos\theta$ and $y(\theta) = \sin\theta$, then for each value of θ we get a point on the curve. When $\theta = 0$ the point is $(1, 0)$, when $\theta = \pi/2$ it is $(0, 1)$, and when $\theta = 3\pi/4$ it is $(-1/\sqrt{2}, 1/\sqrt{2})$. The curve determined by letting θ run through all values from 0 to 2π is a circle of radius 1 centered at the origin.

While the parametric equations $x = \cos\theta$, $y = \sin\theta$ and the equation $x^2 + y^2 = 1$ have the same graph, the parametric equations convey a sense of motion from $\theta = 0$ at $(1, 0)$ counterclockwise around the circle to $(1, 0)$ again when $\theta = 2\pi$. In the curve with the parameter eliminated, $x^2 + y^2 = 1$, we have only a static graph.

This chapter provides an especially good opportunity to exercise the use of a graphing utility, a graphing calculator or a computer with graphic software.

Example 1 Graph the parametric equations

$$x(\theta) = \sin^2\theta,$$

$$y(\theta) = \cos^2\theta,$$

for $0 \le \theta \le 2\pi$.

Solution 1 To eliminate the parameter θ, we note that $x + y = \sin^2\theta + \cos^2\theta = 1$, so that the graph is that of the line segment $x + y = 1$, where $x \ge 0$, $y \ge 0$. The graph is given in Figure 11.1. The curve begins $(\theta = 0)$ at $(0, 1)$, goes down the line $x + y = 1$ to $(1, 0)$ when $\theta = \pi/2$, and returns to $(0, 1)$ when $\theta = \pi$. It repeats the trip for $\pi \le \theta \le 2\pi$.

Solution 2 On some graphing calculators the method of graphing in parametric form is particularly easy. Also, the sense of motion to be conveyed by parametric equations is displayed most vividly. One must enter the equations, as per the instructions in the manual, with the graphic utility in the parametric mode. Sometimes the parameter is called T rather than θ. To get a good viewing "window" one must set ranges for x, y, and T. In this example, $-0.5 \leq x \leq 1.5$, $-0.5 \leq y \leq 1.5$, and $0 \leq T \leq 6.4$ will suffice nicely. The graph is as seen in Figure 11.1.

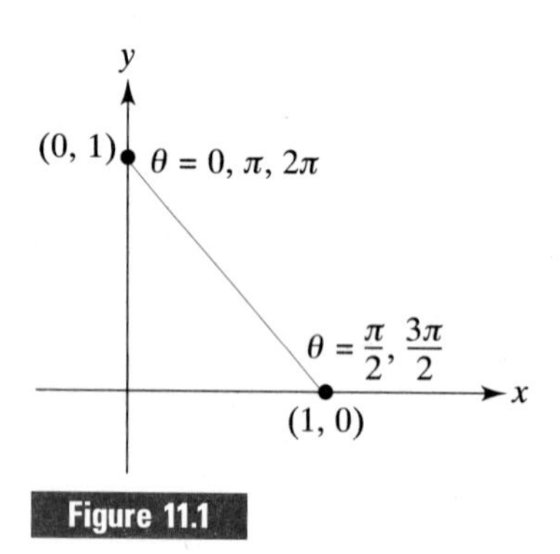

Figure 11.1

If an object is projected in air at an initial velocity v_0 and at an angle of elevation β, then from Figure 11.2 we see that the velocity vector has a horizontal component of $v_0 \cos \beta$ and a vertical component of $v_0 \sin \beta$. From physics we know that at time t, the parametric equations of the point (x, y) on the trajectory of the object are

$$x = (v_0 \cos \beta)t,$$
$$y = (v_0 \sin \beta)t - 16t^2. \quad (1)$$

The trajectory, neglecting wind and air resistance, is a parabola as shown in Figure 11.2.

Example 2 A tank fires a shell with initial velocity of 288 ft/s at a 30° angle of elevation. Where and when does the shell hit the ground?

Solution Setting $v_0 = 288$, $\theta = \pi/6$ in Equations (1) we get the parametric equations

$$x(t) = \left(288 \cos \frac{\pi}{6}\right)t = 249t$$

$$y(t) = \left(288 \sin \frac{\pi}{6}\right)t - 16t^2 = 144t - 16t^2.$$

The shell hits the ground when $y = 0$, hence we must solve

$$144t - 16t^2 = 0.$$

The solutions are $t = 0$ and $t = 9$, so the shell is airborne for 9 seconds; $x(9) = 2245$, so it traveled for 2245 ft. Its maximum height occurs at $t = 4.5$ seconds and is $y(4.5) = 324$ ft.

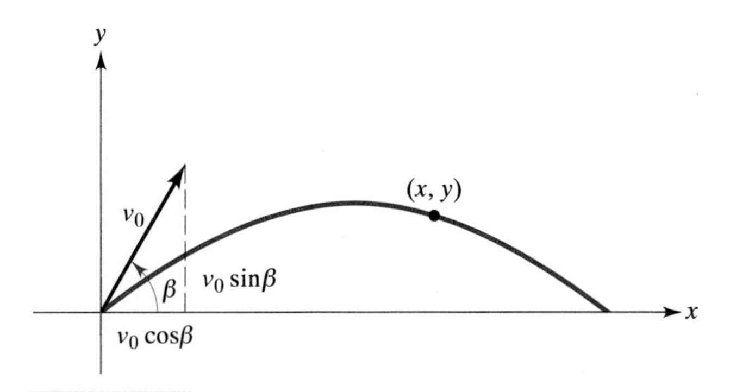

Figure 11.2

Exercise 11.1

In Exercises 1 through 6 convert from parametric form to rectangular form, i.e., eliminate the parameter, to sketch for $0 \le \theta \le 2\pi$.

1. $x = 5 \cos \theta$, $y = 5 \sin \theta$
2. $x = 2 + \cos \theta$, $y = 1 + \sin \theta$
3. $x = 2 \sin^2 \theta$, $y = 3 \cos^2 \theta$
4. $x = \sec^2 \theta$, $y = \tan^2 \theta$
5. $x = 3 \tan^2 \theta$, $y = 2 \sec^2 \theta$
6. $x = 2 \cos \theta$, $y = 3 \sin \theta$

In Exercises 7 through 10 use a graphing calculator to sketch the parametric equations

7. $x = \theta^2$
   $y = \sin\theta$
   $0 \le \theta \le 2\pi$

8. $x = \sec\theta$
   $y = \tan\theta$
   $-\pi/2 < \theta < \pi/2$

9. $x = \theta\sin\theta$
   $y = \cos\theta$
   $0 \le \theta \le 4\pi$

10. $x = \theta\cos\theta$
    $y = \theta\sin\theta$
    $0 \le \theta \le 2\pi$

11. A mortar fires a shell at an angle of 45° with an initial velocity of 320 ft/s. How far does it travel and how high does it go?

11.2 POLAR COORDINATES

We saw in Section 1.5, in the definition of the trigonometric functions, that with any point P in the plane we can associate an ordered pair (x, y) of rectangular coordinates and a pair (r, θ) of coordinates which are the *polar coordinates* of P, as seen in Figure 11.3.

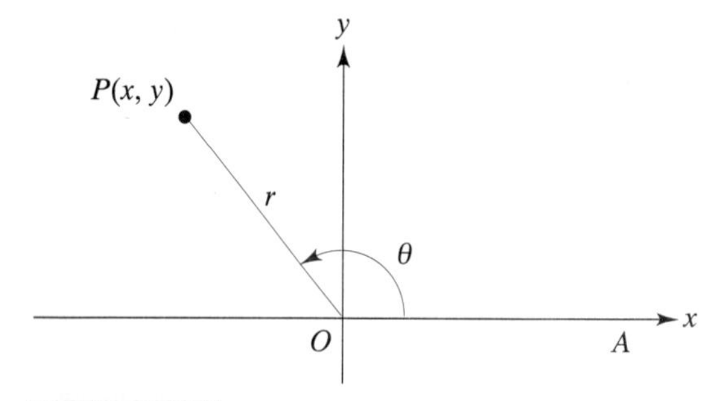

Figure 11.3

More specifically, the distance r from the origin O, now called the *pole,* to P is the *radius vector* while θ, the *vectorial angle,* is positive or negative depending on whether it is measured counterclockwise or clockwise from the *polar axis* (the positive x axis) A. The radius vector r is positive if measured from the pole along the terminal side of θ and is negative if measured along the reflection (extension) of the terminal side through the pole.

Each pair (r, θ) of polar coordinates determines a point in the plane but a given point may have many polar coordinates. For example $(3, \pi/6)$, $(3, -11\pi/6)$, $(-3, 7\pi/6)$, and $(-3, -5\pi/6)$ are all polar coordinates of the same point as seen in Figure 11.4.

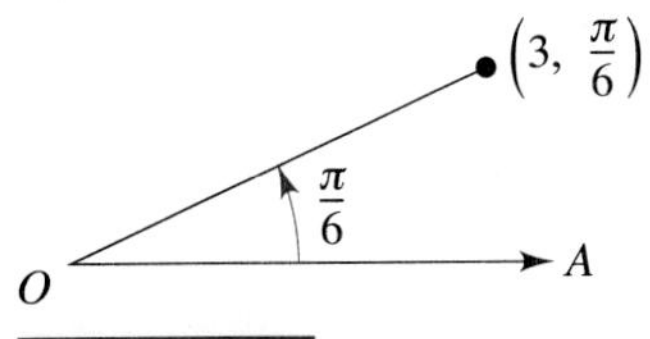

Figure 11.4

We see then that

$$
\begin{aligned}
(r, \theta) &= (-r, \theta + \pi) \\
&= (-r, \theta - \pi) \qquad (2) \\
&= (r, \theta + 2n\pi), \ n \text{ any integer.}
\end{aligned}
$$

Example 1 Plot $(-2, \pi/3)$ in polar coordinates and give three other pairs of polar coordinates for the point.

Solution We extend 2 units through the pole on an angle of $\pi/3$, as indicated in Figure 11.5. From Equations (2) we have

$$
\begin{aligned}
\left(-2, \frac{\pi}{3}\right) &= \left(2, -\frac{2\pi}{3}\right) = \left(2, \frac{4\pi}{3}\right) \\
&= \left(-2, -\frac{5\pi}{3}\right).
\end{aligned}
$$

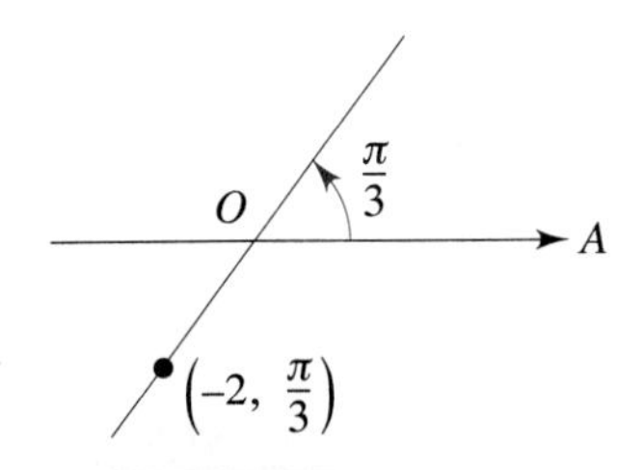

Figure 11.5

From Figure 11.3 it is easy to see that the following relations hold between the rectangular and the polar coordinates of a point. If P has rectangular coordinates (x, y) and polar coordinates (r, θ), then

$$
\begin{aligned}
\mathbf{x} &= \mathbf{r} \cos \boldsymbol{\theta}, \\
\mathbf{y} &= \mathbf{r} \sin \boldsymbol{\theta}. \qquad (3)
\end{aligned}
$$

Also,

$$r^2 = x^2 + y^2,$$

and

$$\tan\theta = \frac{y}{x}, \qquad \text{for } x \neq 0. \tag{4}$$

Unfortunately, these equations do not determine r and θ uniquely. We must, therefore, impose our own conditions on r and θ to make the conversion from rectangular to polar coordinates. We may choose to insist that r be nonnegative. This yields

$$r = \sqrt{x^2 + y^2}. \tag{5}$$

Now if $\tan\theta = y/x$, we get

$$\theta = \text{Tan}^{-1}\frac{y}{x}, \qquad x \neq 0. \tag{6}$$

But the range of Tan^{-1} is $-\pi/2 < \theta < \pi/2$, so that the value of θ from Equation (6), together with Equation (5), will *not* represent any point to the left of the y axis. For example, if $x = -1$, $y = 1$, we would get $r = \sqrt{2}$ and $\theta = \text{Tan}^{-1}(-1) = -\pi/4$, which represents a point in the fourth quadrant. There are several ways out of this dilemma. One way is to allow negative values of r and to keep Equation (6). Another is to keep Equation (5) and instead of Equation (6) to use

$$\theta = \begin{cases} \text{Tan}^{-1}\dfrac{y}{x}, & \text{if } x > 0 \\ \pi + \text{Tan}^{-1}\dfrac{y}{x}, & \text{if } x < 0. \end{cases} \tag{7}$$

Most graphing calculators have the ability to convert from rectangular to polar coordinates. They keep $r \geq 0$ and use the formula

$$\theta = \begin{cases} \text{Tan}^{-1}\dfrac{y}{x} & \text{if } (x, y) \text{ is in quadrants I, IV} \\ \pi + \text{Tan}^{-1}\dfrac{y}{x} & \text{if } (x, y) \text{ is in quadrant II} \\ -\pi + \text{Tan}^{-1}\dfrac{y}{x} & \text{if } (x, y) \text{ is in quadrant III.} \end{cases} \tag{8}$$

Finally, we may keep both r and θ positive by using Equation (5) and

$$\theta = \begin{cases} \text{Tan}^{-1} \dfrac{y}{x} & \text{if } (x, y) \text{ is in quadrant I} \\[2ex] \pi + \text{Tan}^{-1} \dfrac{y}{x} & \text{if } (x, y) \text{ is in quadrants II or III} \\[2ex] 2\pi + \text{Tan}^{-1} \dfrac{y}{x} & \text{if } (x, y) \text{ is in quadrant IV.} \end{cases} \qquad (9)$$

Naturally, in all these expressions if $x = 0$, $\text{Tan}^{-1}\, y/x$ is undefined. So if $x = 0$, we put $r = |y|$ and $\theta = \pi/2$ if $y > 0$, and if $y < 0$, we put $\theta = -\pi/2$ or $\theta = 3\pi/2$.

Example 2 Find the rectangular coordinates of the point with $(-2, \pi/3)$ as its polar coordinates.

Solution Using Equations (3), we have

$$x = r \cos \theta = -2 \cos \frac{\pi}{3} = -1,$$

$$y = r \sin \theta = -2 \sin \frac{\pi}{3} = -\sqrt{3}.$$

The rectangular coordinates are $(-1, -\sqrt{3})$.

Example 3 Find polar coordinates of the point $(-2, -2)$ in rectangular coordinates.

Solution From Equations (4) we have

$$r = \sqrt{x^2 + y^2} = 2\sqrt{2}.$$

And since $x < 0$,

$$\theta = \pi + \text{Tan}^{-1} 1 = \frac{5\pi}{4}.$$

Exercise 11.2

1. Identify each point (r, θ) on the graph by its polar coordinates provided $r > 0$ and $0 \le \theta < 2\pi$.

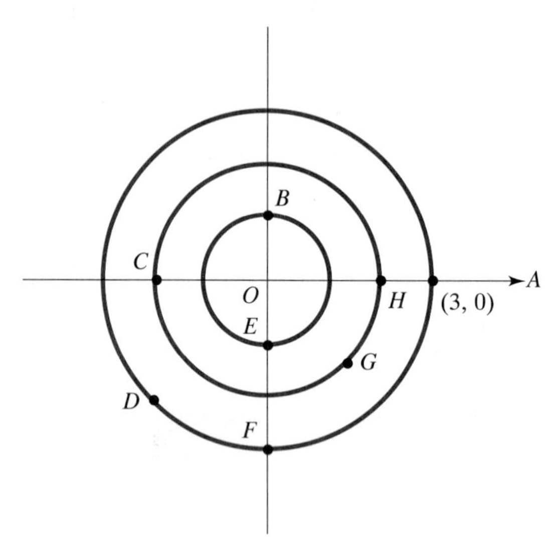

2. Answer Exercise 1 provided $r \ge 0$ and $-\pi \le \theta < \pi$.
3. Answer Exercise 1 with $-\pi/2 < \theta \le \pi/2$ and no restriction on r.

In Exercises 4 through 9 plot the points $P(r, \theta)$ in polar coordinates.

4. $P(2, \pi)$
5. $P(-3, \pi/6)$
6. $P(-1, -\pi/2)$
7. $P(5, -\pi/9)$
8. $P(2, 5\pi/4)$
9. $P(-\sqrt{2}, -7\pi/4)$

In Exercises 10 through 15 write three other pairs of polar coordinates for the given point.

10. $P(-1, \pi/6)$
11. $P(-1, 5\pi/6)$
12. $P(-4, \pi)$
13. $P(-\sqrt{2}, -7\pi/6)$
14. $P(\sqrt{5}, \pi/3)$
15. $P(-2, -\pi/4)$

Find the rectangular coordinates of the point $P(r, \theta)$ in Exercises 16 through 21.

16. $P(-1, \pi/6)$
17. $P(-1, 5\pi/6)$
18. $P(3\sqrt{2}, -3\pi/4)$
19. $P(-\sqrt{2}, -7\pi/6)$
20. $P(-3, \pi/3)$
21. $P(-2, -\pi/4)$

Find nonnegative polar coordinates for the following points in rectangular coordinates.

22. $(0, 5)$
23. $(5, 0)$
24. $(2\sqrt{3}, -2)$
25. $(\sqrt{3}, -1)$
26. $(-4, -3)$
27. $(-5, -12)$

11.3 GRAPHING IN POLAR COORDINATES

We consider now the problem of graphing a function $r = f(\theta)$ in polar coordinates, provided f is a function whose domain includes the interval $0 \leq \theta \leq 2\pi$. The *graph* of an equation $r = f(\theta)$ *in polar coordinates* is the set of all points (r, θ) whose coordinates satisfy the equation.

Example 1 Construct a graph of $r = \sin\theta$.

Solution 1 We assign convenient values of θ, $0 \leq \theta \leq 2\pi$, and plot the corresponding points (r, θ). Doing this we see that the following points are on the graph:

$$(0,0), \left(\frac{1}{\sqrt{2}}, \frac{\pi}{4}\right), \left(1, \frac{\pi}{2}\right), (0, \pi), \left(-1, \frac{3\pi}{2}\right).$$

The graph is a circle as shown in Figure 11.6. It is traced counterclockwise twice from the pole to $(1, \pi/2)$ and back.

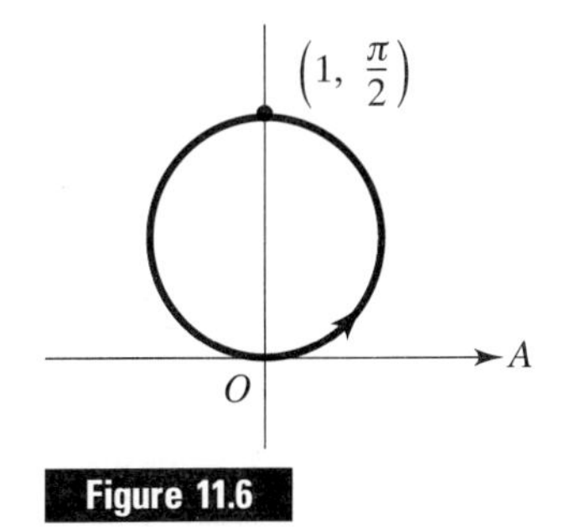

Figure 11.6

Solution 2 Some graphing calculators will graph $r = f(\theta)$ in polar coordinates. All the user needs to do is to enter the function and set the range of values for θ and r. Other graphing calculators convert from polar equations to parametric equations in order to graph $r = f(\theta)$. For them, we use

$$x = r\cos\theta = f(\theta)\cos\theta$$
$$y = r\sin\theta = f(\theta)\sin\theta$$

with $0 \le \theta \le 2\pi$ to sketch $r = f(\theta)$ in parametric form. The cursor traces the curve as θ moves from 0 to 2π.

Example 2 Graph $r = \cos 3\theta$, for $0 \le \theta \le 2\pi$.

Solution 1 We make a table of values showing changes in r as θ changes in steps of length $\pi/6$:

θ	r
$0 \to \pi/6$	$1 \to 0$
$\pi/6 \to \pi/3$	$0 \to -1$
$\pi/3 \to \pi/2$	$-1 \to 0$
$\pi/2 \to 2\pi/3$	$0 \to 1$
$2\pi/3 \to 5\pi/6$	$1 \to 0$
$5\pi/6 \to \pi$	$0 \to -1$

Continuing in this way as θ goes from π to 2π we get the graph in Figure 11.7.

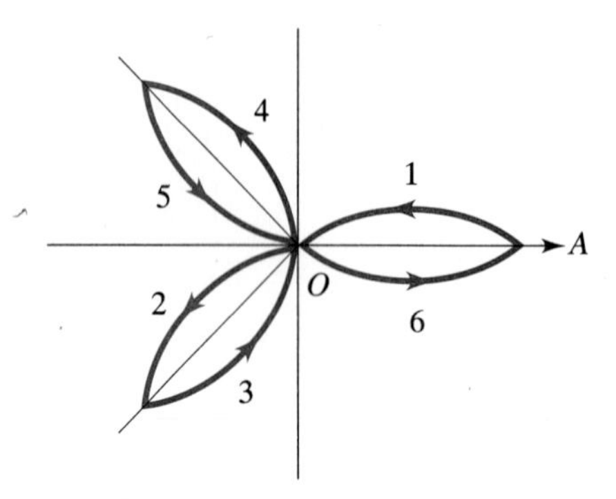

Figure 11.7

Solution 2 Graph $r = \cos 3\theta$ on a graphing calculator either in polar form or in parametric form. The graph, a three-leaf rose, is traced as the arrows indicate in Figure 11.7.

Exercise 11.3

Sketch the graphs in polar coordinates. Use a graphing calculator if one is available.

1. $r = 3$
2. $r = -2$
3. $r = 4/\sin\theta$
4. $r = -1/\cos\theta$
5. $r = 1/(\cos\theta + \sin\theta)$
6. $r = \cos\theta$
7. $r = 2\sin\theta$
8. $r = \sin 2\theta$
9. $r = \cos 2\theta$
10. $r = \sin 3\theta$
11. $r = \sin(\theta/2)$
12. $r = \cos(\theta/2)$
13. $r = \sin(3\theta/2)$
14. $r = 2 + 3\cos\theta$
15. $r = \sin^2\theta + \cos^2\theta$
16. $r = \cos^2\theta - \sin^2\theta$

KEY TERMS

Parametric equations, polar coordinates, radius vector, vectorial angle, polar axis, graph of an equation in polar coordinates

REVIEW EXERCISES

Sketch the graph of the parametric equations for $0 \le \theta \le 2\pi$.

1. $x = 2\sin\theta\cos\theta$
   $y = \cos^2\theta - \sin^2\theta$
2. $x = \cos^2\theta$
   $y = \cos\theta\sin\theta$
3. $x = \tan\theta$
   $y = \sec\theta$
   $0 < \theta < \pi/2$
4. $x = 2 + \cos\theta$
   $y = -3 + \sin\theta$

Convert from rectangular to polar coordinates.

5. $(-2, -2)$
6. $(0, 4)$
7. $(0, -3)$
8. $(\sqrt{3}, \sqrt{2})$
9. $(-\sqrt{2}, 0)$
10. $(-\sqrt{3}, \sqrt{3})$

Graph in polar coordinates, $0 \le \theta \le 2\pi$.

11. $r = 2\cos\theta$
12. $r = -\sqrt{2}$
13. $r = \theta$
14. $r = \sin 2\theta$
15. $r = \cos 2\theta$
16. $r = 2\sin\theta\cos\theta$
17. A pitcher throws a baseball horizontally 90 mi/h. If the point of release is 7 ft above the ground, at what height does it cross home plate 60.5 ft away?

12

Complex Numbers

12.1 COMPLEX NUMBERS

It will be recalled that there are quadratic equations with real coefficients that have no real roots. The simplest such equation is $x^2 + 1 = 0$, which obviously has no real solution, since the square of any real number—whether positive, negative, or zero—is nonnegative. (An analogous linear equation with positive coefficients is $x + 1 = 0$, which necessitates the postulation of negative numbers for its solution—though the negative numbers probably arose from other considerations.) Another example of a quadratic equation with real coefficients but with no real roots is $x^2 - 4x + 13 = 0$. It was for the solution of such equations that the system of *complex numbers* was first investigated, but this remarkable system has proved to be invaluable in many other connections.

The system of complex numbers is really the system of *ordered pairs* of *real* numbers—a first, then a second—(a, b), in which equality, addition, and multiplication are defined in a certain specified way. Usually, however, complex numbers are written as $a + bi$, and in this form the defining properties* are as follows:

Equality: $a + bi = c + di$ **if and only if** $a = c$ **and** $b = d$,

Addition: $(a + bi) + (c + di) = (a + c) + (b + d)i$,

Multiplication: $(a + bi)(c + di) = (ac - bd) + (ad + bc)i$.

It is easy to show that all the ordinary rules for adding and multiplying real numbers—the associative and commutative laws, etc.—carry over to the system of complex numbers. For this reason, we speak of the *field* of complex numbers.

*In terms of the (a, b) notation, which we shall not adopt, these definitions are
Equality: $(a, b) = (c, d)$ if and only if $a = c$ and $b = d$
Addition: $(a, b) + (c, d) = (a + c, b + d)$
Multiplication: $(a, b)(c, d) = (ac - bd, ad + bc)$

There is no need to memorize the definitions of addition and multiplication of complex numbers. Just remember that addition and multiplication are ordinary addition and multiplication, with the single special property that

$$i^2 = -1.$$

There is, however, one property of the real numbers that does not carry over to the complex numbers. *The complex number field is not ordered;* we do not say that one complex number is less than or greater than another.

Complex numbers have been defined as numbers that obey the definitions of equality, addition, and multiplication listed above. This approach is equivalent to the following definition.

A *complex number* is a number of the form $a + bi$, where a and b are real numbers and $i = \sqrt{-1}$.

If $b \neq 0$, the complex number $a + bi$ is called an *imaginary number.*

If $a = 0$ and $b \neq 0$, the complex number $a + bi$ is called a *pure imaginary number.*

If $b = 0$, the complex number $a + bi$ becomes a, a real number.

Hence we see that the field of complex numbers includes all real numbers and all imaginary numbers.

Illustrations *Imaginary numbers:* $3i, 2 + 5i, -7 + 8i, 9 - i, -1 - 6i,$

Pure imaginary numbers: $3i, -4i, \sqrt{-49}, -\sqrt{-2},$

Real numbers: $4, \frac{1}{7}, 0, -\frac{2}{9}, 5, -\sqrt{2}, \pi.$

All these numbers are complex numbers.

Since $\sqrt{a} \cdot \sqrt{a} = a$ (by the definition of square root), we see that if $i = \sqrt{-1}$, then $i^2 = -1$. Moreover, $i^3 = i^2 \cdot i = -i$, $i^4 = (i^2)^2 = (-1)^2 = 1$, $i^5 = i^4 \cdot i = i$, $i^6 = i^4 \cdot i^2 = -1$, $i^{87} = i^{84} \cdot i^3 = (i^4)^{21} i^3 = 1^{21}(-i) = -i$.

In solving the equation $x^2 - 4x + 13 = 0$, we find the roots to be $x = 2 \pm \sqrt{-9}$. Remembering that $i = \sqrt{-1}$, we have $x = 2 \pm 3i$. Notice that if $2 + 3i$ is substituted for x in the equation $x^2 - 4x + 13 = 0$, we get

$$4 + 12i + 9i^2 - 8 - 12i + 13 = 0.$$

If i^2 is replaced by -1, we obtain $17 + 12i - 12i - 17 = 0$. This shows that $2 + 3i$ is a perfectly good root of the equation, provided we understand that i is a number whose square is -1.

The complex numbers $a + bi$ and $a - bi$ are said to be *complex conjugates* of each other. Notice that the roots of $x^2 + 4 = 0$ are $2i$ and $-2i$, which are pure imaginary complex conjugates. The roots of the equation $x^2 - 4x + 13 = 0$ are the conjugate imaginary numbers $2 + 3i$ and $2 - 3i$. It can be shown that if an imaginary number $(a + bi)$ is a root of an equation with *real* coefficients, then the conjugate imaginary $(a - bi)$ is also a root of this equation.

Since i is a number whose square is -1, the best procedure in handling complex numbers is to perform all operations as if i were an ordinary letter and then replace i^2 with -1. It is to be noted that the quotient of two complex numbers is obtained by multiplying numerator and denominator by the conjugate of the denominator. For example,

$$\frac{7 + 5i}{3 - i} = \frac{(7 + 5i)(3 + i)}{(3 - i)(3 + i)} = \frac{21 + 7i + 15i + 5i^2}{9 - i^2}$$

$$= \frac{16 + 22i}{10} = \frac{8}{5} + \frac{11}{5}i.$$

This result can be checked by multiplying $\frac{8}{5} + \frac{11}{5}i$ by $3 - i$. What should the result be?

All complex numbers should first be written in the form $a + bi$. Thus $3 + \sqrt{-49} = 3 + \sqrt{49}\sqrt{-1} = 3 + 7i$. This procedure is suggested to avoid mistakes such as $\sqrt{-5} \cdot \sqrt{-5} = \sqrt{(-5)(-5)} = \sqrt{25} = 5$. This is obviously incorrect because, by the definition of square root, $\sqrt{-5}$ is a number that when multiplied by itself becomes -5. The correct way of handling this is $\sqrt{-5} \cdot \sqrt{-5} = i\sqrt{5} \cdot i\sqrt{5} = 5i^2 = -5$. This result agrees with the definition of square root.

Exercise 12.1

Perform each of the indicated operations and express the result in the form $a + bi$.

1. $(6 - 5i) + (2 + 3i)$
2. $(4 + i) + (5 - 6i) - (1 + 2i)$
3. $(3 + 7i) - (9 + 7i)$
4. $(8 - 5i) - (3 + 4i)$
5. $(4 - i)(5 + 6i)$
6. $(2 + 9i)(3 + 2i)$
7. $(6 + \sqrt{-5})(9 + 2\sqrt{-5})$
8. $(1 + 4\sqrt{-3})(7 - \sqrt{-3})$
9. $(3 - 8i)^2$
10. $(6 + i)^2$

11. i^{1996}
12. $i^7 + i^8 + i^9 + i^{10}$
13. $i^{1776} - i^{1492}$
14. $i^3 - i^{13}$
15. $\dfrac{4 + 5i}{1 - 6i}$
16. $\dfrac{8 - 9i}{7 - 4i}$
17. $\dfrac{5 + 8i}{3 + 2i}$
18. $\dfrac{9 - 2i}{2 + 7i}$
19. $\dfrac{-5 + 6i}{2i}$
20. $\dfrac{7}{8 + 3i}$

Find the values of the real numbers x and y.

21. $6x + 5yi = 3 - 15i$
22. $(x - i)(6 + 7i) = 19 + yi$
23. $(1 + xi)(y + 5i) = 21 + i$
24. $(x + 3i)(4 - 7i) = 41 + yi$
25. Find the value of $x^2 - 8x + 65$ if $x = 4 + 7i$.
26. Show by substitution that $\frac{5}{4} - \frac{9}{4}i$ is a root of the equation $8x^2 - 20x + 53 = 0$.
27. Write a complex number that is not an imaginary number.
28. Prove that the sum of two conjugate imaginary numbers is a real number.
29. Prove that the product of two conjugate imaginary numbers is a positive real number.

12.2 GRAPHICAL REPRESENTATION OF COMPLEX NUMBERS

Let us represent (as in the case of the x axis of a rectangular coordinate system) the real numbers by points on a horizontal directed line (Figure 12.1). Let the vector V represent the directed segment connecting the origin O to the point corresponding to the real number a. Since $ai^2 = -a$, it can be said that multiplying a by $i \cdot i$ is geometrically equivalent to rotating V through 180° about O. Consequently, it is logical to represent the multiplication of a by i as a rotation of V through 90° about O. Accordingly, the number ai will be represented as a point a units above O on the *vertical* line through O. We shall refer to the horizontal axis as the **axis of reals** and the vertical axis as the **axis of (pure) imaginaries.** This

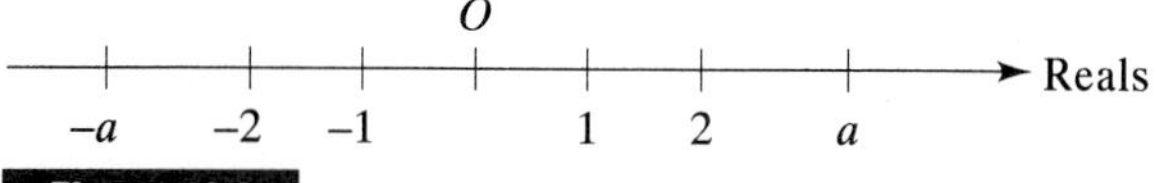

Figure 12.1

system of axes defines a region called the **complex plane.** It is to be noted that, while the unit on the axis of reals is the number 1, the unit on the axis of imaginaries is the imaginary number i. Hence the complex number $a + bi$ is represented by the point a units from the axis of imaginaries and b units from the axis of reals. Figure 12.2 illustrates the graphical representation of complex numbers in the complex plane.

It is often convenient to think of the complex number $a + bi$ as representing the vector OP (Figure 12.2).

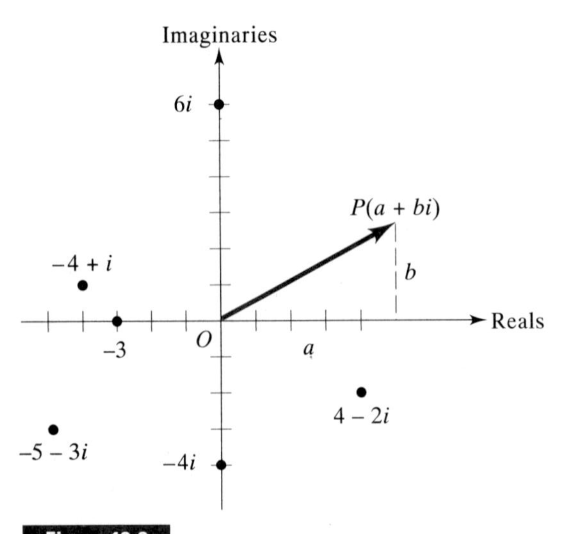

Figure 12.2

Since the sum of $a + bi$ and $c + di$ is $(a + c) + (b + d)i$, we can add the numbers graphically by adding the real parts, a and c, to get the real part of the sum, and adding the imaginary coefficients, b and d, to get the imaginary coefficient. This is illustrated in Figure 12.3. The result is

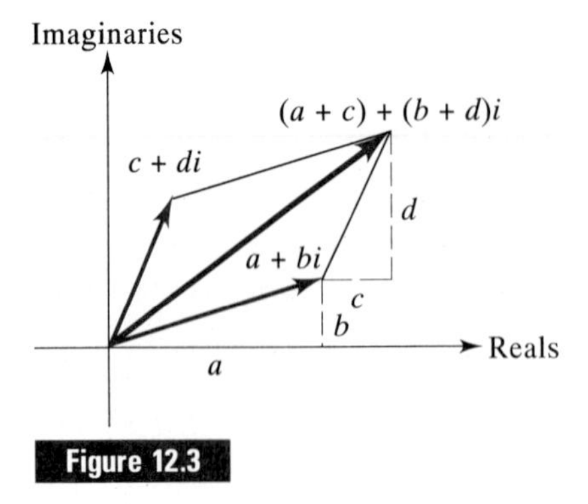

Figure 12.3

exactly the same as if we had applied the parallelogram law to the vectors representing the numbers $a + bi$ and $c + di$. Three complex numbers can be added graphically by first obtaining the sum of two of them and then adding this to the third.

We can subtract $c + di$ from $a + bi$ graphically by adding $a + bi$ to $-c - di$.

12.3 POLAR FORM OF A COMPLEX NUMBER

Let point P in the complex plane represent the complex number $a + bi$. The **absolute value** of $a + bi$ is the distance r from O to P. It is always considered positive. The **amplitude** of $a + bi$ is the angle measured from the positive axis of reals to the line OP. Absolute value is also called *modulus;* amplitude is sometimes called *argument.* From Figure 12.4, it is obvious that

$$r = \sqrt{a^2 + b^2}, \qquad \tan\theta = \frac{b}{a}, \tag{1}$$

and

$$a = r\cos\theta, \qquad b = r\sin\theta. \tag{2}$$

These equations hold regardless of the quadrant in which P lies. If the last equation is multiplied by i and added to the preceding one, we get

$$\boldsymbol{a + bi = r(\cos\theta + i\sin\theta)}. \tag{3}$$

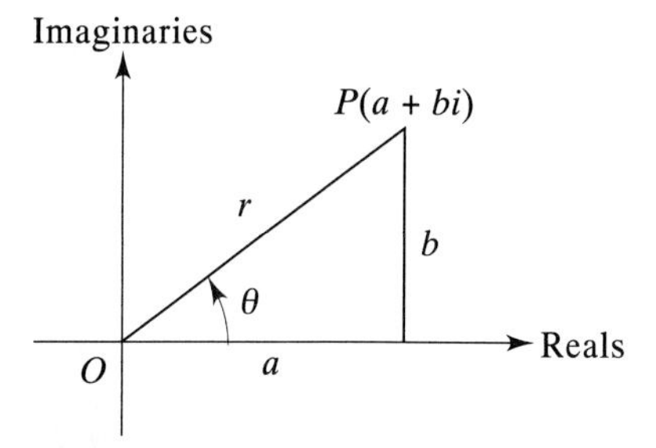

Figure 12.4

The expression $r(\cos\theta + i\sin\theta)$ is called the **polar form** of the complex number while $a + bi$ is called the **algebraic form** of the complex number. The polar form employs the polar coordinates (r, θ) of the point with rectangular coordinates (a, b). Polar form is sometimes called *trigonometric* form and $r(\cos\theta + i\sin\theta)$ is sometimes written in abbreviated form $r \text{ cis } \theta$.

A complex number in polar form is easily converted to algebraic form by using Equations (2) above. As we saw in Section 11.2, the conversion from algebraic (rectangular) form to polar form is not so easy. We will use Equations (9) of Section 11.2 in this conversion to keep $r \geq 0$ and $\theta \geq 0$. Hence the complex number $a + bi$ equals $r(\cos\theta + i\sin\theta)$ if

$$r = \sqrt{a^2 + b^2}$$

and

$$\theta = \begin{cases} \operatorname{Tan}^{-1}\dfrac{b}{a}, & \text{when } (a, b) \text{ is in Q I} \\ \pi + \operatorname{Tan}^{-1}\dfrac{b}{a}, & \text{when } (a, b) \text{ is in Q II or Q III} \\ 2\pi + \operatorname{Tan}^{-1}\dfrac{b}{a}, & \text{when } (a, b) \text{ is in Q IV.} \end{cases}$$

In order to get the correct value of θ, we should plot the point (a, b), i.e., we should plot the point $a + bi$ in the complex plane. The amplitude θ of a real number or of a pure imaginary number can be found by inspection of its location. For instance as we see in Figure 12.2, the amplitude of -3 is π and that of $-4i$ is $3\pi/2$.

Example Express each of the following in trigonometric form:

$$\text{(a) } 3 - 3i, \qquad \text{(b) } -4.$$

Solution (a) Plot the number in the complex plane. Equations (1) give us $r = \sqrt{18} = 3\sqrt{2}$, $\tan\theta = -\frac{3}{3} = -1$. From the last equation, θ could be $3\pi/4$ or $5\pi/4$. From Figure 12.5 we see that θ must be $5\pi/4$. Hence

$$3 - 3i = 3\sqrt{2}\left(\cos\frac{5\pi}{4} + i\sin\frac{5\pi}{4}\right)$$

This result can be checked by replacing $\cos 5\pi/4$ and $\sin 5\pi/4$ with $\sqrt{2}/2$ and $-\sqrt{2}/2$, respectively, and then demonstrating that the right side is actually equal to the left side.

(b) After plotting the number, Figure 12.5, we find by inspection that $r = 4$ and $\theta = \pi$. Hence we can see immediately that

$$-4 = 4(\cos\pi + i\sin\pi)$$

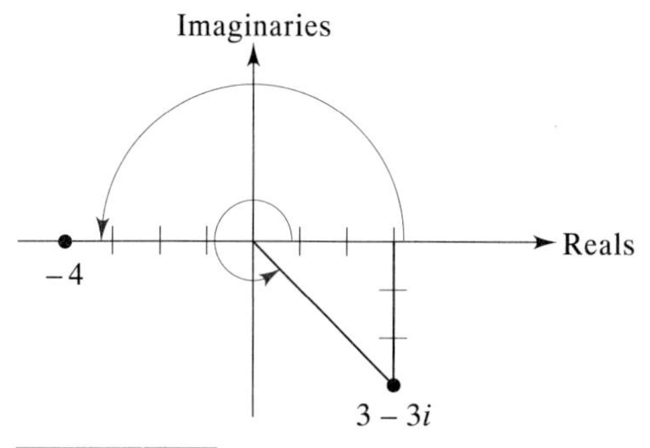

Figure 12.5

It is to be carefully noted that, regardless of the signs of a and b, r *is always positive,* and *the signs in front of* $\cos\theta$ *and* $i \sin\theta$ *are always positive.*

Exercise 12.2

Perform the indicated operations graphically and check the results algebraically.

1. $(5 + 2i) + (-3 + 4i)$
2. $(4 - 3i) + (-2 - i)$
3. $(-6 - 5i) - (1 - 3i)$
4. $(-7 + 2i) + (-1 + 2i)$
5. $(5 + 6i) - (-2 + 3i)$
6. $-7i - (4 - 2i)$
7. $(1 + i) + (3 - 2i) + (2 + 6i)$
8. $(1 - 3i) + (4 - 2i) - (3 - i)$

Plot each of the following complex numbers and then express it in trigonometric form.

9. $-\sqrt{2} + i\sqrt{2}$
10. $-3 - 3i$
11. $4 + 4i$
12. $5 - 5i$
13. 7
14. $-6i$
15. $9i$
16. -2
17. $-8 - 6i$
18. $15 + 8i$
19. $\sqrt{3} - i$
20. $-1 + i\sqrt{3}$

Plot each of the following complex numbers and then express it in algebraic form. Write the result with three-decimal-place accuracy.

21. $6(\cos 5\pi/3 + i \sin 5\pi/3)$
22. $4(\cos 5\pi/6 + i \sin 5\pi/6)$
23. $10(\cos 10\pi/9 + i \sin 10\pi/9)$
24. $8(\cos \pi/18 + i \sin \pi/18)$

25. The complex number $2(\cos \pi/6 - i \sin \pi/6)$ is not in trigonometric form [as defined in Equation (3) of Section 12.3]. (a) Why isn't it? (b) Express the given number in (proper) trigonometric form.
26. On one system of coordinates, plot and label the number $2 + 3i$, its conjugate, and its negative.
27. What is the amplitude (a) of a positive real number? (b) of a negative real number? (c) of bi if $b > 0$? (d) of bi if $b < 0$?
28. Show that the negative of $r(\cos \theta + i \sin \theta)$ is

$$r[\cos(\theta + \pi) + i \sin(\theta + \pi)].$$

29. Show that the conjugate of $r(\cos \theta + i \sin \theta)$ is

$$r[\cos(-\theta) + i \sin(-\theta)].$$

12.4 MULTIPLICATION OF COMPLEX NUMBERS IN POLAR FORM

Finding products, powers, and roots of complex numbers in polar form is especially easy.

Theorem **The absolute value of the product of two complex numbers is the product of their absolute values; the amplitude of the product is the sum of their amplitudes:**

$$\mathbf{r_1(\cos \theta_1 + i \sin \theta_1) \cdot r_2(\cos \theta_2 + i \sin \theta_2) = r_1 r_2[\cos(\theta_1 + \theta_2) + i \sin(\theta_1 + \theta_2)].}$$

Proof Let $r_1(\cos \theta_1 + i \sin \theta_1)$ and $r_2(\cos \theta_2 + i \sin \theta_2)$ be any two complex numbers in polar form. Their product is

$$\begin{aligned} & r_1(\cos \theta_1 + i \sin \theta_1) \cdot r_2(\cos \theta_2 + i \sin \theta_2) \\ &= r_1 r_2(\cos \theta_1 \cos \theta_2 + i \sin \theta_1 \cos \theta_2 \\ &\quad + i \cos \theta_1 \sin \theta_2 + i^2 \sin \theta_1 \sin \theta_2) \\ &= r_1 r_2[(\cos \theta_1 \cos \theta_2 - \sin \theta_1 \sin \theta_2) \\ &\quad + i(\sin \theta_1 \cos \theta_2 + \cos \theta_1 \sin \theta_2)] \\ &= r_1 r_2[\cos(\theta_1 + \theta_2) + i \sin(\theta_1 + \theta_2)]. \end{aligned}$$

Illustration

$$2\left(\cos\frac{3\pi}{8} + i\sin\frac{3\pi}{8}\right)\cdot 3\left(\cos\frac{5\pi}{8} + i\sin\frac{5\pi}{8}\right)$$
$$= 2\cdot 3\left[\cos\left(\frac{3\pi}{8}+\frac{5\pi}{8}\right) + i\sin\left(\frac{3\pi}{8}+\frac{5\pi}{8}\right)\right]$$
$$= 6(\cos\pi + i\sin\pi) = 6(-1 + i\cdot 0) = -6.$$

This theorem can be extended to include the product of any number of complex numbers:

$$r_1(\cos\theta_1 + i\sin\theta_1)\cdot r_2(\cos\theta_2 + i\sin\theta_2)\cdots r_n(\cos\theta_n + i\sin\theta_n)$$
$$= r_1r_2\cdots r_n[\cos(\theta_1+\theta_2+\cdots+\theta_n) + i\sin(\theta_1+\theta_2+\cdots+\theta_n)].$$

12.5 DE MOIVRE'S THEOREM

If n is any real number,

$$[r(\cos\theta + i\sin\theta)]^n = r^n(\cos n\theta + i\sin n\theta).$$

Proof For n a positive integer (by mathematical induction).

Part 1 *Verification*

For $n = 1$ the theorem is true, since

$$[r(\cos\theta + i\sin\theta)]^1 = r(\cos\theta + i\sin\theta).$$

For $n = 2$ the theorem is true, since

$$[r(\cos\theta + i\sin\theta)]^2 = r^2(\cos 2\theta + i\sin 2\theta).$$

Part 2 We will prove that the truth of the theorem for any particular integer k implies its truth for the next integer $k + 1$. Let k represent any particular value of n. Assuming that

$$[r(\cos\theta + i\sin\theta)]^k = r^k(\cos k\theta + i\sin k\theta), \quad (A)$$

we must prove that

$$[r(\cos\theta + i\sin\theta)]^{k+1} = r^{k+1}[\cos(k+1)\theta + i\sin(k+1)\theta]. \quad (B)$$

An examination of the left sides of the equations suggests that we multiply Equation (*A*) by $r(\cos\theta + i\sin\theta)$. Doing this, we obtain

$$[r(\cos\theta + i\sin\theta)]^{k+1} = r^k(\cos k\theta + i\sin k\theta)\cdot r(\cos\theta + i\sin\theta).$$

Applying the theorem above, we get

$$\begin{aligned}[r(\cos\theta + i\sin\theta)]^{k+1} &= r^{k+1}[\cos(k\theta+\theta) + i\sin(k\theta+\theta)]\\ &= r^{k+1}[\cos(k+1)\theta + i\sin(k+1)\theta].\end{aligned}$$

which is identical with Equation (*B*). This proves that if the theorem is true for $n = k$, then it must be true for $n = k + 1$.

Part 3 *Conclusion* The theorem is true for $n = 1$ and 2 (Part 1). Since it is true for $n = 2$, it is true for $n = 3$ (Part 2, where $k = 2$ and $k + 1 = 3$). Since it is true for $n = 3$, it is true for $n = 4$, and so on for all positive integers n.

It can be shown that De Moivre's theorem is true for all real values of n. We shall use it for only two cases: (1) when n is a positive integer, and (2) when n is the reciprocal of a positive integer. The proof of the latter case is omitted in this text.

Example 1 Use De Moivre's theorem to find the value of $(-1 + i)^{10}$.

Solution After plotting $(-1 + i)$ and putting it in trigonometric form, we have

$$-1 + i = \sqrt{2}\left(\cos\frac{3\pi}{4} + i\sin\frac{3\pi}{4}\right).$$

Applying De Moivre's theorem:

$$
\begin{aligned}
(-1 + i)^{10} &= \left[\sqrt{2}\left(\cos \frac{3\pi}{4} + i \sin \frac{3\pi}{4}\right)\right]^{10} \\
&= (\sqrt{2})^{10}\left(\cos 10 \cdot \frac{3\pi}{4} + i \sin 10 \cdot \frac{3\pi}{4}\right) \\
&= 32\left(\cos \frac{30\pi}{4} + i \sin \frac{30\pi}{4}\right) \\
&= 32\left(\cos \frac{3\pi}{2} + i \sin \frac{3\pi}{2}\right) \\
&= -32i.
\end{aligned}
$$

Theorem **The n nth roots of $r(\cos \theta + i \sin \theta)$ are given by the formula**

$$
\sqrt[n]{r}\left(\cos \frac{\theta + k \cdot 2\pi}{n} + i \sin \frac{\theta + k \cdot 2\pi}{n}\right),
$$

where $k = 0, 1, 2, \ldots, n - 1$.

Proof Assuming that De Moivre's theorem is true when n is the reciprocal of a positive integer, we have

$$
\begin{aligned}
\sqrt[n]{r(\cos \theta + i \sin \theta)} &= [r(\cos \theta + i \sin \theta)]^{1/n} \\
&= r^{1/n}\left(\cos \frac{\theta}{n} + i \sin \frac{\theta}{n}\right).
\end{aligned}
$$

Since $\cos \theta$ and $\sin \theta$ are periodic functions (Section 8.1) with a period of 2π, we can say that $\cos \theta = \cos(\theta + k \cdot 2\pi)$ and $\sin \theta = \sin(\theta + k \cdot 2\pi)$, where k is an integer. Hence

$$
\begin{aligned}
&\sqrt[n]{r(\cos \theta + i \sin \theta)} \\
&\qquad = \sqrt[n]{r}\left(\cos \frac{\theta + k \cdot 2\pi}{n} + i \sin \frac{\theta + k \cdot 2\pi}{n}\right).
\end{aligned}
$$

It is easy to show that the right side of this equation takes on n distinct values when k takes on the values $0, 1, 2, \ldots,$ $n - 1$. But if k takes on a value larger than $(n - 1)$, the result is merely a duplication of one of the n roots already found.

Example 2 Find the three cube roots of $-8i$.

Solution After plotting the number and putting it in trigonometric form, we have

$$-8i = 8\left(\cos\frac{3\pi}{2} + i\sin\frac{3\pi}{2}\right).$$

Apply the theorem on roots. The three cube roots of $-8i$ are

$$\sqrt[3]{8}\left(\cos\frac{\frac{3\pi}{2} + k\cdot 2\pi}{3} + i\sin\frac{\frac{3\pi}{2} + k\cdot 2\pi}{3}\right)$$
$$= 2\left[\cos\left(\frac{\pi}{2} + k\cdot\frac{2\pi}{3}\right) + i\sin\left(\frac{\pi}{2} + k\cdot\frac{2\pi}{3}\right)\right].$$

Let the three roots be r_1, r_2, r_3. Then

$$r_1 = 2\left(\cos\frac{\pi}{2} + i\sin\frac{\pi}{2}\right) = 2i, \qquad (k = 0)$$

$$r_2 = 2\left(\cos\frac{7\pi}{6} + i\sin\frac{7\pi}{6}\right) = -\sqrt{3} - i, \qquad (k = 1)$$

$$r_3 = 2\left(\cos\frac{11\pi}{6} + i\sin\frac{11\pi}{6}\right) = \sqrt{3} - i. \qquad (k = 2)$$

The three roots are equally spaced on a circle with radius 2 and center at the origin (Figure 12.6). Notice that for $k = 3$, we obtain r_1 again.

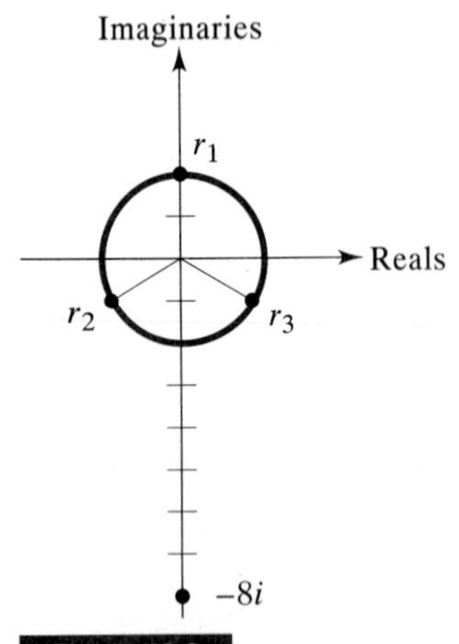

Figure 12.6

Exercise 12.3

Perform the indicated multiplications and then express the results in algebraic form.

1. $5\left(\cos\frac{4\pi}{9} + i\sin\frac{4\pi}{9}\right) \cdot 6\left(\cos\frac{2\pi}{9} + i\sin\frac{2\pi}{9}\right)$

2. $8\left(\cos\frac{7\pi}{8} + i\sin\frac{7\pi}{8}\right) \cdot \left(\cos\frac{3\pi}{8} + i\sin\frac{3\pi}{8}\right)$

3. $3\left(\cos\frac{8\pi}{9} + i\sin\frac{8\pi}{9}\right) \cdot 2\left(\cos\frac{23\pi}{18} + i\sin\frac{23\pi}{18}\right)$

4. $7\left(\cos\frac{5\pi}{9} + i\sin\frac{5\pi}{9}\right) \cdot 4\left(\cos\frac{10\pi}{9} + i\sin\frac{10\pi}{9}\right)$

For each of the following products, (a) express the factors in trigonometric form, (b) find their product trigonometrically, and (c) check your result by finding the product algebraically.

5. $-5i(-\sqrt{3} + i)$
6. $(-4\sqrt{3} - 4i)(-\sqrt{3} + 3i)$
7. $(3\sqrt{2} + 3i\sqrt{2})(6 - 6i)$
8. $(-2 - 2i\sqrt{3})6i$

For each of the following products, (a) express the factors in algebraic form, (b) find their product algebraically, and (c) check your result by finding the product trigonometrically.

9. $8\left(\cos\frac{3\pi}{4} + i\sin\frac{3\pi}{4}\right) \cdot 6\left(\cos\frac{\pi}{4} + i\sin\frac{\pi}{4}\right)$

10. $10\left(\cos\frac{5\pi}{6} + i\sin\frac{5\pi}{6}\right) \cdot 4\left(\cos\frac{5\pi}{6} + i\sin\frac{5\pi}{6}\right)$

11. $2\left(\cos\frac{2\pi}{3} + i\sin\frac{2\pi}{3}\right) \cdot \left(\cos\frac{5\pi}{6} + i\sin\frac{5\pi}{6}\right)$

12. $4\left(\cos\frac{4\pi}{3} + i\sin\frac{4\pi}{3}\right) \cdot 7\left(\cos\frac{3\pi}{2} + i\sin\frac{3\pi}{2}\right)$

Use De Moivre's theorem to find the value of each of the following. Express results in algebraic form.

13. $[2(\cos 108° + i\sin 108°)]^5$
14. $[\sqrt{2}(\cos 85° + i\sin 85°)]^6$
15. $(3 + 3i)^4$
16. $(1 - i)^8$

17. $(-2 + 2i)^5$

18. $(-\sqrt{2} - i\sqrt{2})^{10}$

19. $\left(\frac{\sqrt{3}}{2} + \frac{1}{2}i\right)^{100}$

20. $(\sqrt{3} - 3i)^4$

21. $(\sqrt{3} + i)^7$

22. $(8 + 6i)^4$

Find all the indicated roots of the following complex numbers. Express results in algebraic form. Round off approximate results to three-decimal-place accuracy.

23. The cube roots of $1000\left(\cos \frac{9\pi}{11} + i \sin \frac{9\pi}{11}\right)$
24. The fourth roots of $256\left(\cos \frac{4\pi}{3} + i \sin \frac{4\pi}{3}\right)$
25. The square roots of $8i$
26. The square roots of $-2i$
27. The square roots of $2 - 2i\sqrt{3}$
28. The cube roots of -8
29. The cube roots of $64i$
30. The cube roots of $-4\sqrt{2} + 4i\sqrt{2}$
31. The fourth roots of -256
32. The fourth roots of i
33. The fifth roots of 1
34. The fourth roots of $-8 - 8i\sqrt{3}$

Find all the roots of the following equations.

35. $x^6 - 1{,}000{,}000 = 0$ (*Hint:* The roots of the equation are the six sixth roots of 1,000,000.)
36. $x^{12} - 4096 = 0$
37. $x^{10} + 1 = 0$
38. $x^5 = -32$
39. Prove that

$$\frac{r_1(\cos \theta_1 + i \sin \theta_1)}{r_2(\cos \theta_2 + i \sin \theta_2)} = \frac{r_1}{r_2}[\cos(\theta_1 - \theta_2) + i \sin(\theta_1 - \theta_2)].$$

KEY TERMS

Complex number, real number, pure imaginary number, complex plane, polar form of a complex number, absolute value, amplitude, De Moivre's theorem, roots of a complex number

REVIEW EXERCISES

Perform the indicated operations:

1. $(5 - 2\sqrt{3}i) + (-\pi + i)$
2. $(\sqrt{3} + \sqrt{2}i) + (\sqrt{3} - \sqrt{2}i)$
3. $(\sqrt{3} + \sqrt{2}i)(\sqrt{3} - \sqrt{2}i)$
4. $i^{1994} - i^{1865}$
5. $i^{1993} + i^{1994} + i^{1995} + i^{1996}$
6. $\dfrac{1 + i}{1 - i}$
7. $3\left(\cos \dfrac{2\pi}{3} + i \sin \dfrac{2\pi}{3}\right) \cdot 5\left(\cos \dfrac{\pi}{3} + i \sin \dfrac{\pi}{3}\right)$
8. $\left[\sqrt{10}\left(\cos \dfrac{5\pi}{6} + i \sin \dfrac{5\pi}{6}\right)\right]^6$
9. $\left[2\left(\cos \dfrac{\pi}{15} + i \sin \dfrac{\pi}{15}\right)\right]^5$

Express in polar form:

10. $2 - 2i$
11. $-6i$
12. $\dfrac{1}{2} - \dfrac{\sqrt{3}}{2}i$
13. $1 + \text{i}$

Find the indicated roots:

14. The fourth roots of $1 + i$.
15. The cube roots of $\dfrac{1}{2} - \dfrac{\sqrt{3}}{2}i$.
16. The fifth roots of $32i$.

13

Topics from Analytic Geometry

13.1 STRAIGHT LINES

Suppose we have the equation of a curve in the coordinate plane and wish to find its equation with respect to another pair of axes which are parallel respectively to the original axes. To see how the coordinates are changed we examine Figure 13.1 showing the new origin at (h, k). Let (x, y) be the coordinates of P with respect to the old axes and (x', y') the coordinates of P with respect to the new axes. Then from Figure 13.1 we have

$$x = h + x',$$
$$y = k + y'.$$

These translation equations express the relationship between two sets of coordinates for the plane with the property that the corresponding axes are parallel.

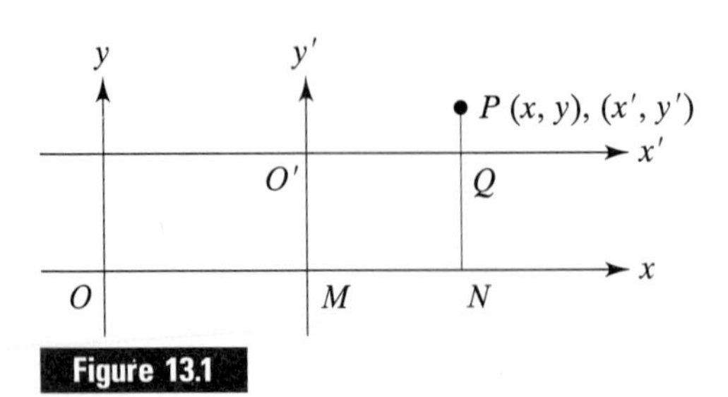

Figure 13.1

Example 1 Find the new equation for $x^2 + y^2 - 6x + 4y - 3 = 0$ if the origin is translated to $(3, -2)$.

Solution The translation equations become $x = 3 + x'$, $y = -2 + y'$. Substitution in the equation yields $(x' + 3)^2 + (y' - 2)^2 - 6(x' + 3) + 4(y' - 2) - 3 = 0$ or, after simplification,

$$x'^2 + y'^2 = 16.$$

This is a circle centered at the point $(3, -2)$ in the original coordinates having a radius of 4. Or we may say it is a circle of radius 4 centered at the origin in the x', y'-coordinate system.

Example 1 illustrates that a change of coordinates to a new system may provide a simpler form for a particular equation.

Example 2 Where should one translate the origin to simplify the equation of the circle

$$x^2 + y^2 + 6x - 10y + 18 = 0?$$

Solution We complete the squares on the x terms and the y terms to get

$$(x + 3)^2 + (y - 5)^2 = 16.$$

So if $x' = x + 3$ and $y' = y - 5$, the equation becomes $x'^2 + y'^2 = 16$. Consequently, if we translate the origin to $(-3, 5)$, the equation of the circle has its simplest form.

The slope of a line is a very important concept in much of mathematics and we are ready to state the definition of this fundamental notion.

The *slope of a nonvertical line* in the plane is the tangent of the angle the line makes with the positive direction of the x axis.

Horizontal lines have slope 0, but vertical lines with equations of the form $x = a$ have *no slope,* i.e., the notion of slope is not applied to vertical lines.

The line $y = x$ has slope 1 since it bisects Q I and intersects the x axis at the origin in an angle of 45° having a tangent of 1.

Example 3 Find the equation of the line through the origin of slope m.

Solution From Figure 13.2 we see that $m = \tan\theta = y/x$ for any point (x, y) on the line. Thus the equation is $y = mx$.

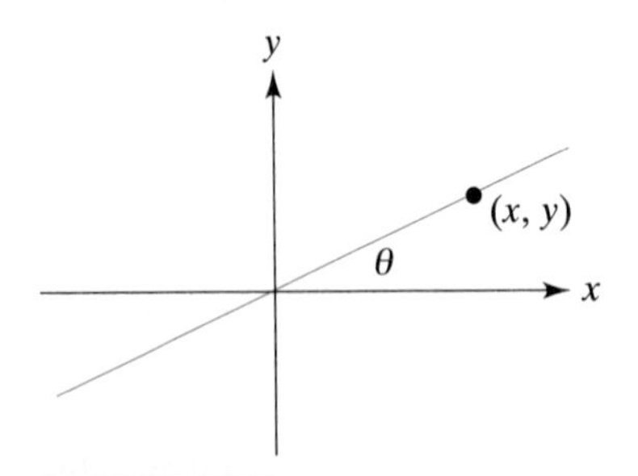

Figure 13.2

Example 4 Find the equation of the line of slope m which intersects the y axis at $(0, b)$.

Solution If we translate the origin to $(0, b)$ the equation becomes $y' = mx'$ where $y' = y - b$ and $x' = x - 0$. Thus the equation is

$$y - b = m(x - 0) \qquad \text{or} \qquad y = mx + b.$$

This is the **slope-intercept form** of the line since it utilizes the slope m and the y intercept b of the line.

If we apply the same argument as in Example 4, we see that the equation of a line through (a, b) of slope m is

$$y - b = m(x - a),$$

which is known as the **point-slope form** of a line.

Finally, we turn our attention to finding the angle between two lines. Toward this end consider lines L_1 of slope m_1 and L_2 of slope m_2 making angles of θ_1 and θ_2 with the positive direction of the x axis, respectively, as seen in Figure 13.3. Since the exterior angle θ_2 of a triangle is equal to the sum of the remote interior angles θ_1 and φ, we have $\varphi = \theta_2 - \theta_1$. Thus

$$\tan \varphi = \tan(\theta_2 - \theta_1) = \frac{\tan \theta_2 - \tan \theta_1}{1 + \tan \theta_2 \tan \theta_1}$$

$$= \frac{m_2 - m_1}{1 + m_1 m_2}.$$

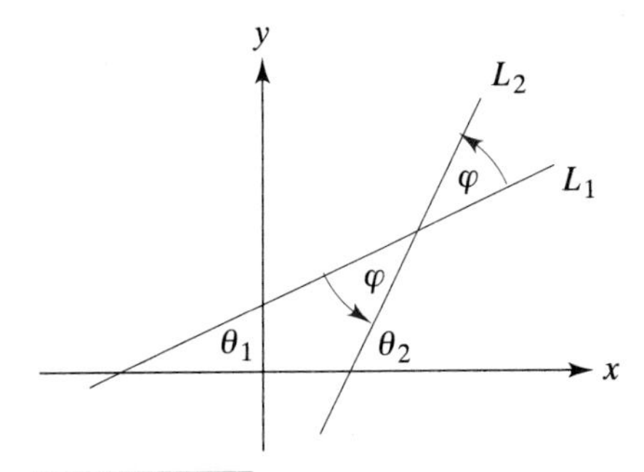

Figure 13.3

If $m_1 = m_2$, $\tan \varphi = 0$ and so $\varphi = 0$ and the lines are parallel. Likewise if $1 + m_1 m_2 = 0$, then $\tan \varphi$ is undefined and so $\varphi = 90°$ making the lines perpendicular. We summarize these last remarks.

Lines L_1 and L_2 of slope m_1 and m_2 are parallel if and only if $m_1 = m_2$, and are perpendicular if and only if $m_1 m_2 = -1$ or $m_1 = -1/m_2$.

Example 5 Find the equation of the line through $(-1, \pi)$ and perpendicular to the line

$$y = -5x + 17.$$

Solution The line $y = -5x + 17$ is in slope intercept form and has a slope of -5, so the line we seek has a slope of $\frac{1}{5}$. Using the point-slope form we have the desired equation

$$y - \pi = \tfrac{1}{5}(x + 1).$$

Exercise 13.1

1. Show $(x - a)^2 + (y - b)^2 = r^2$ is the equation of a circle of radius r with center at (a, b).
2. Find the **two-point form** of a line. That is, find the equation of a line through (a, b) and (c, d).
3. Two lines passing through (e, π) make an angle of 45°. If the slope of one line is 1, find the equation of the other line.
4. Is the triangle with vertices $A(2, 5)$, $B(-5, 7)$ and $C(-2, -9)$ a right triangle? Solve the triangle.

In Exercises 5 through 11, find the equation of the line with the given properties.

5. Through $(1 - e, \pi - 6)$ with slope -2.
6. With slope $-\pi$ and intercept -1.
7. Through $(5, 6)$ and $(-2, 3)$.
8. Horizontal and tangent to

$$(x - 1)^2 + (y + 4)^2 = 9.$$

9. Parallel to $x = -4$ and 8 units away from it.
10. Perpendicular to $y = \frac{1}{7}(x + 4)$ and passing through $(5, -6)$.
11. Parallel to $y = \frac{1}{7}(x + 4)$ and passing through $(5, -6)$.
12. Find the angle between $y = \frac{1}{7}(x + 4)$ and $y + \sqrt{2} = 3(x - 7)$.
13. Find the angle between $2x + 3y = 5$ and $y = \sqrt{2}\,x$.

13.2 THE PARABOLA

A **conic,** or **conic selection,** is any curve in the plane having an equation of the form

$$Ax^2 + Bxy + Cy^2 + Dx + Ey + F = 0,$$

where $A, B, \ldots, F$ are constants. We continue our study of the conics—the circle is a conic as is a straight line—by introducing the **parabola,** a curve with many applications in physics, astronomy, economics, solar energy, and satellite communication.

A *parabola* is the set of points in the plane that are equidistant from a fixed point, called the *focus* of the parabola, and a fixed line, called the *directrix* of the parabola.

Suppose the focus is at the point $(0, a)$, $a > 0$, and the directrix is the line $y = -a$. Then the **vertex,** the point on the parabola midway between the focus and directrix, is at the origin. The line through the vertex and focus and perpendicular to the directrix is the **axis** of the parabola.

As we see from Figure 13.4, if the point $P(x, y)$ is on the parabola, its distance to $(0, a)$ is $\sqrt{x^2 + (y - a)^2}$ and its distance to $y = -a$ is $y + a$.

Hence we have

$$\sqrt{x^2 + (y - a)^2} = y + a,$$
$$x^2 + (y - a)^2 = (y + a)^2,$$
$$x^2 = 4ay,$$

where a is the distance from the vertex to the focus.

We get exactly the same equation if $a < 0$, i.e., if the focus is below the x axis and the directrix is above it.

Similarly, if the focus is at $(a, 0)$, $a > 0$, on the x axis and the directrix is $x = -a$, we get $y^2 = 4ax$ for the equation of the parabola opening to the right.

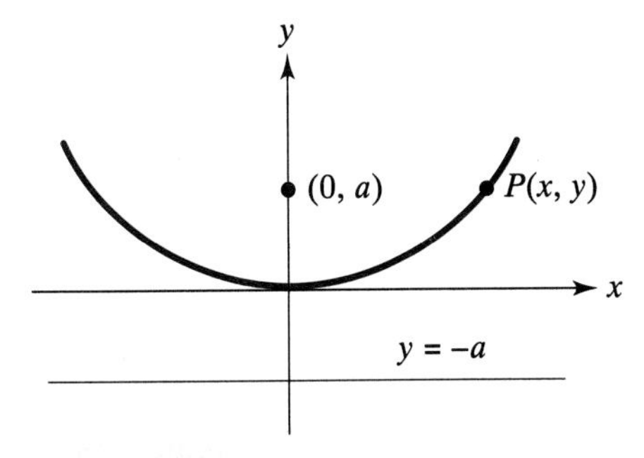

Figure 13.4

The equations for a parabola with vertex at the origin can yield, by using the translation equations of Section 13.1, **standard forms** for the equations of a parabola with vertex at (h, k):

$$(y - k)^2 = 4a(x - h)$$

is the equation of a parabola with vertex (h, k), focus $(h + a, k)$, and that opens to the right if $a > 0$ or to the left if $a < 0$.

$$(x - h)^2 = 4a(y - k)$$

is the equation of a parabola with vertex (h, k), focus $(h, k + a)$, and that opens upward if $a > 0$ or downwards if $a < 0$. Parabolas opening upward have a minimum value at their vertex, while those opening downward have a maximum value at their vertex.

Example 1 Find the equation of the parabola with focus at $(3, -2)$ and vertex at $(3, 1)$.

Solution Since the vertex is 3 units above the focus, the parabola opens downward and $a = -3$. We have

$$(x - 3)^2 = -12(y - 1)$$

or

$$(x - 3)^2 = 12 - 12y.$$

Example 2 Sketch a graph of the parabola with equation

$$y^2 - 6y - 8x = 31,$$

identifying the vertex and focus.

Solution Completing the square on the y terms we have

$$y^2 - 6y + 9 = 8x + 31 + 9$$
$$(y - 3)^2 = 8(x + 5),$$

so that the vertex is at $(-5, 3)$, $a = 2$, and the parabola opens to the right. The focus is at $(-3, 3)$ and $x = -7$ is the directrix, as we see in Figure 13.5.

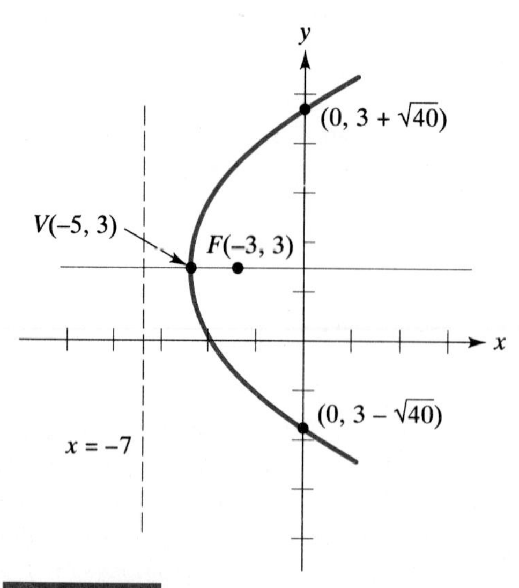

Figure 13.5

Exercise 13.2

Find the equation of the parabola that satisfies the conditions in Exercises 1 through 5.

1. Vertex at $(4, 1)$, $x = 2$ is the directrix.
2. Vertex at $(3, 2)$, focus at $(3, 4)$.
3. Focus at $(2, -3)$, $x = 6$ is directrix.
4. Focus at $(-2, 2)$, $y = 4$ is directrix.
5. Vertex at $(-1, -2)$, axis is vertical, passes through $(3, 6)$.

In Exercises 6 through 9, write the equation for the parabola.

6.

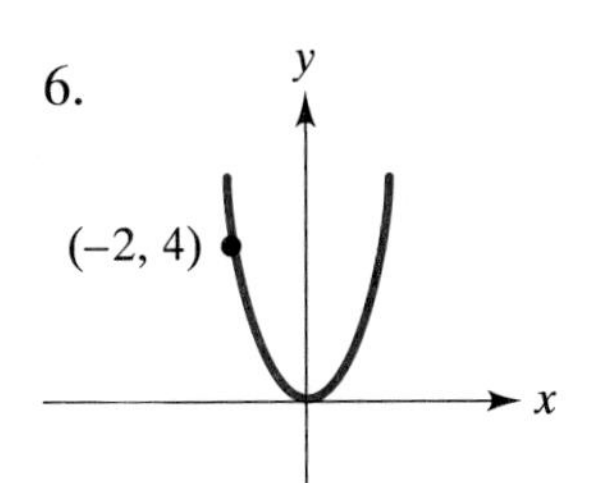

7.

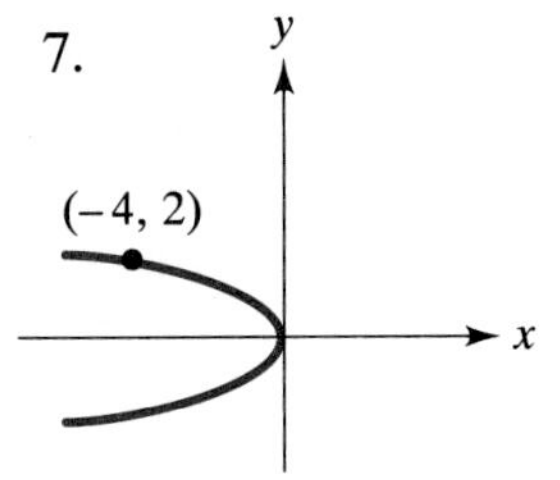

8.

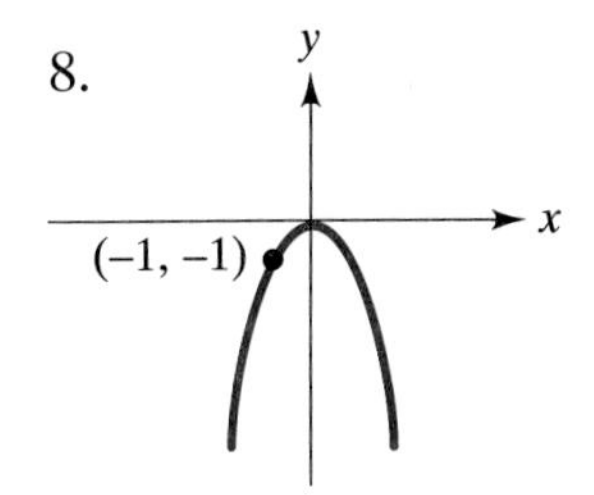

9.

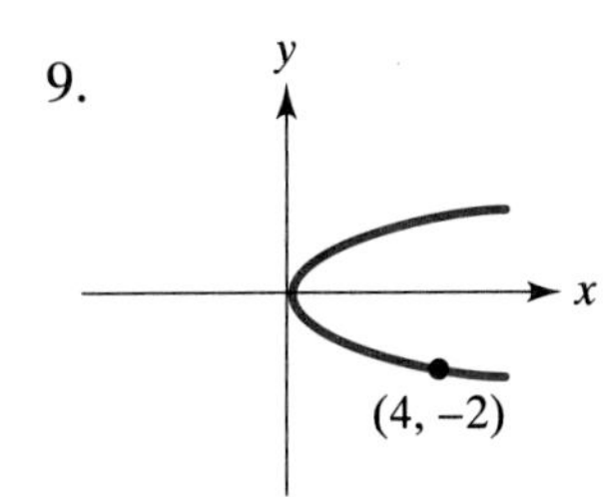

Express each equation in standard form, give the coordinates of the focus and vertex, and sketch a graph for Exercises 10 through 14.

10. $y^2 + 8x + 8 = 0$
11. $y^2 - 12x - 48 = 0$
12. $x^2 + 4x + 16y + 4 = 0$
13. $y^2 - 4y + 8x - 28 = 0$
14. $x^2 - 8x - 6y - 8 = 0$
15. Galveston is 140 mi east of and 115 mi north of Corpus Christi. A ship leaves each port at noon, one headed due east from Corpus Christi at 15 mi/h and one due south from Galveston at 10 mi/h. When is the distance between the ships a minimum (equivalently, when is the square of the distance a minimum)? How far apart are they then? What is the bearing of one ship from the other?

16. A TV-satellite dish has a parabolic cross section and its receiver, located at the focus, is 24 in from the vertex. If the origin is at the vertex, find the equation of the parabola.

13.3 THE ELLIPSE

The next conic to be examined is the ellipse, a closed oval figure that, among other things, describes the path of the planets around the sun.

An *ellipse* is the set of points P in the plane with the property that the sum of the distances from P to two fixed points is a constant. The two fixed points are the *foci*, plural of focus, of the ellipse.

We locate the origin midway between the foci so that one focus is at $F(c, 0)$ and the other at $F'(-c, 0)$. If $2a$ is the constant sum of the distances PF and PF' and if P has coordinates (x, y), we have

$$PF + PF' = 2a,$$

$$\sqrt{(x-c)^2 + y^2} + \sqrt{(x+c)^2 + y^2} = 2a.$$

Transposing a radical, squaring both sides, then squaring again, we get

$$(a^2 - c^2)x^2 + a^2y^2 = a^2(a^2 - c^2). \qquad (1)$$

In Figure 13.6 we see that $FF' = 2c$ and $PF + PF' = 2a > 2c$, so that $a^2 > c^2$. Letting $b^2 = a^2 - c^2$ and simplifying Equation 13.1 we get

$$\frac{x^2}{a^2} + \frac{y^2}{b^2} = 1.$$

An examination of Figure 13.6 reveals that the ellipse is symmetric with respect to its **major axis** from the vertex $V'(-a, 0)$ to the vertex $V(a, 0)$. It is also symmetric with respect to its **minor axis** BB' which intersects the major axis at the **center** of the ellipse midway between the foci.

Had we placed the foci at $(0, c)$ and $(0, -c)$ on the y axis, then an analysis similar to the one above would yield the equation

$$\frac{y^2}{a^2} + \frac{x^2}{b^2} = 1,$$

where $2a$ is the length of the major axis—along the y axis now—and $2b$ is the length of the minor axis.

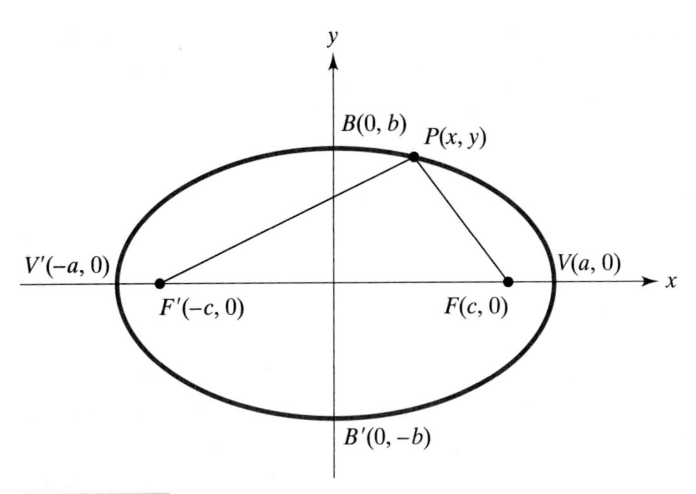

Figure 13.6

Example 1 Find the equation of the ellipse with foci at $(0, 3)$ and $(0, -3)$ and a vertex at $(0, 5)$. Sketch the ellipse.

Solution The center is at $(0, 0)$ and the major axis lies along the y axis. Note that $c = 3$ and $a = 5$, so that $b^2 = a^2 - c^2 = 16$. The equation is

$$\frac{y^2}{25} + \frac{x^2}{16} = 1$$

and the graph is in Figure 13.7.

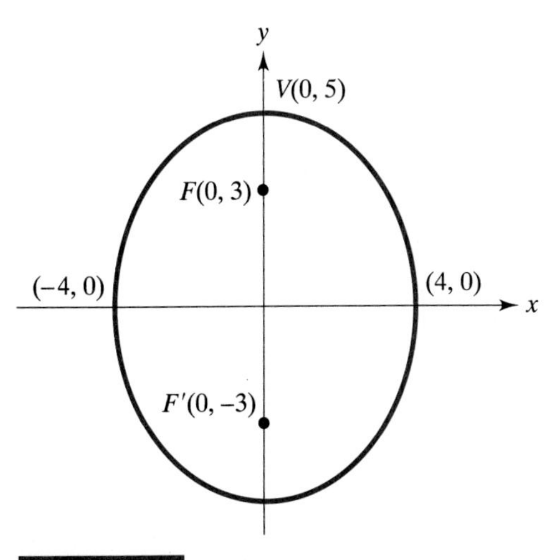

Figure 13.7

Just as we saw with the parabola, we may write the equation of an ellipse with its center at (h, k) instead of the origin. If a new coordinate system has its origin at (h, k), then in that system the equation of the ellipse is

$$\frac{x'^2}{a^2} + \frac{y'^2}{b^2} = 1$$

or

$$\frac{y'^2}{a^2} + \frac{x'^2}{b^2} = 1.$$

Since $x' = x - h$ and $y' = y - k$ we get the **standard form** of the equation of an ellipse with center at (h, k) to be

$$\frac{(x-h)^2}{a^2} + \frac{(y-k)^2}{b^2} = 1,$$

or, if the major axis is vertical,

$$\frac{(y-k)^2}{a^2} + \frac{(x-h)^2}{b^2} = 1,$$

where the quantities a, b and c have the same relationships to an ellipse as they did when the center was at $(0, 0)$.

Example 2 Put $x^2 + 4y^2 + 6x + 16y + 21 = 0$ into standard form. Identify the foci, center and ends of the axes. Sketch the graph.

Solution Completing the square on the x terms and y terms yields

$$(x^2 + 6x + 9) + 4(y^2 + 4y + 4) = -21 + 9 + 16$$

or

$$\frac{(x+3)^2}{4} + \frac{(y+2)^2}{1} = 1.$$

The center is $(-3, -2)$, $a = 2$, $b = 1$, and $c = \sqrt{3}$. The foci are $(-3 \pm \sqrt{3}, -2)$, the vertices are $(-3 \pm 2, -2)$ and the ends of the minor axis are $(-3, -2 \pm 1)$, as seen in Figure 13.8.

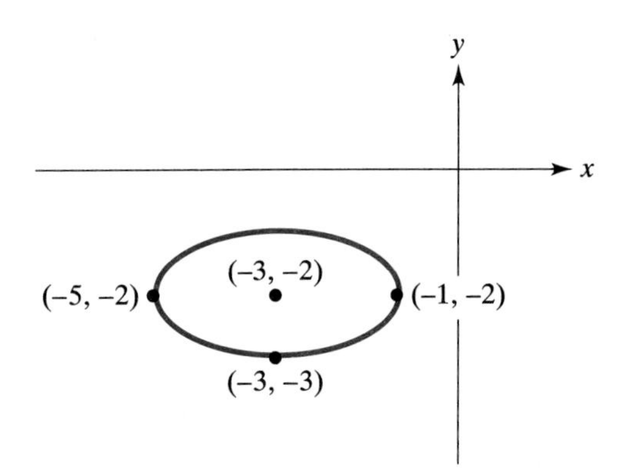

Figure 13.8

Exercise 13.3

In Exercises 1 through 8 find the coordinates of the foci, the center, and the vertices. Sketch a graph.

1. $\dfrac{x^2}{9} + \dfrac{y^2}{4} = 1$
2. $\dfrac{x^2}{4} + \dfrac{y^2}{9} = 1$
3. $\dfrac{(x-3)^2}{16} + \dfrac{(y-2)^2}{9} = 1$
4. $\dfrac{(x+3)^2}{25} + \dfrac{(y-1)^2}{9} = 1$
5. $\dfrac{(x-1)^2}{9} + \dfrac{(y+3)^2}{16} = 1$
6. $4x^2 + y^2 + 8x - 4y - 8 = 0$
7. $4y^2 + 9x^2 - 24y - 72x + 144 = 0$
8. $4x^2 + 8y^2 - 4x - 24y - 13 = 0$

Write the equation of the ellipse satisfying the stated conditions and sketch a graph of the curve in Exercises 9 through 15.

9. Center at $(5, 1)$, vertex at $(5, 4)$, end of the minor axis at $(3, 1)$.
10. Vertex at $(6, 3)$, foci at $(\pm 4, 3)$.
11. Ends of minor axis at $(-1, 2)$ and $(-1, -4)$, focus at $(1, -1)$.
12. Center at the origin, vertex at $(0, -6)$, end of minor axis at $(4, 0)$.
13. Center at $(5, 4)$, length of major axis is 16, length of minor axis is 6, major axis is horizontal.

14. Foci at $(\pm 5, 0)$, length of major axis is 18.
15. Center at $(3, 2)$, focus at $(3, 7)$, vertex at $(3, -5)$.
16. The perimeter of a triangle is 40 while $(0, 5)$ and $(0, -5)$ are two of its vertices. Find the locus (graph) of the third vertex.
17. Carry out all the steps in deriving Equation (1).
18. Graph $r = 15/(3 - 2\cos\theta)$, $0 \le \theta \le 2\pi$, in *polar* coordinates, using a graphing calculator if one is available.

13.4 THE HYPERBOLA

The final conic to be studied here is the hyperbola, a curve whose definition is strikingly similar to that of the ellipse.

> **A *hyperbola* is the set of points *P* in the plane with the property that the *difference* of the distances from P to two fixed points, the foci, is a constant.**

As before, we locate the foci at $F(c, 0)$ and $F'(-c, 0)$ and let $P(x, y)$ be a point on the hyperbola. If $2a$ is the difference of the distances PF and PF', then as we see in Figure 13.9

$$PF' - PF = 2a,$$

or

$$PF' - PF = -2a,$$

depending on whether P is to the right or left of the y axis. From the distance formula we have

$$\sqrt{(x + c)^2 + y^2} - \sqrt{(x - c)^2 + y^2} = \pm 2a.$$

Transposing one radical, squaring, simplifying, and squaring again we get

$$(c^2 - a^2)x^2 - a^2y^2 = a^2(c^2 - a^2).$$

Set $b^2 = c^2 - a^2 > 0$ and divide by a^2b^2 to get

$$\frac{x^2}{a^2} - \frac{y^2}{b^2} = 1. \tag{2}$$

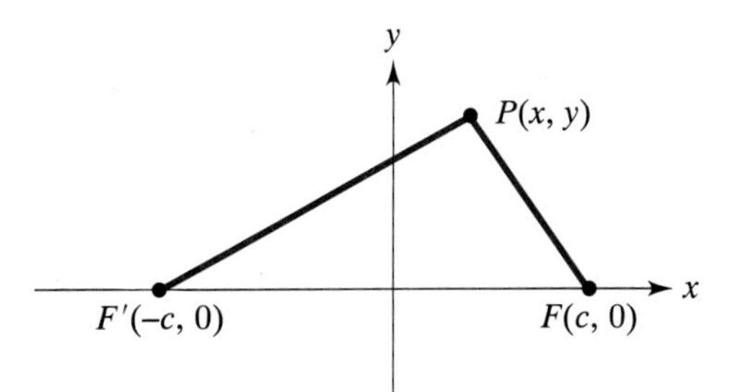

Figure 13.9

The points $(a, 0)$ and $(-a, 0)$ are the vertices, the center is midway along the axis between the foci, while the segment $(0, b)$ to $(0, -b)$ is the **conjugate axis.** The hyperbola is seen in Figure 13.10, opening right and left.

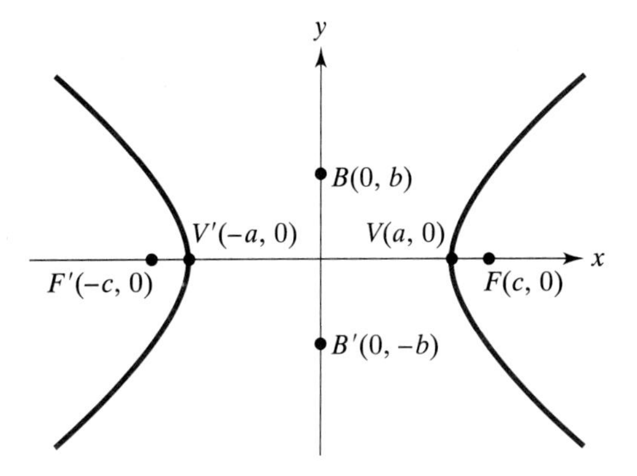

Figure 13.10

Had the roles of x and y been reversed so that the foci were $F(0, c)$ and $F'(0, -c)$ we would have obtained the equation for a hyperbola opening up and down

$$\frac{y^2}{a^2} - \frac{x^2}{b^2} = 1 .$$

Of course, if the center is (h, k) rather than the origin, the **standard form** of the equations becomes

$$\frac{(x-h)^2}{a^2} - \frac{(y-k)^2}{b^2} = 1, \tag{3}$$

and

$$\frac{(y-k)^2}{a^2} - \frac{(x-h)^2}{b^2} = 1. \tag{4}$$

The hyperbola has two **asymptotes** passing through its center (h, k). The hyperbola represented by Equation (3), opening to the right and left, has asymptotes

$$y - k = \frac{b}{a}(x - h)$$

and
$$y - k = -\frac{b}{a}(x - h),$$

while the asymptotes of the hyperbola opening up and down represented by Equation (4) are

$$y - k = \frac{a}{b}(x - h)$$

and
$$y - k = -\frac{a}{b}(x - h).$$

Example Sketch the curve, showing its center, foci, vertices and asymptotes:

$$x^2 - 2y^2 + 6x + 4y + 5 = 0.$$

Solution The standard form of the equation is

$$\frac{(x + 3)^2}{2} - \frac{(y - 1)^2}{1} = 1,$$

so that the center is $(-3, 1)$, $a = \sqrt{2}$ and $b = 1$. $c^2 = a^2 + b^2 = 3$, so $c = \sqrt{3}$. The foci are $(-3 \pm \sqrt{3}, 1)$, the vertices are $(-3 \pm \sqrt{2}, 1)$ and the asymptotes are

$$y - 1 = \frac{1}{\sqrt{2}}(x + 3)$$

and
$$y - 1 = -\frac{1}{\sqrt{2}}(x + 3).$$

The curve is graphed in Figure 13.11.

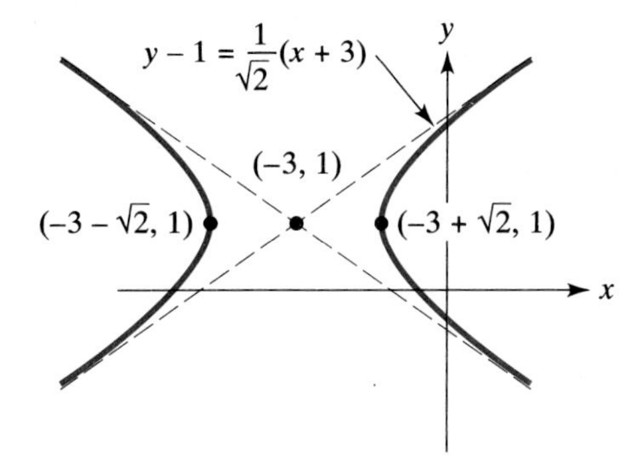

Figure 13.11

Exercise 13.4

In Exercises 1 through 7 find the coordinates of the foci, the vertices, the center, and the equations of the asymptotes of the hyperbola. Sketch the curve showing all of the points and asymptotes.

1. $\frac{x^2}{4} - \frac{y^2}{16} = 1$
2. $\frac{y^2}{16} - \frac{x^2}{4} = 1$
3. $\frac{(y - 2)^2}{5} - \frac{(x + 1)^2}{9} = 1$
4. $\frac{(x - \pi)^2}{e^2} - \frac{(y - e)^2}{\pi^2} = 1$
5. $9x^2 - 4y^2 + 90x + 189 = 0$
6. $36x^2 - 64y^2 = 2304$
7. $4y^2 - 9x^2 + 8y - 54x - 81 = 0$
8. Prove the details in the derivation of Equation (2).

In Exercises 9 through 11 write the equation of the hyperbola satisfying the given conditions.

9. Center at $(0, 0)$, a vertex at $(1, 0)$, focus at $(\sqrt{2}, 0)$.
10. Center at $(-2, 2)$, a focus at $(6, 2)$, a vertex at $(4, 2)$.
11. Center at $(3, -1)$, a focus at $(3, 3)$, a vertex at $(3, 2)$.
12. Graph $r = 4/(2 + 3 \cos \theta)$, $0 \le \theta \le 2\pi$, in *polar* coordinates.

13.5 ROTATION OF AXES

In Section 13.1 we examined the effect of translating axes to a new system with the new axes parallel to the old ones. We now will examine a transformation of axes in which the origin remains fixed but the new axes are rotated through an acute angle about the origin as seen in Figure 13.12.

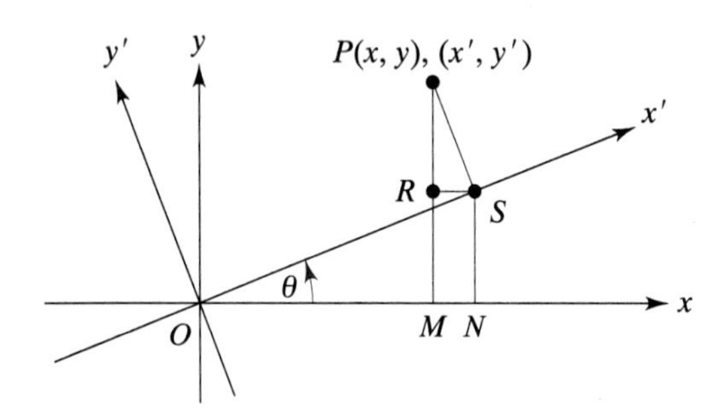

Figure 13.12

If P has coordinates (x, y) in the original system and (x', y') in the rotated system then from Figure 13.12 we see that

$$x = 0M, \qquad y = MP, \qquad x' = 0S, \qquad \text{and} \qquad y' = SP,$$

where RS and NS are parallel to the x and the y-axis respectively. Now

$$x = 0M = 0N - MN = 0N - RS = x' \cos\theta - y' \sin\theta,$$
$$y = MP = MR + RP = NS + RP = x' \sin\theta + y' \cos\theta.$$

We have derived the transformation formulas for a **rotation of axes** through the angle θ:

$$x = x' \cos\theta - y' \sin\theta,$$
$$y = x' \sin\theta + y' \cos\theta.$$

The main use of these rotation formulas is to remove the xy term from a general conic section

$$Ax^2 + Bxy + Cy^2 + Dx + Ey + F = 0.$$

Indeed, if we substitute the rotation equations in this equation and collect terms to get

$$A'x'^2 + B'x'y' + C'y'^2 + D'x' + E'y' + F' = 0,$$

we will find that

$$\begin{aligned}
A' &= A\cos^2\theta + B\sin\theta\cos\theta + C\sin^2\theta,\\
B' &= B\cos 2\theta - (A - C)\sin 2\theta,\\
C' &= A\sin^2\theta - B\sin\theta\cos\theta + C\cos^2\theta,\\
D' &= D\cos\theta + E\sin\theta,\\
E' &= E\cos\theta - D\sin\theta,\\
F' &= F.
\end{aligned}$$

We seek a value for θ so that $B' = 0$, hence we find

$$\tan 2\theta = \frac{B}{A - C}.$$

if $A \neq C$. If $A = C$, $B' = B\cos 2\theta = 0$ holds provided $\theta = 45°$.

We have discovered that if we rotate axes through the angle θ for which

$$\tan 2\theta = \frac{B}{A - C}, \qquad \text{if } A \neq C,$$

or if $A = C$, if we set $\theta = 45°$, then the conic section $Ax^2 + Bxy + Cy^2 + Dx + Ey + F = 0$ will have no $x'y'$ term.

Example 1 Rotate axes through an angle θ to remove the xy term from

$$x^2 - 2xy + y^2 - 8\sqrt{2}x - 8 = 0.$$

Solution We have $A = C = 1$, so that $\theta = 45°$ and the rotation formulas are

$$x = \frac{1}{\sqrt{2}}x' - \frac{1}{\sqrt{2}}y', \quad y = \frac{1}{\sqrt{2}}x' + \frac{1}{\sqrt{2}}y'.$$

Substitution in the equation yields

$$\tfrac{1}{2}(x'^2 - 2x'y' + y'^2) - (x'^2 - y'^2) + \tfrac{1}{2}(x'^2 + 2x'y' + y'^2) - 8(x' - y') - 8 = 0,$$

or

$$y'^2 + 4y' = 4x' + 4,$$

which in standard form is

$$(y' + 2)^2 = 4(x' + 2).$$

This is a parabola with vertex at $(-2, -2)$ opening to the right in the $x'y'$ coordinate system, as we see in Figure 13.13.

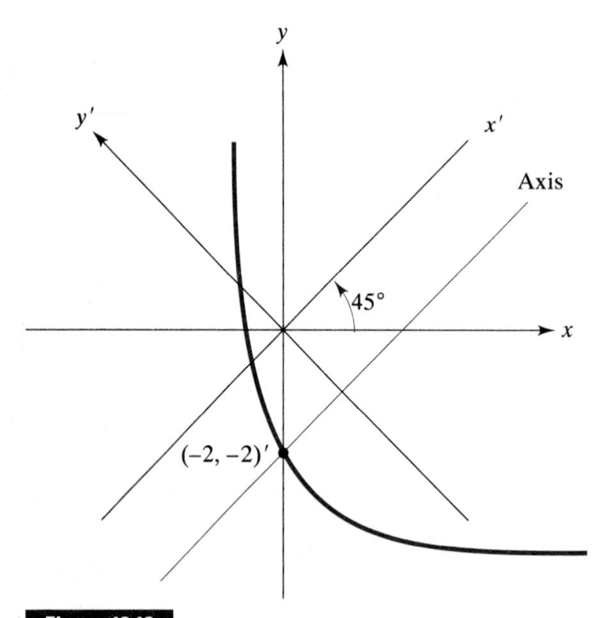

Figure 13.13

Example 2 Rotate axes through an angle θ to remove the xy term from

$$x^2 + 4xy + 4y^2 + 2\sqrt{5} - \sqrt{5}y = 0.$$

Solution We have $\tan 2\theta = -\frac{4}{3}$, and we need $\sin \theta$ and $\cos \theta$. From $\tan 2\theta = -\frac{4}{3}$, we find $\cos 2\theta = -\frac{3}{5}$. Using the half-angle formulas of Section 6.5 we have

$$\sin\theta = \sqrt{\frac{1-\cos 2\theta}{2}} = \frac{2}{\sqrt{5}},$$

$$\cos\theta = \sqrt{\frac{1+\cos 2\theta}{2}} = \frac{1}{\sqrt{5}}.$$

The rotation formulas are

$$x = \frac{1}{\sqrt{5}}(x' - 2y'),$$

$$y = \frac{1}{\sqrt{5}}(2x' + y'),$$

and when substituted in the equation we get

$$\tfrac{1}{5}(x'^2 - 4x'y' + 4y'^2) + \tfrac{4}{5}(2x'^2 - 3x'y' - 2y'^2)$$
$$+ \tfrac{4}{5}(4x'^2 + 4x'y' + y'^2) + 2(x' - 2y') - (2x' + y') = 0.$$

This simplifies to $x'^2 = y'$, the equation of a parabola with vertex at the origin opening up in the rotated coordinates, as seen in Figure 13.14.

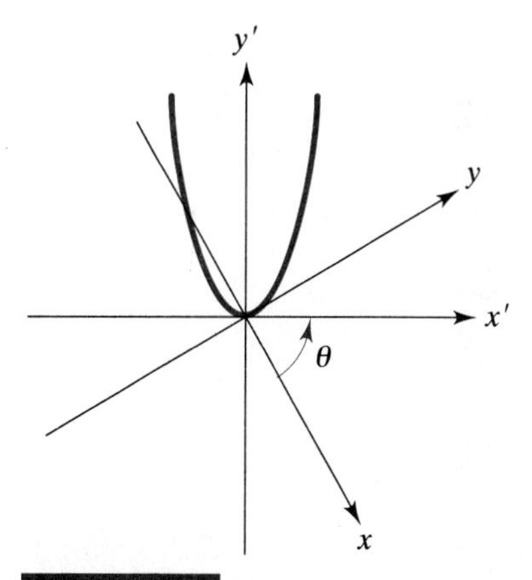

Figure 13.14

Exercise 13.5

Rotate coordinate axes through the appropriate angle θ to remove the xy term. Find the transformed equation and sketch a graph, showing both coordinate systems.

1. $xy = 7$
2. $3x^2 + 2\sqrt{3}xy + y^2 - 2x - 2\sqrt{3}y - 2 = 0$
3. $xy - x + y = 0$
4. $9x^2 - 6xy + y^2 + x + 1 = 0$
5. $x^2 - \sqrt{3}xy + 2y^2 - 2 = 0$
6. $x^2 + xy + y^2 - 1 = 0$
7. $73x^2 - 72xy + 52y^2 + 100x - 200y + 100 = 0$
8. $3x^2 - 10xy + 3y^2 + 22x - 26y + 43 = 0$.
9. Carefully calculate the new coefficients $A', B', \ldots, F'$ as given above to verify their accuracy.

KEY TERMS

Translation equations, slope of a line, intercept of a line, parabola, focus, vertex, directrix, axis, ellipse, foci, vertices, hyperbola, asymptotes

REVIEW EXERCISES

1. Find the line through (5, 6) and perpendicular to $3x - 2y = 19$.
2. Find the new equation for

$$x^2 + y^2 - 2x + 4y = 0,$$

   if the origin is translated to $(1, -2)$.
3. Write the equation of the parabola with focus at $(2, -3)$ and $x = 5$ as its directrix.
4. Sketch a graph of $(x - 2)^2/4 + (y + 1)^2/9 = 1$.
5. Put the equation in standard form and sketch its graph:

$$12y^2 - 4x^2 + 72y + 16x + 44 = 0$$

6. Rotate axes to remove the xy term and sketch a graph:

$$13x^2 - 10xy + 13y^2 - 72 = 0.$$

APPENDIX

A.1 CIRCULAR AND EXPONENTIAL FUNCTIONS

In calculus it is shown that the exponential function e^θ and the circular functions $\cos\theta$ and $\sin\theta$ can be represented by *convergent infinite series.* These are

$$e^\theta = 1 + \theta + \frac{\theta^2}{2!} + \frac{\theta^3}{3!} + \frac{\theta^4}{4!} + \frac{\theta^5}{5!} + \cdots + \frac{\theta^n}{n!} + \cdots,$$

$$\cos\theta = 1 - \frac{\theta^2}{2!} + \frac{\theta^4}{4!} - \frac{\theta^6}{6!} + \cdots + (-1)^n\frac{\theta^{2n}}{(2n)!} + \cdots,$$

$$\sin\theta = \theta - \frac{\theta^3}{3!} + \frac{\theta^5}{5!} - \frac{\theta^7}{7!} + \cdots + (-1)^n\frac{\theta^{2n+1}}{(2n+1)!} + \cdots.$$

Speaking loosely, as we include more and more terms of a series, their sum more closely approaches the expression on the left side. In the case of these three series, this statement is true for all values of θ.

Notice that $\cos\theta$, an *even* function, is expressed as a series of even powers of θ. Also, the *odd* function $\sin\theta$ is expressed as a series of odd powers of θ.

It is series such as these that are used in constructing a table like that appearing at the back of this book. For example, with $\theta = 0.178$, and using only two terms of the cosine and sine series, we have

$$\cos 0.178 \doteq 1 - \frac{(0.178)^2}{2} \doteq 1 - 0.0158 = 0.9842,$$

$$\sin 0.178 \doteq 0.178 - \frac{(0.178)^3}{6} \doteq 0.178 - 0.0009 = 0.1771,$$

which agree, to four decimal places, with the corresponding entries in the table. These same series are also used by calculators in approximating trigonometric function values, but many more terms of the series are used.

Regardless of how these series were derived, we can take them as alternative definitions of functions that we shall designate by e^θ, $\cos\theta$, and $\sin\theta$, respectively—a far cry from triangles.

Except for the alternating signs in the series for $\sin\theta$ and $\cos\theta$, the sum of these series is the same as the series for e^θ. This prompts us to substitute $i\theta$ for θ in the series for e^θ, which yields

$$\begin{aligned}
e^{i\theta} &= 1 + (i\theta) + \frac{(i\theta)^2}{2!} + \frac{(i\theta)^3}{3!} + \frac{(i\theta)^4}{4!} + \frac{(i\theta)^5}{5!} + \cdots + \frac{(i\theta)^m}{m!} + \cdots \\
&= 1 + i\theta - \frac{\theta^2}{2!} - i\frac{\theta^3}{3!} + \frac{\theta^4}{4!} + i\frac{\theta^5}{5!} + \cdots \\
&\quad + (-1)^n\frac{\theta^{2n}}{(2n)!} + i(-1)^n\frac{\theta^{2n+1}}{(2n+1)!} + \cdots \\
&= \left[1 - \frac{\theta^2}{2!} + \frac{\theta^4}{4!} - \cdots + (-1)^n\frac{\theta^{2n}}{(2n)!} + \cdots\right] \\
&\quad + i\left[\theta - \frac{\theta^3}{3!} + \frac{\theta^5}{5!} - \cdots + (-1)^n\frac{\theta^{2n+1}}{(2n+1)!} + \cdots\right] \\
&= \cos\theta + i\sin\theta\,.
\end{aligned}$$

Thus in strictly nonrigorous fashion, but with heuristic intent, we have obtained *Euler's formula:*

$$\boldsymbol{e^{i\theta} = \cos\theta + i\sin\theta\,.}$$

An important special case, $\theta = \pi$, yields the unusual relation

$$e^{i\pi} + 1 = 0\,,$$

which involves 0, the real unit 1, the imaginary unit i, and two famous irrational numbers, e and π. Moreover, since $e^{\pi i} = -1$, it follows that $\ln(-1) = \pi i$.

Raising both sides of Euler's formula to the power n, we get $(e^{i\theta})^n = (\cos\,\theta + i\,\sin\,\theta)^n$. But $(e^{i\theta})^n = e^{in\theta} = \cos\,n\theta + i\,\sin\,n\theta$. Hence we have an alternative derivation of De Moivre's theorem:

$$(\cos\,\theta + i\,\sin\,\theta)^n = \cos\,n\theta + i\,\sin\,n\theta\,.$$

Exercise A.1

1. Use the series definition of $\cos\,\theta$ to prove that $\cos(-\theta) = \cos\,\theta$.
2. Use the series definition of $\sin\,\theta$ to prove that $\sin(-\theta) = -\sin\,\theta$.
3. Use the series for $\sin\,\theta$ to compute $\sin\,0.1$ correct to eight decimal places.
4. Use Euler's formula to derive the formulas for $\sin(A + B)$ and $\cos(A + B)$. *Hint:* Replace θ with $A + B$; notice that $e^{A+B} = e^A e^B$; apply the definition of equality of complex numbers.
5. Use Euler's formula to prove that

   (a) $\cos\,\theta = \dfrac{e^{i\theta} + e^{-i\theta}}{2}$,

   (b) $\sin\,\theta = \dfrac{e^{i\theta} - e^{-i\theta}}{2i}$.

6. Use formula (a) of Problem 5 to prove that

$$\cos\,5\theta = 16\,\cos^5\,\theta - 20\,\cos^3\,\theta + 5\,\cos\,\theta\,.$$

   [*Hint:* Express the right side in terms of exponential functions; use the binomial formula to expand; simplify to $\frac{1}{2}(e^{5i\theta} + e^{-5i\theta})$, which is $\cos\,5\theta$.]

A.2 HYPERBOLIC TRIGONOMETRIC FUNCTIONS

Recalling the definition of the circular trigonometric functions from Section 4.4, which employed the circle $x^2 + y^2 = 1$, we see in Figure A.1 that if θ is the length of the arc BP, then θ is the radian measure of the angle BOP and, from Figure A.1, θ is the measure of the area of the sector $AOPB$. Under these conditions for $P(x, y)$, $x = \cos\,\theta$ and $y = \sin\,\theta$. Borrowing the area aspect of this definition and using the hyperbola $x^2 - y^2 = 1$ in place

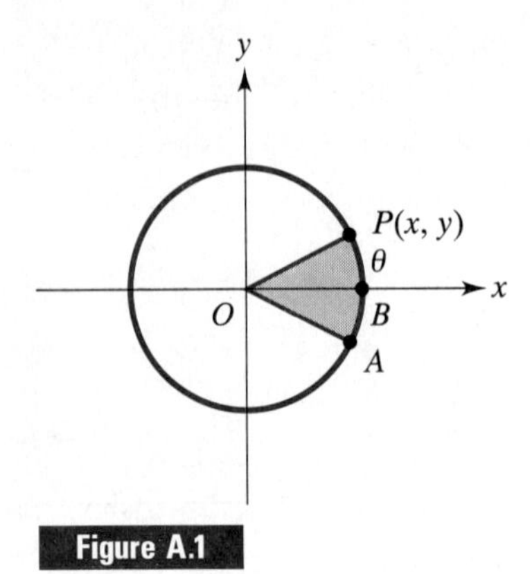

Figure A.1

of the circle, as we see in Figure A.2, if θ is the measure of the shaded area then $x = \cosh\theta$ and $y = \sinh\theta$ are defined to be the **hyperbolic cosine** and the **hyperbolic sine** of θ. It is possible to derive the following expressions for $\sinh\theta$ and $\cosh\theta$ in terms of e^{θ}:

$$\sinh\theta = \frac{e^{\theta} - e^{-\theta}}{2},$$

$$\cosh\theta = \frac{e^{\theta} + e^{-\theta}}{2}.$$

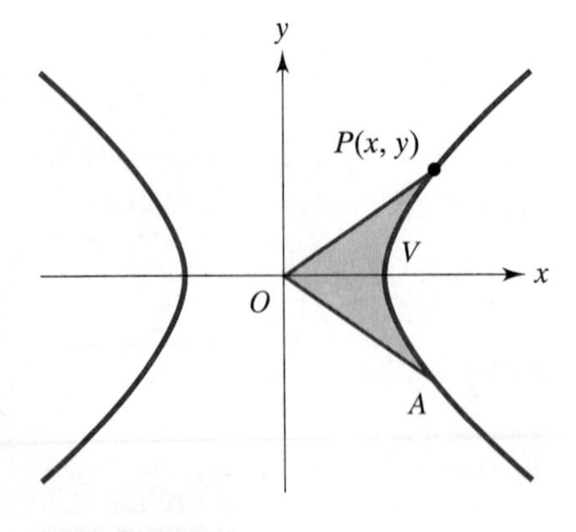

Figure A.2

Example Sketch a graph of $y = \sinh\theta$.

Solution For $\theta > 0$, e^{θ} grows rapidly, while $e^{-\theta}$ decreases, so $\sinh \theta$ grows as θ increases. Similarly if $\theta < 0$, $-e^{-\theta}$ dominates e^{θ} in size and so $\sinh \theta$ is negative as shown in Figure A.3.

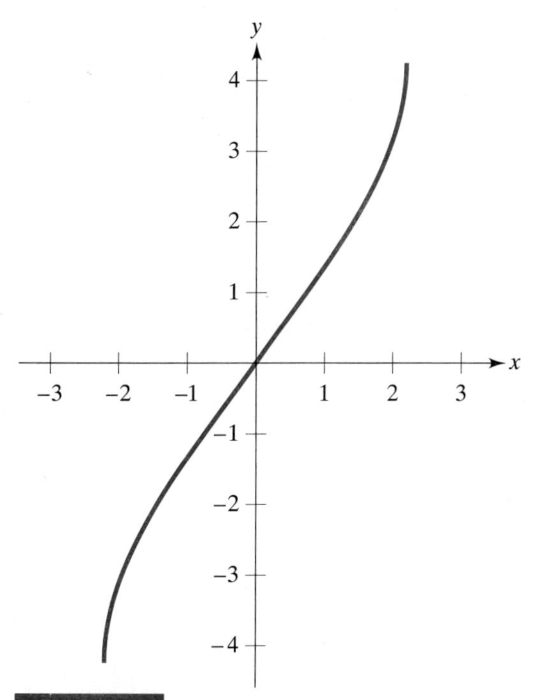

Figure A.3

Exercise A.2

1. Is $y = \sinh \theta$ an odd or an even function?
2. Is $y = \cosh \theta$ an odd or an even function?
3. Graph $y = \cosh \theta$.
4. Prove $\sinh \theta + \cosh \theta = e^{\theta}$, for any θ.
5. Prove $\cosh^2 \theta - \sinh^2 \theta = 1$, for any θ.
6. Use the series on page 237 to show

$$\cosh \theta = 1 + \frac{\theta^2}{2!} + \frac{\theta^4}{4!} + \cdots + \frac{\theta^{2n}}{(2n)!} + \cdots.$$

7. Derive a series expression for $\sinh \theta$.

A.3 CHANGE OF BASE OF LOGARITHMS

If we know the logarithm of a number to the base a, we can find the logarithm of that number to the base b by using

$$\log_b N = \frac{\log_a N}{\log_a b} = (\log_b a)(\log_a N). \tag{1}$$

To prove this, let

$$\log_b N = y.$$

Then

$$N = b^y.$$

Take the logarithm of each side to the base a:

$$\begin{aligned} \log_a N &= \log_a b^y \\ &= y \log_a b \\ \log_a N &= \log_b N \log_a b. \end{aligned}$$

Hence

$$\boldsymbol{\log_b N = \frac{\log_a N}{\log_a b}}.$$

If $N = a$,

$$\log_b a = \frac{1}{\log_a b}.$$

Therefore

$$\log_b N = (\log_b a)(\log_a N).$$

In calculus and higher mathematics, the most suitable system of logarithms is the *natural,* or *napierian,* system, which employs the base e, where e is approximately 2.71828. If $a = 10$ and $b = e$, Equation (1) becomes

$$\boldsymbol{\ln N = \log_e N = \frac{\log_{10} N}{0.43429} = 2.3026 \log_{10} N = 2.3026 \log N.}$$

Exercise A.3

Evaluate each of the following:

1. $\log e^2$
2. $\log 1/e$
3. $\ln 100$
4. $\ln 2^{10}$
5. $\ln \frac{1}{100}$
6. $\ln 10^{-10}$
7. $\ln 1000$
8. $\log_{100} e$
9. $\log_{1000} e$
10. $\log_{1/100} e$

A.4 USING A TABLE OF TRIGONOMETRIC FUNCTIONS

The table at the end of the book lists, to four decimal places (for numbers less than 1) or four significant figures (for numbers greater than 1), the sine, cosine, tangent, and cotangent for acute angles at intervals of one-tenth of a degree. For angles less than 45°, find the name of the function at the *top* of the column, then read *down* until the angle is found at the *left*. For angles greater than 45°, find the name of the function at the *bottom* of the column, then read *up* until the angle is found at the *right*.

Remember that results obtained using a four-place table will not have more than four-figure accuracy no matter how high the degree of accuracy of the given data.

Example 1 Find cos 82.6°.

Solution In the table, in the column with *cos* at its *foot*, move up to the number in line with 82.6°. Thus cos 82.6° = 0.1288.

Example 2 Find tan 27.3°.

Solution In the table, in the column with *tan* at its *head*, come down to the number in line with 27.3°. Hence tan 27.3° = 0.5161.

Example 3 Find θ if $\sin \theta = 0.9385$.

Solution Since sines are found in column 3 reading *down* and in column 6 reading *up,* we must search through these two columns for the number 0.9385. It appears in the sixth column, which has *sin* at its *foot.* This column contains the sines of the angles in the *right* column. On a line with 0.9385, we find in the *right* column the angle 69.8°. Hence,

if $$\sin\theta = 0.9385,$$

then $$\theta = 69.8°.$$

Example 4 Find θ if $\cot\theta = 1.638$.

Solution We search the two cotangent columns, the fourth going up and the fifth going down, and find 1.638 in the fifth column. Since this column has *cot* at its *head,* we associate 1.638 with the angle at the *left.* Hence,

if $$\cot\theta = 1.638,$$

then $$\theta = 31.4°.$$

The student should guard against writing $\cot\theta = 1.638 = 31.4°$. The second equality sign is used incorrectly, because 1.638 does *not* equal 31.4° and $\cot\theta$ does *not* equal 31.4°.

Exercise A.4

Use the four-place table to find the value of each of the following to four decimal places if the number is less than 1 and to four significant digits if the number is greater than or equal to 1.

1. sin 73.2°
2. cot 56.7°
3. cos 22.6°
4. tan 7.9°
5. cot 18.5°
6. sin 40.3°
7. tan 83.0°
8. cos 61.4°
9. cos 49.8°
10. tan 66.1°
11. sin 37.8°
12. cot 24.6°

Use the four-place table to find θ, to the nearest tenth of a degree, from each of the following functions of θ.

13. $\tan\theta = 0.6720$
14. $\cos\theta = 0.7181$
15. $\cot\theta = 0.5658$
16. $\sin\theta = 0.9191$

17. $\cos\theta = 0.1925$
18. $\tan\theta = 9.845$
19. $\sin\theta = 0.0436$
20. $\cot\theta = 5.050$
21. $\sin\theta = 0.0122$
22. $\cot\theta = 3.582$
23. $\cos\theta = 0.6074$
24. $\tan\theta = 1.558$

A.5 INTERPOLATION

When a sports announcer says, "The ball is on the 27-yard line," most football fans realize that the announcer estimates that the ball is $\frac{2}{5}$ of the way from the 25-yard line to the 30-yard line. This process of literally "reading between the lines" is called interpolation. Another example is: "Interpolate to approximate the value of $\sqrt{8}$." Knowing $\sqrt{4} = 2$ and $\sqrt{9} = 3$, we conclude that $\sqrt{8}$ is a number between 2 and 3. Moreover, 8 is $\frac{4}{5}$ of the way from 4 to 9. Assume that for a small increase in a number N, the change in $\sqrt{N}$ is proportional to the change in N. Then $\sqrt{8}$ would lie $\frac{4}{5}$ of the way from 2 to 3. Since $\frac{4}{5}$ of 1 is 0.8, we conclude that $\sqrt{8}$ is approximately 2.8. This result is correct to only one decimal place. The process of interpolation is important in all work involving the use of tables.

We already know that the trigonometric functions do not change uniformly with the change in the angle (if an angle is doubled, its sine does not double). But if the angle is changed by only a few hundredths of a degree, the change in the function is very nearly proportional to the change in the angle.

Example 1 Find sin 65.84°, using the table.

Solution Here we must interpolate between 65.80° and 65.90°.

$$\left.\begin{array}{l}\left.\begin{array}{l}\sin 65.80^\circ \\ \sin 65.84^\circ\end{array}\right\} 4 \\ \sin 65.90^\circ\end{array}\right\} 10 \qquad \left.\begin{array}{l} = 0.9121 \\ = \\ = 0.9128\end{array}\right\} 7$$

As the angle increases 0.10° (from 65.80° to 65.90°), its sine increases 7 ten-thousandths. Our angle is $\frac{4}{10}$ of the way from 65.80° to 65.90°. Hence the sine of our angle should be $\frac{4}{10}$ of the way from 0.9121 to 0.9128. But $\frac{4}{10}(7) = 2\frac{4}{5} \rightarrow 3$. (Round off to 3 because $2\frac{4}{5}$ is closer to 3 than it is to 2.) Since the sine is increasing, *add* the 3 to 0.9121 to get

$$\sin 65.84^\circ = 0.9124.$$

Example 2 Find cos 15.17°, using the Table.

Solution

$$\begin{aligned} \cos 15.10^\circ &= 0.9655 \\ \cos 15.17^\circ &= \\ \cos 15.20^\circ &= 0.9650 \end{aligned}$$

(differences: 7, 10; −5)

An *increase* of 0.10° in the angle produces a *decrease* of 5 ten-thousandths in the cosine. Our angle is $\frac{7}{10}$ of the way from 15.10° to 15.20°. Hence we want $\frac{7}{10}$ of the decrease of 5. But $\frac{7}{10}(5) = 3\frac{1}{2}$. *Subtracting* this number from 0.9655 gives the halfway number 0.96515, which we round off to the larger number. Therefore

$$\cos 15.17^\circ = 0.9652\,.$$

Example 3 Find θ if $\tan\theta = 0.6206$.

Solution

$$\begin{aligned} \tan 31.80^\circ &= 0.6200 \\ \tan\theta &= 0.6206 \\ \tan 31.90^\circ &= 0.6224 \end{aligned}$$

(differences: 10; 6, 24)

Our number 0.6206 is $\frac{6}{24}$ of the way from 0.6200 to 0.6224. Hence θ should be $\frac{6}{24}$ of the way from 31.80° to 31.90°. But $\frac{6}{24}(10) = 2\frac{1}{2}$. Therefore θ is halfway between 31.82° and 31.83°. Choosing the larger number, we find

$$\theta = 31.83^\circ.$$

Example 4 Find θ if $\cos\theta = 0.2810$.

Solution

$$\begin{aligned} \cos 73.60^\circ &= 0.2823 \\ \cos\theta &= 0.2810 \\ \cos 73.70^\circ &= 0.2807 \end{aligned}$$

(differences: 10; 13, 16)

Our number 0.2810 is $\frac{13}{16}$ of "the way down." Hence θ is $\frac{13}{16}(10) = 8\frac{1}{8} \rightarrow 8$ away from 73.60°. Therefore

$$\theta = 73.68°.$$

Note that, in this three-line method of interpolation, *the small angle is always written on top. All the differences are measured from the small angle and its function.*

Exercise A.5

Use the four-place table. Interpolate to find the value of each of the following.

1. sin 69.84°
2. cos 38.52°
3. tan 76.39°
4. cot 22.68°
5. cot 13.09°
6. sin 72.36°
7. cos 23.84°
8. tan 81.72°
9. tan 21.07°
10. cot 55.94°
11. sin 12.48°
12. cos 48.23°
13. cos 65.11°
14. tan 34.13°
15. cot 57.75°
16. sin 29.97°

Interpolate to find θ to the nearest hundredth of a degree.

17. $\tan \theta = 1.593$
18. $\cot \theta = 6.538$
19. $\sin \theta = 0.9828$
20. $\cos \theta = 0.7383$
21. $\cos \theta = 0.9804$
22. $\tan \theta = 3.429$
23. $\cot \theta = 1.932$
24. $\sin \theta = 0.8997$
25. $\cot \theta = 0.4571$
26. $\sin \theta = 0.6058$
27. $\cos \theta = 0.0240$
28. $\tan \theta = 0.3029$
29. $\sin \theta = 0.3742$
30. $\cos \theta = 0.4806$
31. $\tan \theta = 0.8886$
32. $\cot \theta = 0.3318$

ANSWERS

EXERCISE 1.1 PAGES 5–6

1. 10, 7, $\sqrt{29}$, 6
2. 17, 1, $\sqrt{13}$, 3
3. 25, 4, $\sqrt{34}$, 7
5. -8, $-\sqrt{5}$, -10
6. -12, -5, 0
7. -4, 10, 0
9. (b) I and IV
10. (c) II and IV
11. 0, 0
13. $\sqrt{68}$
14. $\sqrt{a^2 + 6a + 58}$
15. 0, 4

EXERCISE 1.2 PAGES 7–8

1. 430°, $-290°$
2. 470°, $-250°$
3. 630°, $-90°$
5. 190°, $-170°$
6. 340°, $-20°$
7. 160°, $-200°$
9. 50°
10. 352°
11. 212°
13. 145°
14. 256°
15. 30°
17. 9600°
18. 102°, 225°, 8640°

EXERCISE 1.3 PAGES 13–14

1. All θ except angles coterminal with 90° or with 270°
2. $\sec\theta \le -1$ or $1 \le \sec\theta$
3. Fourth
5. Third
6. Second
7. 0.98, 0.17, 5.8
9. -0.94, 0.34, -2.8
10. 0.34, -0.94, -0.36
11. -0.64, -0.77, 0.83
13. 0, 0.17, 0.34, 0.50, 0.64, 0.77, 0.87, 0.94, 0.98, 1
14. 1, 0.98, 0.94, 0.87, 0.77, 0.64, 0.50, 0.34, 0.17, 0
15. 0, 1, 0, does not exist, 1, does not exist*
17. -1, 0, does not exist, 0, does not exist, -1
18. 0, -1, 0, does not exist, -1, does not exist

*Answers are in this order: $\sin\theta$, $\cos\theta$, $\tan\theta$, $\cot\theta$, $\sec\theta$, $\csc\theta$.

EXERCISE 1.4 PAGES 18–19

1. $\frac{8}{17}, \frac{15}{17}, \frac{8}{15}, \frac{15}{8}, \frac{17}{15}, \frac{17}{8}$*
2. $\frac{3}{5}, \frac{-4}{5}, \frac{-3}{4}, \frac{-4}{3}, \frac{-5}{4}, \frac{5}{3}$
3. $\frac{-12}{13}, \frac{5}{13}, \frac{-12}{5}, \frac{-5}{12}, \frac{13}{5}, \frac{-13}{12}$
5. $\frac{-\sqrt{10}}{10}, \frac{-3\sqrt{10}}{10}, \frac{1}{3}, 3, \frac{-\sqrt{10}}{3}, -\sqrt{10}$
6. $\frac{-8\sqrt{89}}{89}, \frac{5\sqrt{89}}{89}, \frac{-8}{5}, \frac{-5}{8}, \frac{\sqrt{89}}{5}, \frac{-\sqrt{89}}{8}$
7. $\frac{4\sqrt{97}}{97}, \frac{9\sqrt{97}}{97}, \frac{4}{9}, \frac{9}{4}, \frac{\sqrt{97}}{9}, \frac{\sqrt{97}}{4}$
9. $-\frac{\sqrt{65}}{9}, \frac{4}{9}, -\frac{\sqrt{65}}{4}, -\frac{4\sqrt{65}}{65}, \frac{9}{4}, -\frac{9\sqrt{65}}{65}$
10. $\frac{-3}{10}, \frac{-\sqrt{91}}{10}, \frac{3\sqrt{91}}{91}, \frac{\sqrt{91}}{3}, \frac{-10\sqrt{91}}{91}, \frac{-10}{3}$
11. $\frac{\sqrt{3}}{4}, \frac{-\sqrt{13}}{4}, \frac{-\sqrt{39}}{13}, \frac{-\sqrt{39}}{3}, \frac{-4\sqrt{13}}{13}, \frac{4\sqrt{3}}{3}$

13. IV **14.** III **15.** III
17. II **18.** IV **19.** I
21. Impossible **22.** Impossible **23.** Possible
25. Possible **26.** Possible **27.** Impossible
29. Close to 0° **30.** Close to 90°
31. Close to 0° **33.** Close to 90°
34. Close to 0°

EXERCISE 1.5 PAGES 20–21

1. $\sin\theta = \frac{2\sqrt{13}}{13}$, $\cos\theta = -\frac{3\sqrt{13}}{13}$, $\cot\theta = -\frac{3}{2}$, $\sec\theta = -\frac{\sqrt{13}}{3}$, $\csc\theta = \frac{\sqrt{13}}{2}$
2. $\cos\theta = -\frac{\sqrt{35}}{6}$, $\tan\theta = \frac{\sqrt{35}}{35}$, $\cot\theta = \sqrt{35}$, $\sec\theta = -\frac{6\sqrt{35}}{35}$, $\csc\theta = -6$

*Answers are in this order: $\sin\theta$, $\cos\theta$, $\tan\theta$, $\cot\theta$, $\sec\theta$, $\csc\theta$.

3. $\sin\theta = -\frac{\sqrt{55}}{8}$, $\tan\theta = -\frac{\sqrt{55}}{3}$, $\cot\theta = -\frac{3\sqrt{55}}{55}$, $\sec\theta = \frac{8}{3}$, $\csc\theta = -\frac{8\sqrt{55}}{55}$

5. $\cos\theta = \frac{7}{25}$, $\tan\theta = \frac{24}{7}$, $\cot\theta = \frac{7}{24}$, $\sec\theta = \frac{25}{7}$, $\csc\theta = \frac{25}{24}$

6. $\sin\theta = \frac{4}{5}$, $\tan\theta = -\frac{4}{3}$, $\cot\theta = -\frac{3}{4}$, $\sec\theta = -\frac{5}{3}$, $\csc\theta = \frac{5}{4}$

7. $\sin\theta = -\frac{5}{13}$, $\cos\theta = -\frac{12}{13}$, $\cot\theta = \frac{12}{5}$, $\sec\theta = -\frac{13}{12}$, $\csc\theta = -\frac{13}{5}$

9. $\sin\theta = -\frac{5\sqrt{29}}{29}$, $\tan\theta = \frac{5}{2}$, $\cot\theta = \frac{2}{5}$, $\sec\theta = -\frac{\sqrt{29}}{2}$, $\csc\theta = -\frac{\sqrt{29}}{5}$

10. $\sin\theta = -\frac{1}{3}$, $\cos\theta = \frac{2\sqrt{2}}{3}$, $\cot\theta = -2\sqrt{2}$, $\sec\theta = \frac{3\sqrt{2}}{4}$, $\csc\theta = -3$

11. $\cos\theta = -\frac{\sqrt{10}}{5}$, $\tan\theta = -\frac{\sqrt{6}}{2}$, $\cot\theta = -\frac{\sqrt{6}}{3}$, $\sec\theta = -\frac{\sqrt{10}}{2}$, $\csc\theta = \frac{\sqrt{15}}{3}$

13. $\sin\theta = -\frac{9}{41}$, $\cos\theta = \frac{40}{41}$, $\tan\theta = -\frac{9}{40}$, $\cot\theta = -\frac{40}{9}$, $\csc\theta = -\frac{41}{9}$

14. $\sin\theta = \frac{\sqrt{2}}{10}$, $\cos\theta = \frac{7\sqrt{2}}{10}$, $\tan\theta = \frac{1}{7}$, $\sec\theta = \frac{5\sqrt{2}}{7}$, $\csc\theta = 5\sqrt{2}$

15. $\sin\theta = \frac{a\sqrt{1+a^2}}{1+a^2}$, $\cos\theta = \frac{\sqrt{1+a^2}}{1+a^2}$, $\cot\theta = \frac{1}{a}$, $\sec\theta = \sqrt{1+a^2}$, $\csc\theta = \frac{\sqrt{1+a^2}}{a}$

EXERCISE 1.6 PAGE 22

1. $\sin\alpha = \frac{20}{101}$, $\cos\alpha = \frac{99}{101}$, $\tan\alpha = \frac{20}{99}$, $\sin\beta = \frac{99}{101}$, $\cos\beta = \frac{20}{101}$, $\tan\beta = \frac{99}{20}$

2. $\sin\theta = \frac{39}{89}$, $\cos\theta = \frac{80}{89}$, $\tan\theta = \frac{39}{80}$, $\sin\phi = \frac{80}{89}$, $\cos\phi = \frac{39}{89}$, $\tan\phi = \frac{80}{39}$

3. $\sin\alpha = \frac{g}{h}$, $\cos\alpha = \frac{f}{h}$, $\tan\alpha = \frac{g}{f}$, $\sin\beta = \frac{f}{h}$, $\cos\beta = \frac{g}{h}$, $\tan\beta = \frac{f}{g}$

EXERCISE 1.7 PAGE 26

1. $\frac{1}{2}$
2. $\frac{7}{8}$
3. $-\frac{1}{2}$
5. $-\frac{1}{16}$
6. $2\sqrt{6}$
7. $\frac{13}{16}$
9. True
10. False
11. True
13. False
14. True
15. True
17. False
18. True
19. True

REVIEW EXERCISES PAGES 26–27

2. 5, 3, $\sqrt{34}$
3. -1, 9
5. $\dfrac{-5\sqrt{61}}{61}, \dfrac{-6\sqrt{61}}{61}, \dfrac{5}{6}, \dfrac{6}{5}, \dfrac{-\sqrt{61}}{6}, \dfrac{-\sqrt{61}}{5}$, III
6. 30°, 100°
7. All angles θ except those coterminal with 0° or with 180°.
9. 0
10. $\dfrac{\sqrt{3}-1}{2}$

EXERCISE 2.1 PAGE 31

1. (a) 0.7182, (b) 0.718, (c) 0.72
2. (a) 8.917, (b) 8.92, (c) 8.9
3. (a) 45.74, (b) 45.7, (c) 46
5. (a) 30.63°, (b) 30.6°, (c) 31°
6. (a) 22.95°, (b) 22.9°, (c) 23°
7. (a) 9.649°, (b) 9.65°, (c) 9.6°
9. $42.65 \le x < 42.75$
10. $261.5 \le x < 262.5$
11. $0.07575 \le x < 0.07585$
13. $5999.5 \le x < 6000.5$
14. $0.19445 \le x < 0.19455$
15. $2.8165 \le x < 2.8175$

EXERCISE 2.2 PAGES 34–35

2. 78.2328°
3. 19.3219°
5. 213.5192°
6. 0.9573
7. 0.6569
9. 0.1391
10. 2.978
11. 0.6472
13. 0.4784
14. 0.6444
15. 2.468
17. 3.219
18. 83.4°
19. 44.1°
21. 66.8°
22. 78.9°
23. 54.7°
25. 14.5°
26. 130.38
27. 0.63817
29. 4.4306
30. 1.0000

EXERCISE 2.3 PAGES 38–39

1. $\sin 240° = \frac{-\sqrt{3}}{2}$, $\cos 240° = -\frac{1}{2}$

2. $\sin 315° = -\frac{\sqrt{2}}{2}$, $\cos 315° = \frac{\sqrt{2}}{2}$

3. $\sin 150° = \frac{1}{2}$, $\cos 150° = -\frac{\sqrt{3}}{2}$

5. $\sin 330° = -\frac{1}{2}$, $\cos 330° = \frac{\sqrt{3}}{2}$

6. $\sin 210° = -\frac{1}{2}$, $\cos 210° = -\frac{\sqrt{3}}{2}$

7. $\sin 225° = -\frac{\sqrt{2}}{2}$, $\cos 225° = -\frac{\sqrt{2}}{2}$

9. $\sin 495° = \frac{\sqrt{2}}{2}$, $\cos 495° = -\frac{\sqrt{2}}{2}$

10. $\sin 480° = \frac{\sqrt{3}}{2}$, $\cos 480° = -\frac{1}{2}$

11. $\sin 1020° = -\frac{\sqrt{3}}{2}$, $\cos 1020° = \frac{1}{2}$

13. False
14. True
15. True
17. True
18. False
19. False
21. 27°, 153°, 207°
22. 80°, 260°, 280°
23. 136°, 224°, 316°
25. −1.1504
26. −0.27564
27. −1.9626
29. −0.95931
30. −2.7034
31. −0.30736
33. −0.78348
34. 0.98788
35. −0.64479

EXERCISE 2.4 PAGE 41

1. $\sin(-30°) = -\frac{1}{2}$, $\cos(-30°) = \frac{\sqrt{3}}{2}$, $\tan(-30°) = -\frac{\sqrt{3}}{3}$

2. $\sin(-45°) = -\frac{\sqrt{2}}{2}$, $\cos(-45°) = \frac{\sqrt{2}}{2}$, $\tan(-45°) = -1$

3. $\sin(-90°) = -1$, $\cos(-90°) = 0$, $\tan(-90°)$ does not exist

5. False
6. False
7. True
9. True
10. True
11. False
13. False
14. False
15. True

18. (b), (e), (g) even; (a), (d), (f) odd; (c), (h) neither

REVIEW EXERCISES PAGE 42

1. 4.8377, 4.9400, 1.0211
2. (a) 156.5958° (b) 0.39721, −0.91773, −0.43282 (c) 23.404°
3. (a) 25.43°, (b) 2.13°, (c) 45.29°
5. (a) $-\frac{1}{2}, -\frac{\sqrt{3}}{2}$ (b) $\frac{\sqrt{3}}{2}, \frac{1}{2}$ (c) $\frac{\sqrt{3}}{2}, -\frac{1}{2}$
6. (a) Even (b) Odd (c) Neither (d) Odd

EXERCISE 3.1 PAGES 47–48

1. $A = 72°, b = 18, c = 58$
2. $B = 21°, b = 1.5, c = 4.1$
3. $B = 17.5°, a = 63.7, c = 66.8$
5. $B = 37.75°, a = 78.63, b = 60.88$
6. $A = 49.2°, a = 0.0504, b = 0.0435$
7. $A = 21.5°, B = 68.5°, c = 957$
9. $A = 41.4°, B = 48.6°, b = 45.4$
10. $A = 40.08°, B = 49.92°, a = 6338$
11. $A = 53.15°, B = 36.85°, a = 66.72$
13. $A = 54.46°, B = 35.54°, c = 3420$
14. $A = 39°, B = 51°, c = 94$
15. $A = 67°, a = 41, b = 18$
17. $A = 55.4°, B = 34.6°, a = 0.661$
18. $B = 10.00°, a = 8462, c = 8592$
19. $A = 53.88°, b = 4317, c = 7324$

EXERCISE 3.2 PAGES 51–53

1. 800 ft
2. 20 m
3. 555 ft
5. 1:54 P.M.
6. 1053 ft
7. 1130 m
9. 110 mi, 130 mi
10. N 70° E, S 70° W
11. S 55° E, N 55° W
13. N 16° E
14. 2063 ft
15. 3.14143 m, 3.14159 m
17. 62.1 km/h
18. 51.11°
19. N 5.0° E
25. (a) 2.25 m, (b) 9.74 m, (c) 1.23 m
26. 35.26°

EXERCISE 3.3 PAGES 56–58

1. 10:38 A.M.
2. 4:23 P.M.
3. 492 km
5. (a) 13.31°, (b) 1637 lb
6. 7561 lb
7. (a) 90 lb, (b) 31 lb
9. 40.6 km/h, 234 km/h
10. 222°, 92 km/h
11. 284.7°, 145 mi/h
13. 2.50 min
14. S 36.9° W, 300 ft/min
15. 54 mi/h, S 23° W
17. No, its speed was 28 mi/h
18. 18.2°
19. S 7.82°, 54.10 lb

EXERCISE 3.4 PAGE 63

1. $B = 83.3°$, $a = 168$, $c = 129$
2. $C = 30.3°$, $a = 689$, $b = 890$
3. $A = 75°$, $b = 3.8$, $c = 5.7$
5. $B = 59°$, $a = 0.95$, $b = 0.82$
6. $C = 60°$, $b = 9.0$, $c = 7.9$
7. $A = 100.0°$, $a = 75.6$, $c = 50.1$
9. $A = 42.17°$, $b = 39.87$, $c = 43.43$
10. $C = 37.23°$, $a = 1745$, $c = 1137$
11. $C = 55.13°$, $a = 4828$, $b = 6810$
13. 209.4
14. 810 km, 420 km
15. 420 km/h, 430 km/h

EXERCISE 3.5 PAGE 67

1. $B = 40°$, $C = 111°$, $c = 8.7$
   $B' = 140°$, $C' = 11°$, $c' = 1.8$
2. $B = 53°$, $C = 51°$, $c = 46$
3. No triangle
5. $B = 43.80°$, $C = 32.65°$, $c = 55.18$
6. No triangle
7. $B = 74.3°$, $C = 52.5°$, $c = 36.8$
   $B' = 105.7°$, $C' = 21.1°$, $c' = 16.7$
9. No triangle
10. $B = 47.1°$, $C = 99.6°$, $c = 269$
    $B' = 132.9°$, $C' = 13.8°$, $c' = 65.2$
11. $B = 61.20°$, $C = 57.60°$, $c = 58.97$
13. N 62° W
14. There are two Jacksons that satisfy the conditions of the problem. Jackson, Tennessee, is 240 mi from Vincennes; Jackson, Mississippi, is 440 mi from Vincennes.
15. 63,000,000 mi or 117,000,000 mi

EXERCISE 3.6 PAGES 72–73

1. $b = 33, A = 32°, C = 42°$
2. $a = 6.45, B = 89.6°, C = 42.1°$
3. $c = 494, A = 46.8°, B = 61.8°$
5. $A = 73.1°, B = 62.6°, C = 44.4°$
6. $A = 34°, B = 102°, C = 44°$
7. $A = 131°, B = 29°, C = 20°$
9. $c = 7.082, A = 95.03°, B = 26.87°$
10. $b = 59.12, A = 111.36°, C = 23.65°$
11. $A = 119.74°, B = 45.01°, C = 15.26°$
13. 9.3 km
14. S 82° E, N 36° E
15. S 16.6° W
17. 55°, 65° or 305°, 295°
18. $\sqrt{91}$ lb

EXERCISE 3.7 PAGES 75–76

1. 33
2. 98
3. 48
5. 1188 (exactly)
6. $10\sqrt{66} \doteq 81$
7. 193
9. 0.834
10. 151.6
11. 24.05
13. 18.06° or 161.94°
14. 6.16 acres

REVIEW EXERCISES PAGES 76–78

1. 4.8
2. 22°
3. 72°
5. No triangle
6. 99.8°
7. 2.81
9. 7811
10. 25
11. 180 ft
13. N 71° W, S 30° W
14. 230 mi, 330 mi
15. N 24° E, 430 km
17. 1012 ft
18. 11°
19. $x = \dfrac{a \sin \delta \sin \gamma}{\sin \alpha \sin \phi}$
21. 6.12 km or 1.14 km
23. $5\sqrt{31}$ (exactly)
25. 481.3 ft, 755.7 ft, 13.11 acres

EXERCISE 4.1 PAGES 82–84

1. 90°
2. 60°
3. 45°
5. 36°
6. −140°
7. 378°
9. 200°
10. 810°
11. −900°
13. 337.5°
14. 153°
15. 130°

17. 326.59° **18.** 171.89° **19.** 49.62° **21.** $\frac{4\pi}{3}$

22. $\frac{3\pi}{4}$ **23.** $\frac{\pi}{6}$ **25.** $-\frac{10\pi}{3}$ **26.** $\frac{35\pi}{18}$

27. $\frac{7\pi}{4}$ **29.** $\frac{13\pi}{4}$ **30.** 20π **31.** $\frac{40\pi}{9}$

33. 4.7084 **34.** 0.3526 **35.** 6.0903 **37.** $\frac{\sqrt{3}}{2}$

38. $-\frac{\sqrt{2}}{2}$ **39.** $-\frac{\sqrt{3}}{2}$ **41.** $-\frac{\sqrt{2}}{2}$ **42.** $-\frac{1}{2}$

43. $-\frac{\sqrt{3}}{2}$ **45.** $-\sqrt{3}$ **46.** $-\sqrt{3}$ **47.** 1

49. -1 **50.** Does not exist **51.** 1

53. 0.5402 **54.** 0.9893 **55.** 0.3879 **57.** $\frac{5\pi}{3}$, 14π

58. $\frac{2\pi}{3}, \frac{2\pi}{3}, \frac{11\pi}{15}$ **59.** $\frac{17\pi}{30}$

EXERCISE 4.2 PAGES 89–92

1. (a) 2.61 cm, (b) 20.1 in.
2. (a) 1.80 m, (b) 0.419 m, (c) 10.2 m
3. 2.71 radians, 155.3° **5.** 17.2°
7. 3000 mi, 3300 mi **9.** 5100 mi
10. 19° N **11.** 39° N
13. 109° E **14.** 13,000 km
15. 115 m **17.** 864,000 mi
18. Yes, $\theta > 0.17°$ **19.** 4.92
21. 23.2 cm **22.** 33.0 cm

EXERCISE 4.3 PAGES 93–94

1. 220 radians/s **2.** 79.6 rpm **3.** 32.0 in.
5. 758 mi/h **6.** 8.38 ft/s, 32 rpm **7.** 2.72 rpm
9. 16.2 cm **10.** 13.1 mi/h **11.** 102 meters/min, 812 rpm
13. 17,800 mi/h **14.** 202 mi

REVIEW EXERCISES PAGE 95

1. $\frac{3\pi}{2}, \frac{-5\pi}{3}$
2. 220°, 96.9°
3. $\frac{\sqrt{3}}{2}, \frac{\sqrt{2}}{2}, \frac{\sqrt{3}}{3}$
5. 2300 miles, 4000 miles, 1200 miles
6. 1.6 ft/sec, 0.7 ft/sec.

EXERCISE 5.1 PAGES 100–101

1. $\cot 2B$ **2.** $\tan \frac{2\pi}{5}$ **3.** $\cos 3C$ **5.** $-\cos \frac{7\pi}{10}$
6. $\pm\sec A$ **7.** $-\cot \frac{6\pi}{5}$ **9.** $\sin \frac{7\pi}{5}$ **10.** $\csc \frac{17\pi}{10}$
11. False **13.** False **14.** False **15.** True
17. True **18.** True **19.** False **21.** True
22. True **23.** True **25.** True **26.** False
27. False **29.** False **30.** True **31.** True

EXERCISE 5.2 PAGE 103

1. $\sin \theta + \cos \theta$ **2.** 1
3. $3 \sec \theta$ **5.** $\sin \theta$
6. $\pm\sin A$ **7.** 8
9. $\sin \frac{\pi}{17} = \sqrt{1 - \cos^2 \frac{\pi}{17}}$ **10.** $\csc \frac{6\pi}{5} = -\sqrt{1 + \cot^2 \frac{6\pi}{5}}$
11. $\sin 88° = \frac{1}{\csc 88°}$ **13.** $\tan 2B = \pm\sqrt{\sec^2 2B - 1}$
14. $\tan 3A = \frac{1}{\cot 3A}$
15. $\sin \theta = \pm\frac{\tan \theta}{\sqrt{1 + \tan^2 \theta}}$, $\cos \theta = \pm\frac{1}{\sqrt{1 + \tan^2 \theta}}$, $\cot \theta = \frac{1}{\tan \theta}$,
$\sec \theta = \pm\sqrt{1 + \tan^2 \theta}$, $\csc \theta = \pm\frac{\sqrt{1 + \tan^2 \theta}}{\tan \theta}$

EXERCISE 5.3 PAGES 109–112

33. Not an identity. Left side = $\sin\theta\cos\theta \neq 1$. Not true for any value of θ.
34. An identity.
35. Not an identity. Left side = $\pm\sin\theta$. Hence the equation holds true for $0 \leq \theta \leq \pi$, but it is false for $\pi < \theta < 2\pi$. Therefore, it does not hold true for all permissible values of θ.
37. An identity.
38. Not an identity. Left side = $4\sec^2\theta - 2 \neq 4\sec^2\theta + 2$. Not true for any value of θ.
39. Not an identity. Left side = $\cos^2\theta - \sin^2\theta \neq \cos^2\theta - \sin\theta$. True only if $\sin^2\theta = \sin\theta$, that is, $\theta = 0$ or $\pi/2$ and angles coterminal with them. Observe that 0 and π are nonpermissible values of θ, because cot 0 and cot π do not exist.
45. $\frac{\pi}{2}, \frac{3\pi}{2}$
46. $A = \frac{3\pi}{2}$; $B = \frac{\pi}{2}, \frac{3\pi}{2}$
47. $A = 90°. 270°$; $B = 90°, 135°, 270°, 315°$

REVIEW EXERCISES PAGE 112

1. No
2. No
3. Yes
5. No
6. No
7. No

EXERCISE 6.1 PAGES 118–120

1. $\cos 255° = \dfrac{\sqrt{2} - \sqrt{6}}{4} \doteq -0.259$
2. $\cos 345° = \dfrac{\sqrt{2} + \sqrt{6}}{4} \doteq 0.966$
3. $\sin 165° = \dfrac{\sqrt{6} - \sqrt{2}}{4} \doteq 0.259$
6. $\frac{1}{2}$
7. -1
9. $-\cos 4\theta$
17. $\cos 100° = \cos 80° \cos 20° - \sin 80° \sin 20°$
18. $\sin 38° = \sin 11° \cos 27° + \cos 11° \sin 27°$
21. $\dfrac{\sqrt{5} + 4\sqrt{2}}{9} \doteq 0.877$
22. $\frac{117}{125}$
23. (a) $-\frac{156}{205}$, (b) $\frac{133}{205}$, (c) Q IV

EXERCISE 6.2 PAGES 123–125

1. $\tan 75° = 2 + \sqrt{3} \doteq 3.732$ **2.** $\tan 165° = -2 + \sqrt{3} \doteq -0.268$

3. $\sin 195° = \dfrac{\sqrt{2} - \sqrt{6}}{4} \doteq -0.259$

5. -1 **6.** 0 **7.** $-\dfrac{1}{\sqrt{3}}$

13. (a) $\frac{44}{125}$, (b) $-\frac{117}{125}$, (c) $-\frac{4}{3}$, (d) $-\frac{44}{117}$

14. (a) $-\frac{21}{221}$, (b) $-\frac{220}{221}$, (c) $-\frac{171}{140}$, (d) $\frac{21}{220}$

15. -3

17. $\cos 50° = \cos 70° \cos 20° + \sin 70° \sin 20°$

18. $\tan 127° = \dfrac{\tan 40° + \tan 87°}{1 - \tan 40° \tan 87°}$

19. $\tan 6° = \dfrac{\tan 8° - \tan 2°}{1 + \tan 8° \tan 2°}$

37. $17 \sin(\theta + 28.1°) = 17 \sin(\theta + 0.490)$

38. $41 \sin(\theta + 257.3°) = 41 \sin(\theta + 4.491)$

39. $\sqrt{2} \sin(\theta + 135°) = \sqrt{2} \sin\left(\theta + \dfrac{3\pi}{4}\right)$

41. $-6\sqrt{3} \sin \theta + 6 \cos \theta$ **42.** $2\sqrt{2} \sin \theta - 2\sqrt{2} \cos \theta$

EXERCISE 6.3 PAGES 129–132

3. $\sin 157\frac{1}{2}° = \frac{1}{2}\sqrt{2 - \sqrt{2}}$, $\cos 157\frac{1}{2}° = -\frac{1}{2}\sqrt{2 + \sqrt{2}}$

5. $\pm\sin 7B$ **6.** $\cos 111°$ **7.** $5 \cos \dfrac{\theta}{2}$

9. $9 \cos 2C$ **10.** 1 **11.** $\frac{1}{2} \sin 36°$

13. $\sin \dfrac{A}{2} = \dfrac{5\sqrt{34}}{34}$, $\sin 2A = \dfrac{240}{289}$ **14.** $\cos \dfrac{C}{4} = \dfrac{\sqrt{6}}{6}$, $\cos C = -\dfrac{1}{9}$

15. $\cos \dfrac{B}{2} = -\dfrac{\sqrt{13}}{4}$, $\cos 2B = -\dfrac{7}{32}$

17. $\cos 10D = 1 - 2 \sin^2 5D$

18. $\sin B = 2 \sin \dfrac{B}{2} \cos \dfrac{B}{2}$ **19.** $\sin 3C = \pm\sqrt{\dfrac{1 - \cos 6C}{2}}$

21. $\cos 100° = -\sqrt{\dfrac{1 + \cos 200°}{2}}$ **22.** $\sin 66° = \sqrt{\dfrac{1 - \cos 132°}{2}}$

23. $\sin 320° = -2 \sin 160°\sqrt{1 - \sin^2 160}$

25. True
26. True
27. True
29. True
30. True
31. False
33. False
34. False
35. True

EXERCISE 6.4 PAGE 134

1. $3 \cos 4\theta - 3 \cos 14\theta$
2. $6 \cos 7\theta + 6 \cos 5\theta$
3. $\sin 4\theta - \sin \theta$
5. $50 \cos 100° + 50 \cos 50°$
6. $8 \sin 66° + 8 \sin 44°$
7. $2 \sin \frac{9\theta}{2} \cos \frac{\theta}{2}$
9. $-2 \cos 15A \sin 2A$
10. $\sqrt{3} \cos 40°$ or $\sqrt{3} \sin 50°$

REVIEW EXERCISES PAGE 135

6. $13 \sin(\theta + 5.1068)$
7. $\dfrac{\tan \frac{\theta}{2} - \tan \frac{\theta}{3}}{1 + \tan \frac{\theta}{2} \tan \frac{\theta}{3}}$
9. $-2 \sin 12A \sin 3A$

EXERCISE 7.1 PAGES 141–142

1. $\frac{13\pi}{18}, \frac{11\pi}{9}$
2. $\frac{7\pi}{30}, \frac{37\pi}{30}$
3. $\frac{\pi}{6}, \frac{5\pi}{6}, \frac{7\pi}{6}, \frac{11\pi}{6}$
5. $\frac{\pi}{6}, \frac{5\pi}{6}, \frac{7\pi}{6}, \frac{11\pi}{6}$
6. $\frac{\pi}{3}, \frac{2\pi}{3}, \frac{4\pi}{3}, \frac{5\pi}{3}$
7. $\frac{3\pi}{4}, \frac{5\pi}{4}, \frac{7\pi}{4}$
9. $\frac{\pi}{3}, \frac{5\pi}{4}, \frac{5\pi}{3}, \frac{7\pi}{4}$
10. $\frac{2\pi}{3}, \frac{5\pi}{4}, \frac{5\pi}{3}, \frac{7\pi}{4}$
11. $0, \frac{\pi}{3}, \pi, \frac{4\pi}{3}$
13. $\frac{\pi}{6}, \frac{5\pi}{6}, \frac{3\pi}{2}$
14. $0, \frac{7\pi}{6}, \frac{11\pi}{6}$
15. No solution
17. $0, \frac{5\pi}{6}, \pi, \frac{7\pi}{6}$
18. $\frac{\pi}{3}, \frac{\pi}{2}, \frac{2\pi}{3}, \frac{3\pi}{2}$

19. $\frac{\pi}{3}, \frac{5\pi}{3}$

21. No solution

22. $\frac{\pi}{3}, \frac{5\pi}{3}$

23. $\frac{\pi}{3}, \frac{2\pi}{3}, \frac{4\pi}{3}, \frac{5\pi}{3}$

25. $\frac{\pi}{4}, \frac{\pi}{2}, \frac{3\pi}{4}, \frac{3\pi}{2}$

26. No solution

27. $\frac{\pi}{2}, \pi, \frac{3\pi}{2}$

29. $\frac{2\pi}{9}, \frac{4\pi}{9}, \frac{8\pi}{9}, \frac{10\pi}{9}, \frac{14\pi}{9}, \frac{16\pi}{9}$

30. $0, \frac{\pi}{4}, \frac{\pi}{2}, \frac{3\pi}{4}, \pi, \frac{5\pi}{4}, \frac{3\pi}{2}, \frac{7\pi}{4}$

31. True for all permissible values of θ, i.e., all θ **except** $0, \frac{\pi}{2}, \pi, \frac{3\pi}{2}$.

33. $\frac{7\pi}{6}, \frac{11\pi}{6}$

34. $\frac{2\pi}{3}$

35. $\frac{\pi}{6}, \frac{3\pi}{2}$

37. $\frac{2\pi}{3}, \pi$

38. $\frac{\pi}{6}, \frac{3\pi}{2}$

39. $\frac{7\pi}{6}, \frac{11\pi}{6}$

41. $\frac{\pi}{4}, \frac{5\pi}{4}$, 0.98, 4.12

42. 0.10, 3.24

43. 1.18, 1.97, 4.32, 5.11

45. 0.77, 5.52

46. 0.21, 2.15, 3.35, 5.29

47. 0.89, 1.87, 4.03, 5.01

49. 2.74, 4.83

50. 1.17, 3.26

51. 5.20

53. 1.37, 4.92

54. $\frac{\pi}{18}, \frac{7\pi}{18}, \frac{13\pi}{18}, \frac{5\pi}{6}, \frac{19\pi}{18}, \frac{25\pi}{18}, \frac{31\pi}{18}, \frac{11\pi}{6}$

55. $0, \frac{\pi}{6}, \frac{\pi}{2}, \frac{5\pi}{6}, \pi, \frac{7\pi}{6}, \frac{3\pi}{2}, \frac{11\pi}{6}$

57. $0, \pi$

58. $0, \frac{3\pi}{4}, \pi, \frac{7\pi}{4}$

59. No solution

61. $\frac{3\pi}{4}, \frac{5\pi}{4}$

62. $\frac{\pi}{4}, \frac{5\pi}{4}$, 2.04, 5.18

REVIEW EXERCISES PAGE 142

1. $\frac{\pi}{6}, \frac{5\pi}{6}, \frac{7\pi}{6}, \frac{11\pi}{6}$

2. No solution

3. $0, \frac{2\pi}{3}$

5. π

6. 1.1071, 1.8925, 4.2487, 5.0341

7. 0.61, 1.66, 2.70, 2.88, 3.75, 4.80, 5.85, 6.02

EXERCISE 8.1 PAGES 150–151

1. See Figure 8.2.

2. See Figure 8.3.

3. See Figure 8.7.

5. See Figure 8.9.

6. See Figure 8.10.

7. $y = -\sin\theta$

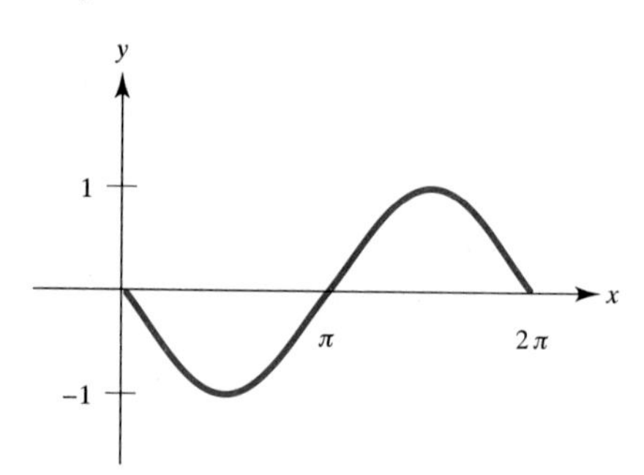

9. π

11. $\frac{\pi}{2}, \frac{3\pi}{2}$

13. $\frac{\pi}{2}$

14. $\frac{\pi}{3}, \frac{5\pi}{3}$

15. $\frac{\pi}{6}, \frac{5\pi}{6}$

17. $\frac{2\pi}{3}, \frac{4\pi}{3}$

18. $\frac{5\pi}{4}, \frac{7\pi}{4}$

19. $\frac{5\pi}{6}, \frac{11\pi}{6}$

21. $\frac{\pi}{6}, \frac{11\pi}{6}$

22. $\frac{\pi}{6}, \frac{7\pi}{6}$

23. No solution

25. $\frac{3\pi}{4}, \frac{7\pi}{4}$

26. π

27. $\frac{5\pi}{6}, \frac{7\pi}{6}$

29. $\frac{\pi}{3}, \frac{2\pi}{3}$

30. $\frac{\pi}{4}, \frac{5\pi}{4}$

31. 0.99, 2.16

33. 1.64, 4.78

34. 0.03, 3.17

35. 1.69, 4.83

37. 0.65, 5.63

38. 3.45, 5.98

39. (a) 13, 7 (b) −6, −16 (c) 12, 11

41. True

42. True

43. True

45. True

EXERCISE 8.2 PAGES 157–159

1. $\frac{\pi}{3}$, 1

2. $\frac{2\pi}{3}$, 4

3. 12, $\frac{1}{3}$

5. $\frac{3}{2}$, ∞

6. 12, 5

7. $\frac{16\pi}{7}$, 4

9. $\frac{2\pi}{3}$, 5, 2 units right

10. $\frac{5\pi}{2}$, 7.5, $\frac{5\pi}{4}$ units left

11. 4, 11, $\frac{2}{3}$ unit left

13. $y = 5 \sin 3x$

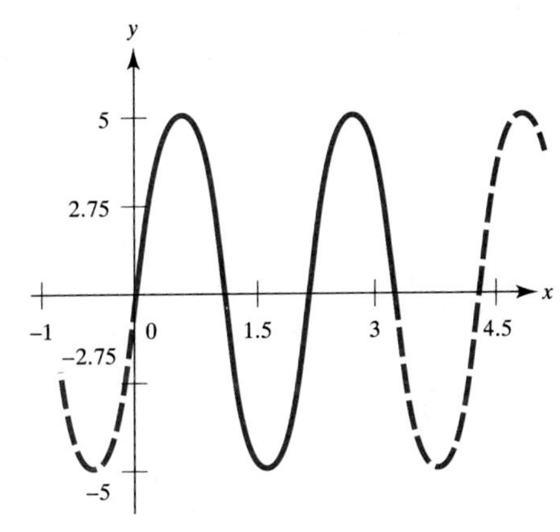

14. $y = 1.5 \cos 2x$

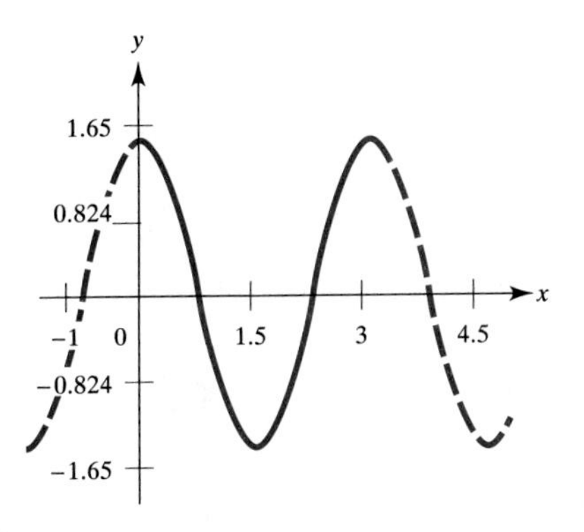

15. $y = \cos \frac{\pi x}{4}$

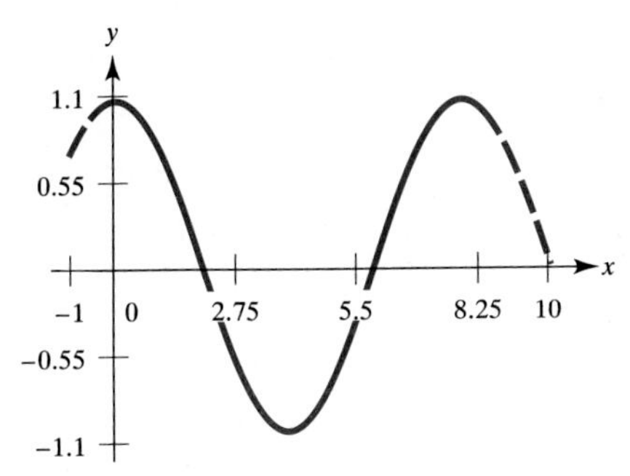

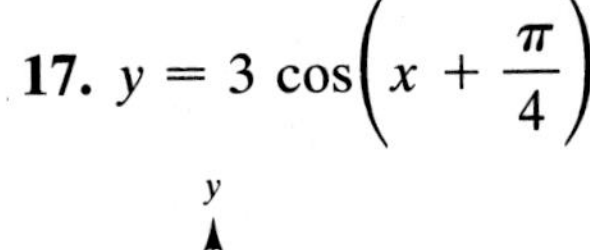

17. $y = 3 \cos\left(x + \frac{\pi}{4}\right)$

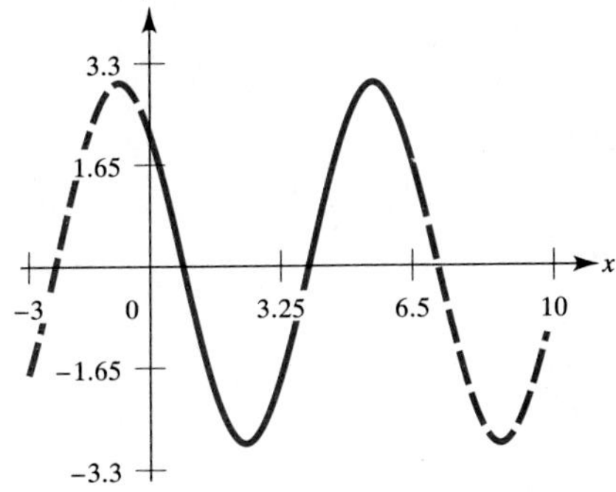

18. $y = \sin(x + \pi)$

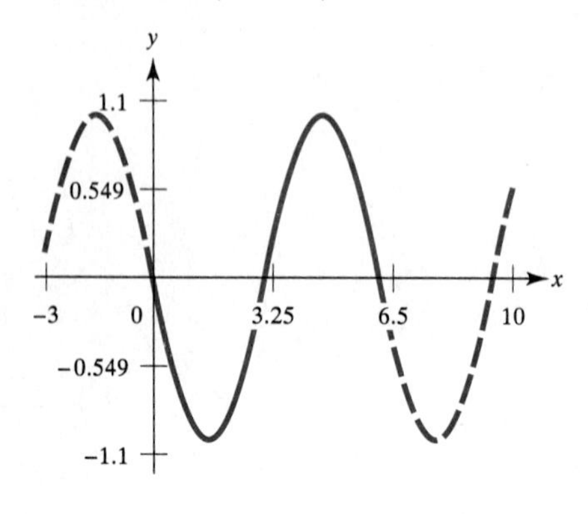

19. $y = 5\sin\left(x - \frac{\pi}{6}\right)$

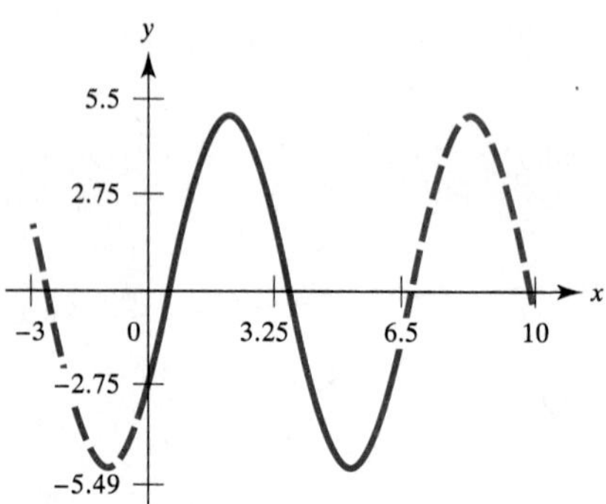

21. $y = 6\sin\left(\frac{x}{3} + \frac{\pi}{12}\right)$

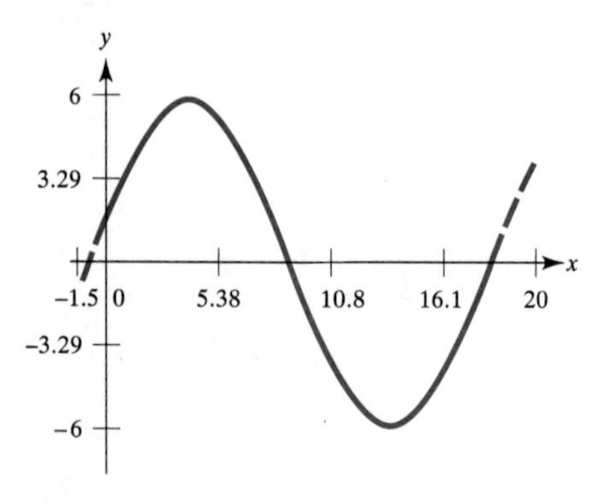

22. $y = 3\sin\left(\frac{x}{2} - \frac{\pi}{6}\right)$

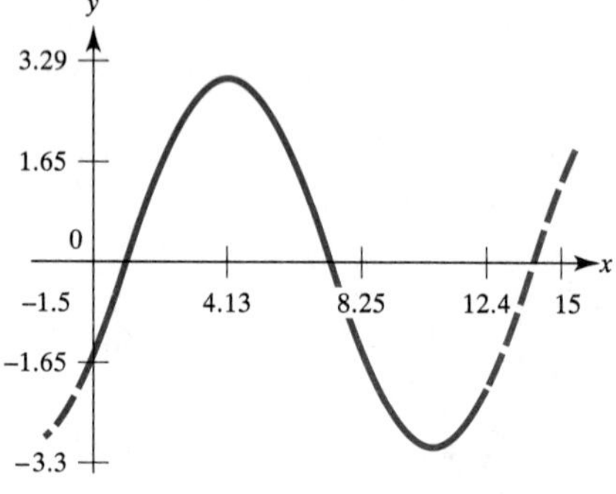

23. $y = 4\sin\left(3x - \frac{\pi}{3}\right)$

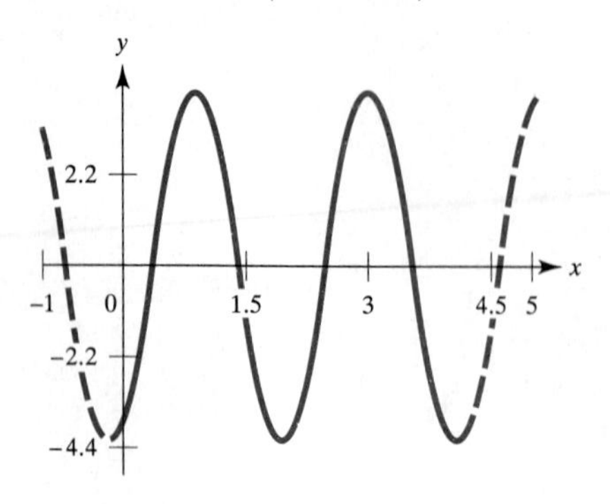

25. $y = 5\sin\left(2x + \frac{\pi}{3}\right)$

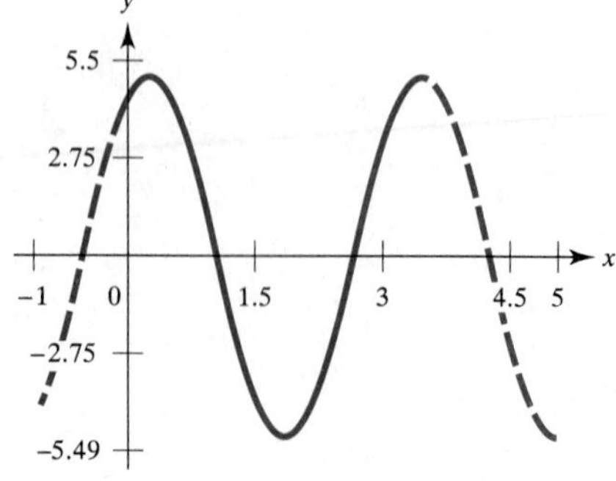

26. $y = -2 \sin x$

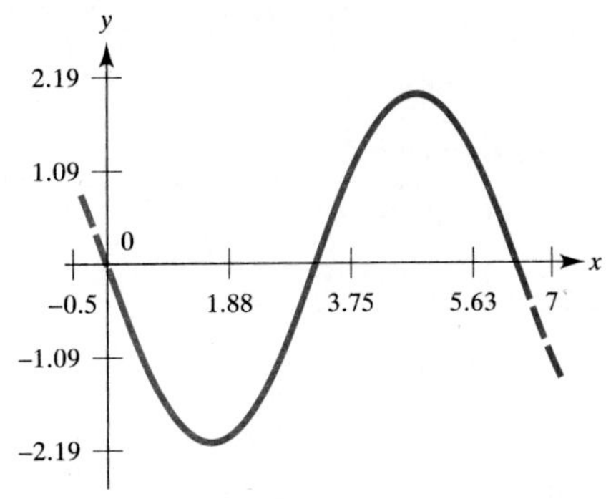

27. $y = -3 \cos 2x$

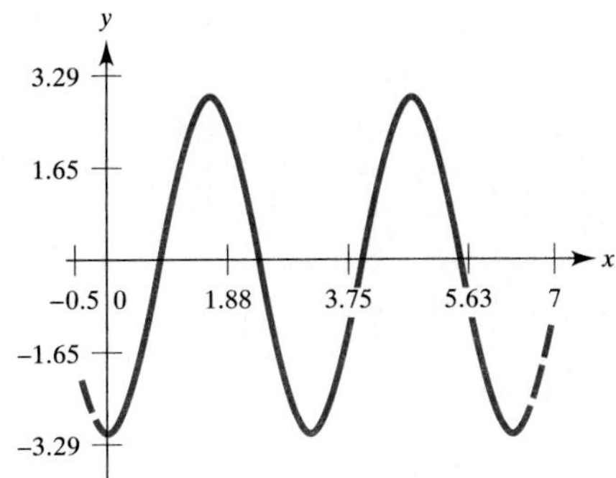

29. $y = 1 + \cos x$

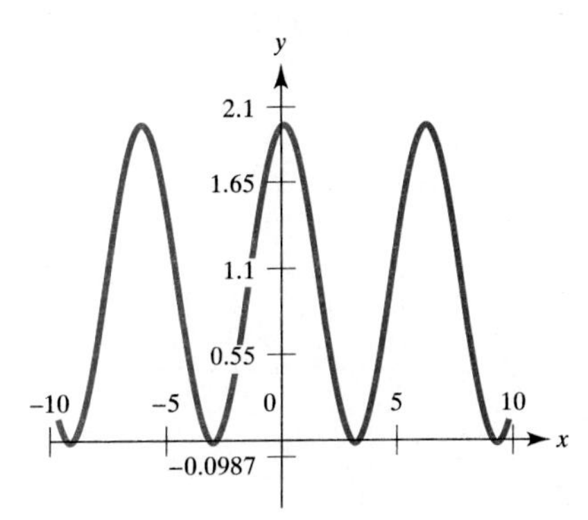

30. $y = -x + \sin x$

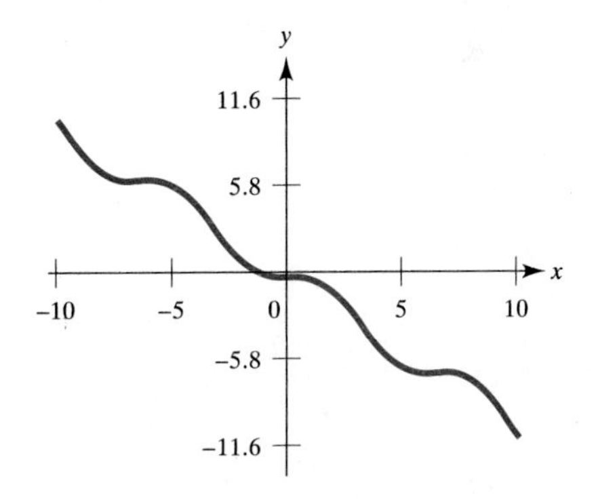

31. $y = \dfrac{x}{\pi} + \cos x$

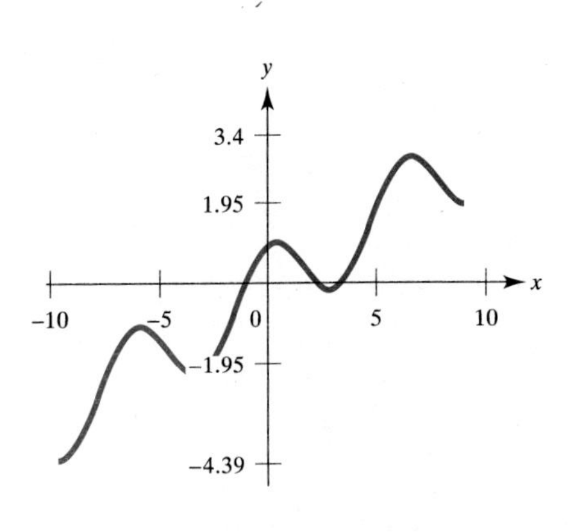

33. $y = \cos x + \cos 2x$

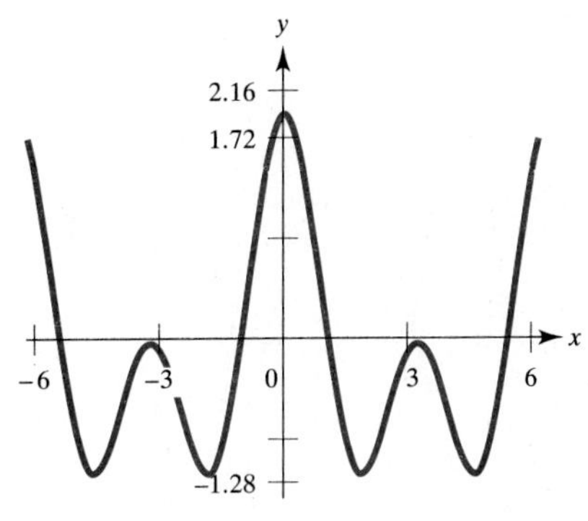

34. $y = \sin 2x + 2 \sin x$

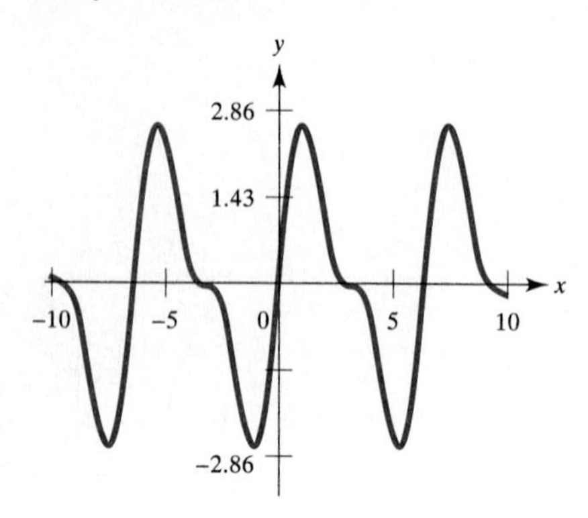

35. $y = \frac{1}{2} \cos 2x + \frac{1}{2}$

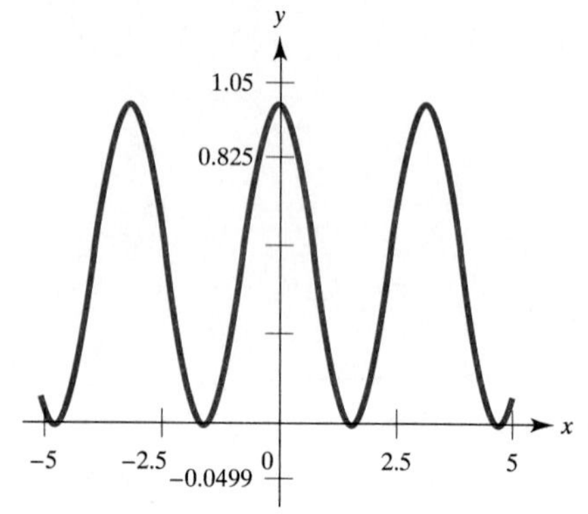

37. $y = \sin x + \sqrt{3} \cos x$

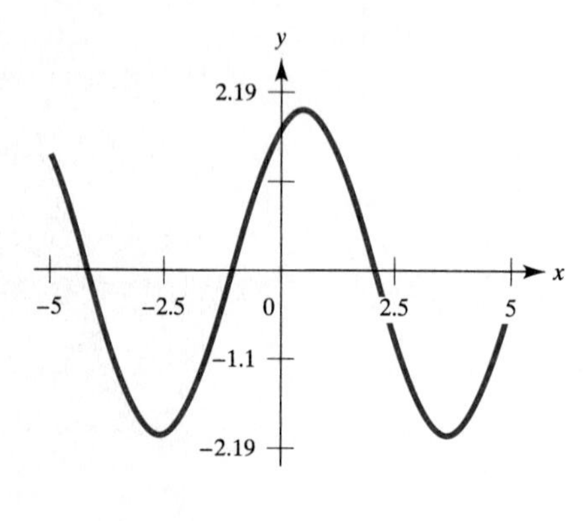

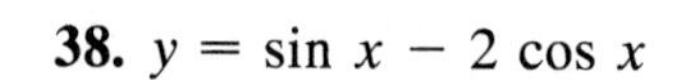

38. $y = \sin x - 2 \cos x$

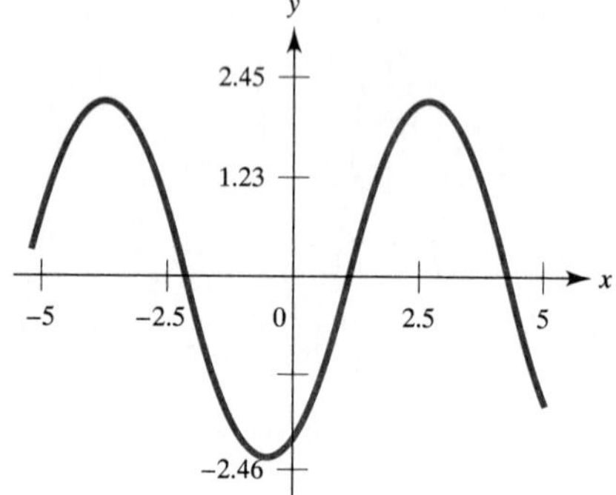

39. $y = \sin x + \cos x$

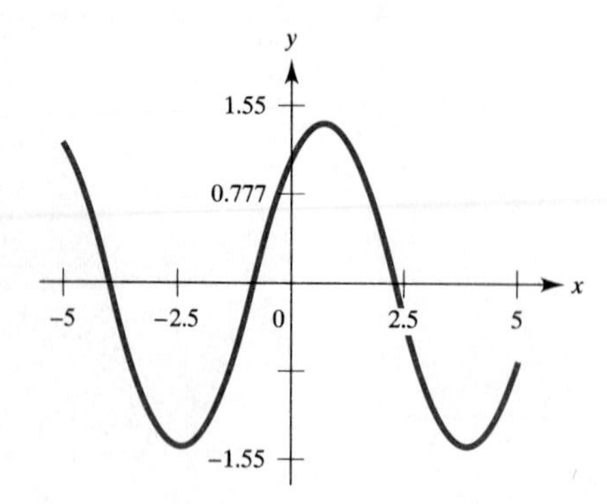

EXERCISE 8.3 PAGE 161

1. $y = 3.7\sin(x - 2.9) + \cos 3x$

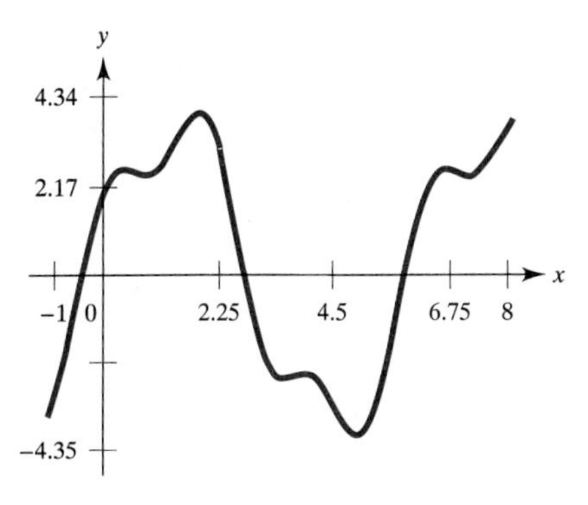

2. $y = \dfrac{x^2}{2} + 6\sin 3x$

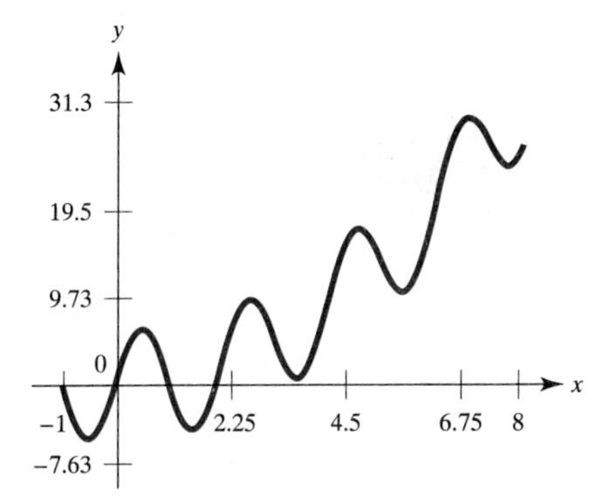

3. $y = \sin^3 x$

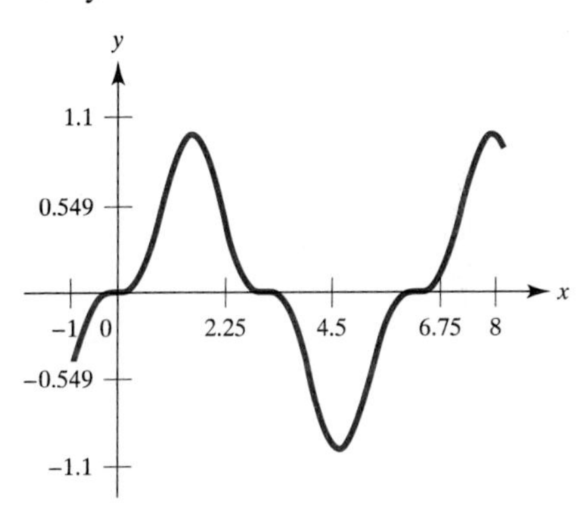

5. $y = -x + \sin 2x - \sin 4x$

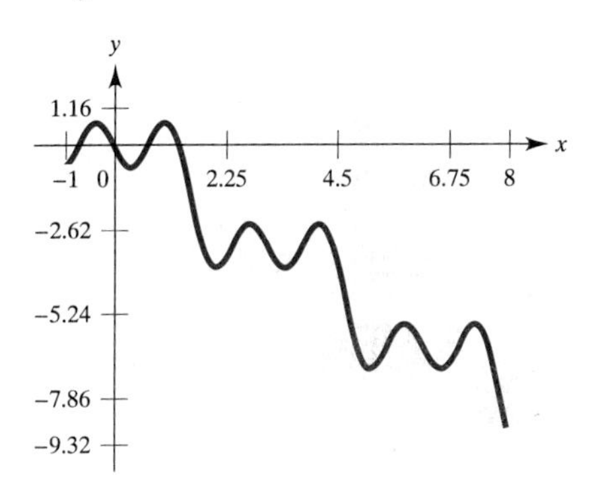

6. $y = \sin^3 x + \cos^3 x - \sin 3x$

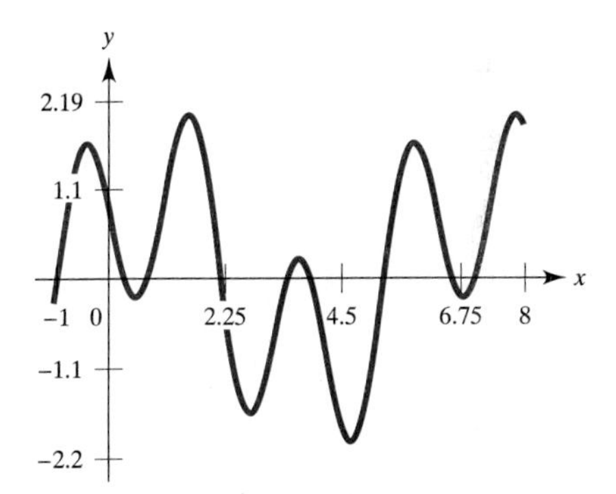

7. 0.5236, 2.6180, 4.7124

9. 0.6981, 1.3963, 2.7925, 3.4907, 4.8870, 5.5851

10. 4.9742, 6.0213

11. 0.8744

13. (a) 1.4669 is the only solution as seen on a graphing calculator.
(b) 0 is the only solution, but most calculators do not show this well.

REVIEW EXERCISES PAGES 161–162

2. $y = -2\sin\left(3\theta - \frac{\pi}{2}\right)$

Amp. 2, period $\frac{2\pi}{3}$,

phase shift is $\frac{\pi}{6}$.

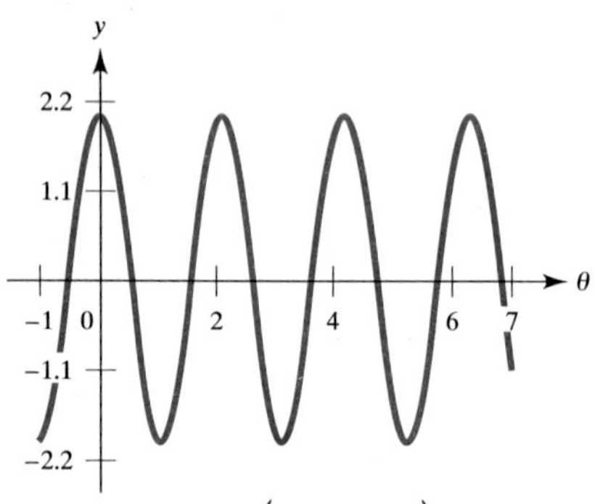

3. $y = -3\sin 2\theta$

Amp. 3, period π.

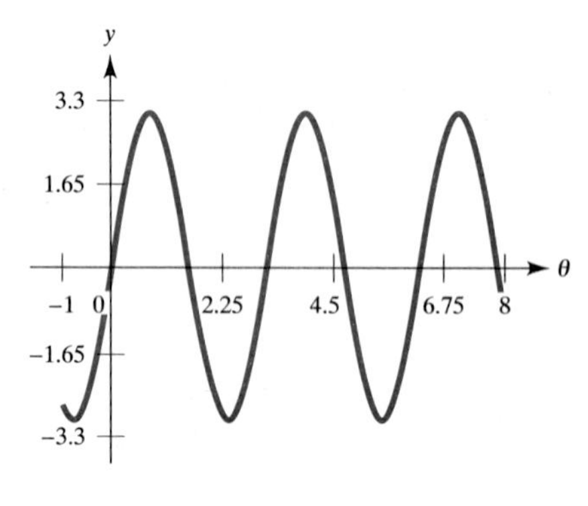

5. $y = -\cos\left(2\theta - \frac{\pi}{4}\right)$

Amp. 1, period π,

phase shift is $\frac{\pi}{8}$.

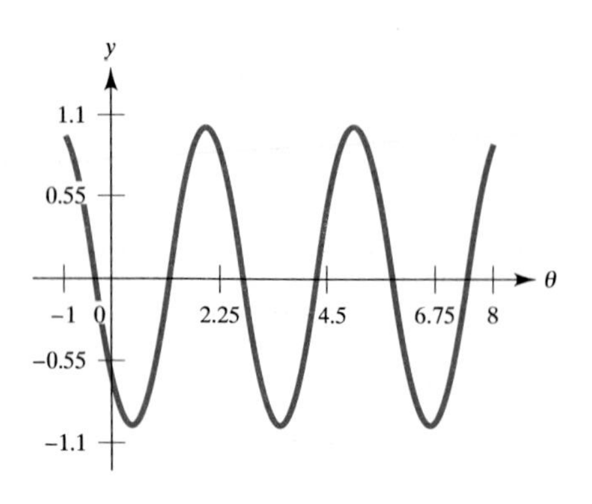

6. $y = -x + \cos x$

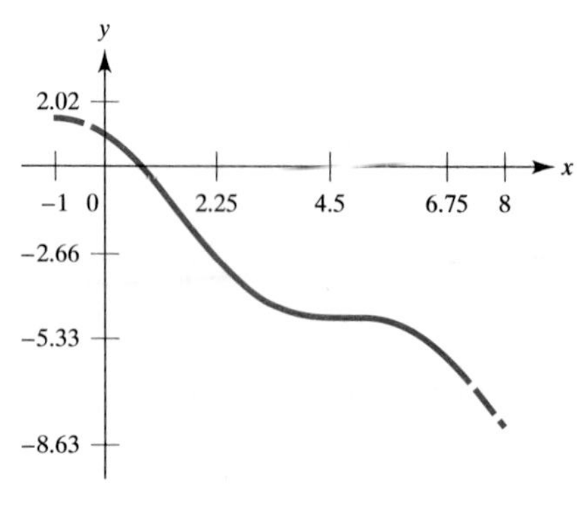

7. $y = 3 \sin x + 4 \cos x$

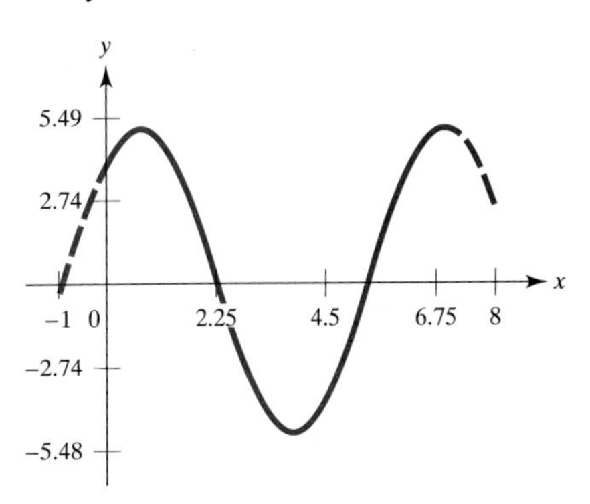

9. $y = 1 - 2 \sin^2 x$

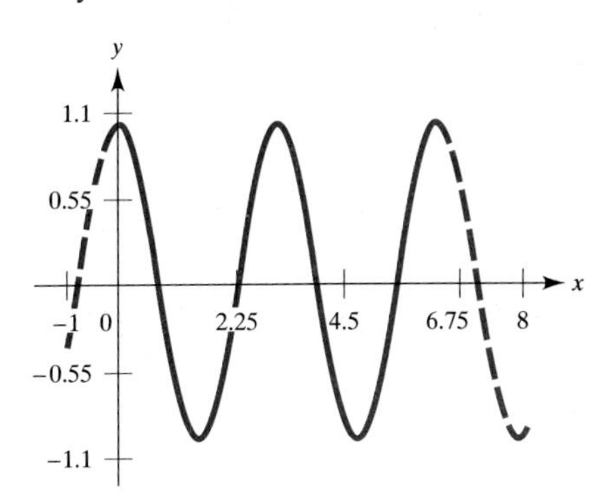

10. 0.9288, 0, −0.9288

11. 0.8246, −0.8246

EXERCISE 9.1 PAGE 164

1. $g(v) = \dfrac{9v - 8}{7}$

2. $g(v) = \dfrac{7v + 3}{v - 2}$

3. $g(v) = v^2 - 4v + 7,\ v \leq 2$

EXERCISE 9.2 PAGE 170

1. $\dfrac{\pi}{3}$ **2.** $-\dfrac{\pi}{6}$ **3.** $\dfrac{\pi}{4}$ **5.** π

6. $\dfrac{\pi}{6}$ **7.** $-\dfrac{\pi}{4}$ **9.** $\dfrac{\pi}{4}$ **10.** $-\dfrac{\pi}{2}$

11. $\dfrac{2\pi}{3}$ **13.** $\dfrac{5\pi}{6}$ **14.** $\dfrac{5\pi}{4}$ **15.** $\dfrac{\pi}{6}$

17. $\dfrac{\pi}{4}$ **18.** $\dfrac{\pi}{3}$ **19.** 0

21. 2.7122 **22.** −0.4555 **23.** −1.5010

25. 1.0681 **26.** 1.3771 **27.** 2.3038

EXERCISE 9.3 PAGES 173–176

1. $\dfrac{1}{u}$ **2.** $\sqrt{1 - u^2}$ **3.** $\dfrac{\sqrt{1 - u^2}}{u}$

5. $\dfrac{\sqrt{15}}{4}$ **6.** $-\dfrac{12}{5}$ **7.** $-\dfrac{3}{\sqrt{13}}$

9. $v\sqrt{1-u^2} - u\sqrt{1-v^2}$
10. $\sqrt{1-u^2}\sqrt{1-v^2} + uv$
11. $uv - \sqrt{1-u^2}\sqrt{1-v^2}$
13. $-\dfrac{\sqrt{11}}{6}$
14. $\dfrac{3 + 2\sqrt{30}}{14}$
15. $-\dfrac{4\sqrt{3} + \sqrt{2}}{10}$
17. $2u^2 - 1$
18. $2u\sqrt{1-u^2}$
19. $2u\sqrt{1-u^2}$
21. $2u$
22. $32u^2 - 1$
23. $\dfrac{5\pi}{6}$
25. $-\dfrac{\pi}{7}$
26. $-\dfrac{2\pi}{5}$
27. $\dfrac{\pi}{3}$
29. $\dfrac{\pi}{9}$
30. $-\dfrac{\pi}{2}$
31. $u, \sqrt{1-u^2}$
33. $\dfrac{u+1}{\sqrt{u^2+2u+5}}, \dfrac{u+1}{2}$
34. $\dfrac{6}{\sqrt{u^2-36}}, \dfrac{u}{6}$
45. True
46. True
47. False
49. True
50. True
51. True
53. False
54. False
55. True
57. $\dfrac{1}{\sqrt{13}}$
58. 0, 1
59. $\dfrac{\sqrt{3}}{2}$
61. $\dfrac{\sqrt{3}}{2}$
62. $\frac{1}{2}$, 1
63. The equation is an identity. It holds true for all permissible values of x, that is, all x except 1 and -1.

REVIEW EXERCISES PAGE 176

1. $g(u) = \dfrac{3-u}{2}$
2. $-0.5692, 1.5703$
3. $-\dfrac{\pi}{4}, \dfrac{3\pi}{4}$
5. $-\dfrac{\pi}{14}, 2v^2 - 1$
6. $v\sqrt{1-u^2} + u\sqrt{1-v^2}$
7. Yes

EXERCISE 10.1 PAGES 180–181

1. 0
2. 5
3. 1
5. $-\frac{1}{4}$
6. -2
7. 1492
9. 10
10. e
11. $\frac{1}{4}$
13. 1776
14. e
15. 0
17. $\ln r = q$
18. $\log 0.001 = -3$
19. $\log_2 1024 = 10$

21. $a^{-1/2} = \frac{1}{3}$
22. $e^x = N$
23. $10^{-4} = 0.0001$
25. $\log(x^2 - 49)$
26. $\log \dfrac{a^3b^2}{\sqrt{c}}$
27. $\log\left(2\pi\sqrt{\dfrac{n}{m}}\right)$
29. True
30. True
31. False
33. False
34. True
35. True
37. True
38. True

EXERCISE 10.2 PAGES 184–185

1. 1
2. 0.41078
3. 0.36788
5. 0.9168
6. e
7. 4.8105
9. 1.1475
10. −1.1475
11. 0.79898
13. e
14. 1
15. $y = e^{\sin x}$

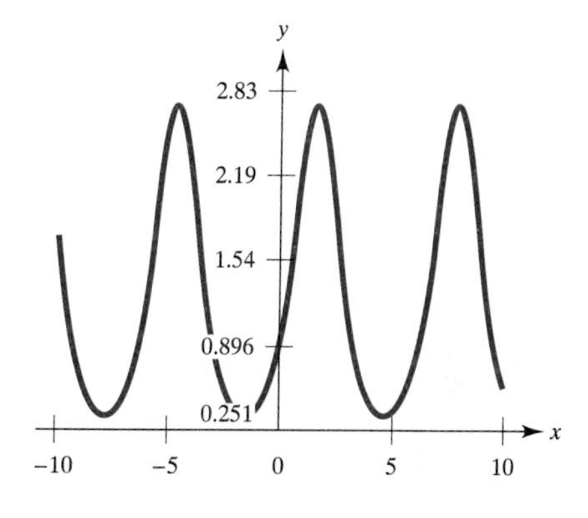

17. $y = \sin e^x$

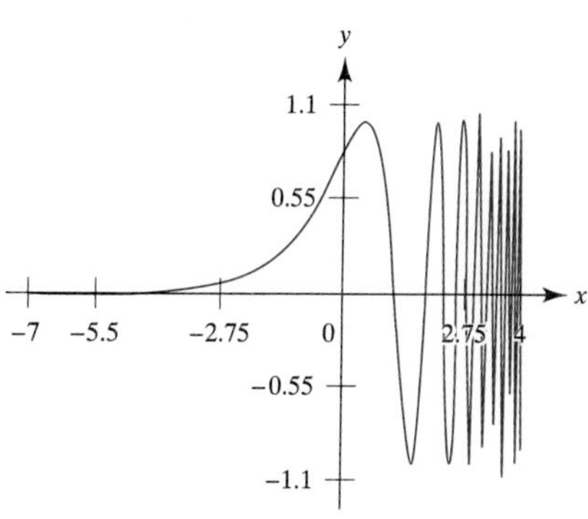

18. $y = \cos e^x$

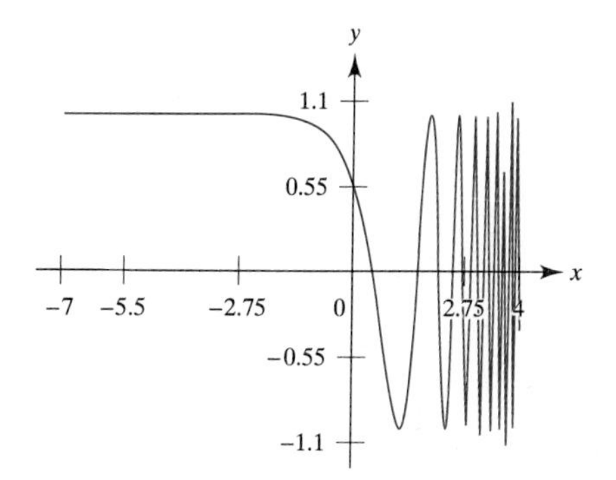

19. $y = -e^{-x}$

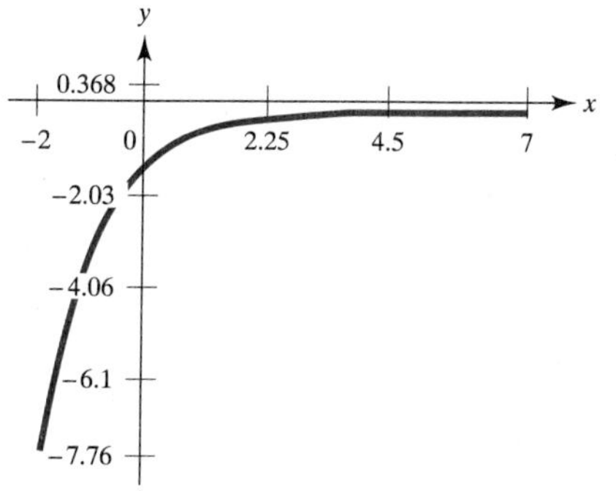

21. (a) 92.2 mg (b) 259
22. (a) 3030 (b) 3.170

EXERCISE 10.3 PAGES 187–188

1. -1 **2.** $3, -1$ **3.** No solution

5. $7, 9$ **6.** 2 **7.** 25

9. $\dfrac{3 \log a + 5 \log c}{4 \log c - 2 \log a}$

10. $\dfrac{b^{2y} - 1}{b^{2y} + 1}$ **11.** $x = 10^{10^y}$

13. $\ln \pi$ **14.** $\dfrac{\ln 0.482}{\ln 5.37} = -0.434$

15. $y = \ln x$
$y = e^x$

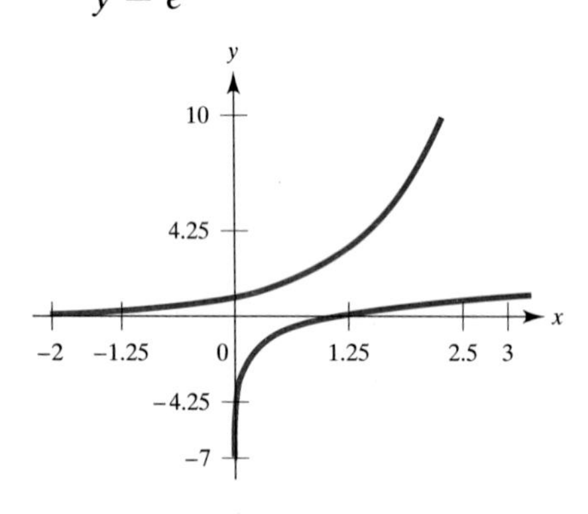

17. $y = e^{-x}$
$y = e^{x^2}$

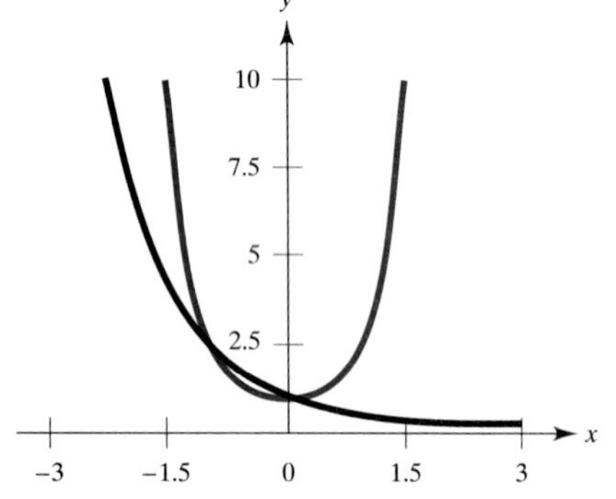

18. $y = e^x + \cos x$
$y = e^x - \ln x$

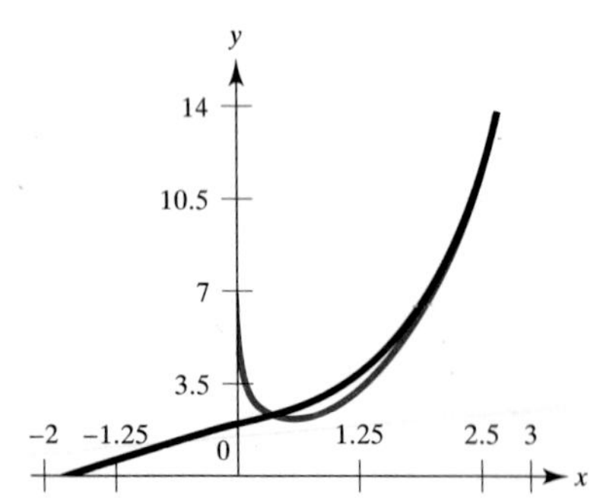

REVIEW EXERCISES PAGE 188

1. $\frac{1}{9}$ **2.** -4

3. 0.0001 **5.** 2.032

6. 4.702

7. $y = e^{-x}$
$y = -e^{x}$

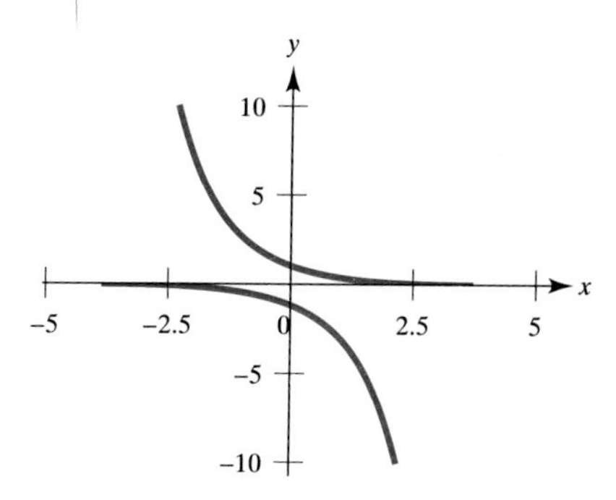

EXERCISE 11.1 PAGES 191–192

1. $x^2 + y^2 = 25$

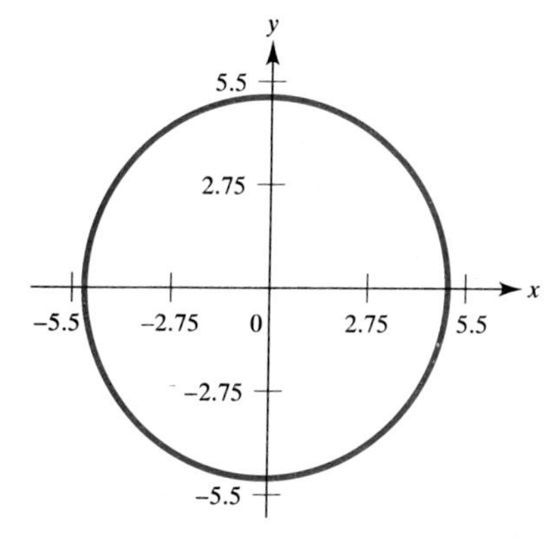

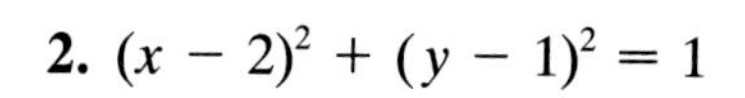

2. $(x - 2)^2 + (y - 1)^2 = 1$

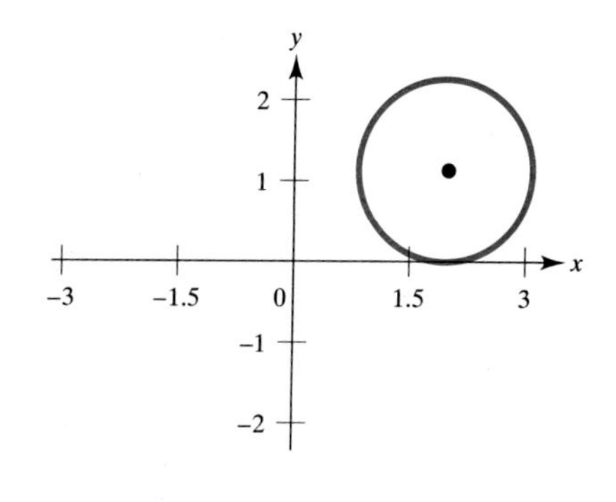

3. $\dfrac{x}{2} + \dfrac{y}{3} = 1$

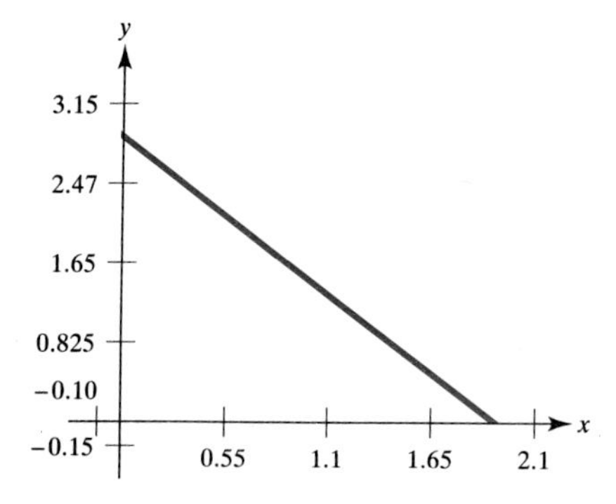

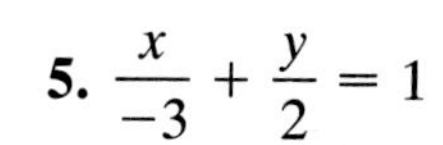

5. $\dfrac{x}{-3} + \dfrac{y}{2} = 1$

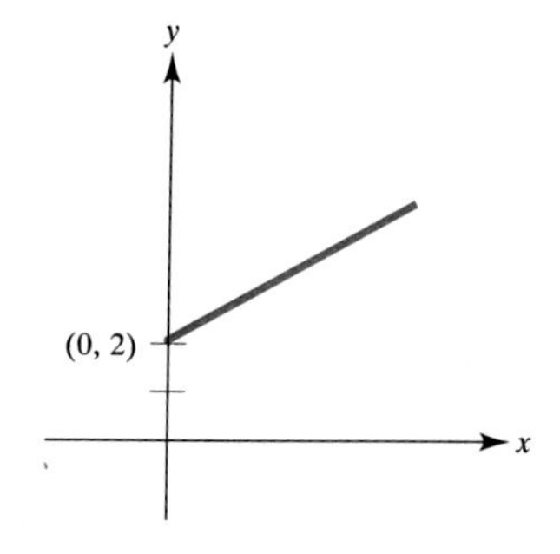

6. $\dfrac{x^2}{4} + \dfrac{y^2}{9} = 1$

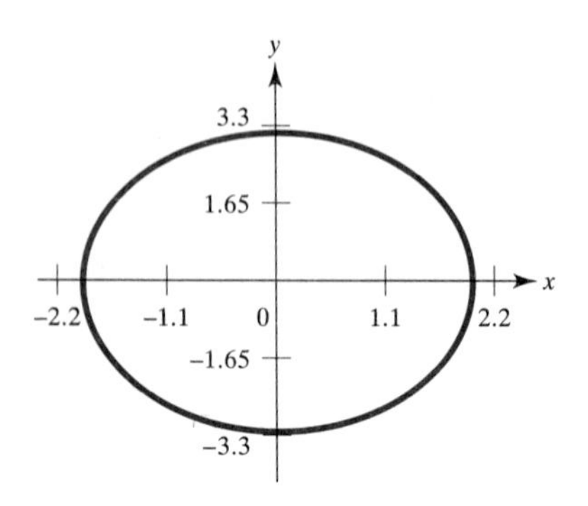

7. $x = \theta^2, y = \sin\theta$
$\theta \le 0 \le 2\pi$

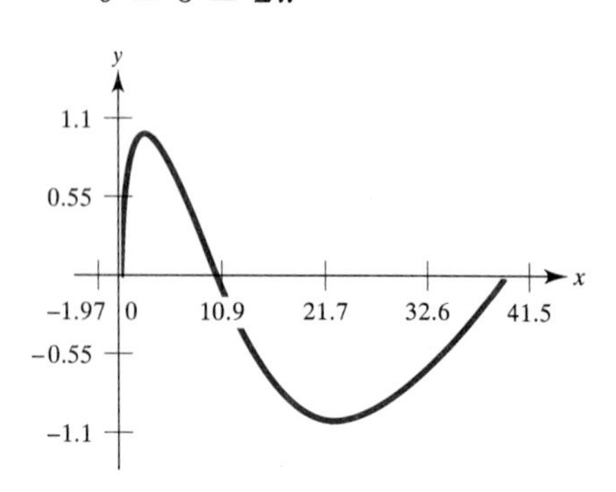

9. $x = \theta \sin\theta, y = \cos\theta$
$0 \le \theta \le 4\pi$

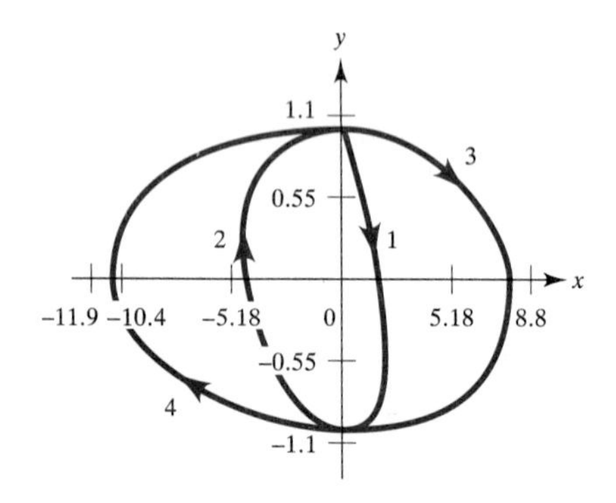

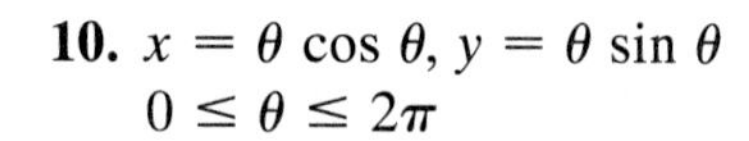

10. $x = \theta \cos\theta, y = \theta \sin\theta$
$0 \le \theta \le 2\pi$

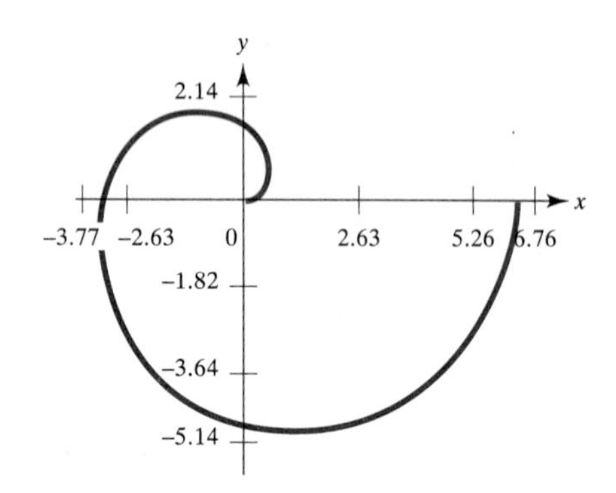

11. It travels 3200 ft, reaching a height of 800 ft.

EXERCISE 11.2 PAGES 196–197

1. $B\left(1, \dfrac{\pi}{2}\right)$, $C(2, \pi)$, $D\left(3, \dfrac{5\pi}{4}\right)$, $E\left(1, \dfrac{3\pi}{2}\right)$, $F\left(3, \dfrac{3\pi}{2}\right)$, $G\left(2, \dfrac{7\pi}{4}\right)$, $H(2, 0)$.

2. $B\left(1, \dfrac{\pi}{2}\right)$, $C(2, -\pi)$, $D\left(3, -\dfrac{3\pi}{4}\right)$, $E\left(1, -\dfrac{\pi}{2}\right)$, $F\left(3, -\dfrac{\pi}{2}\right)$, $G\left(2, -\dfrac{\pi}{4}\right)$, $H(2, 0)$.

3. $B\left(1, \frac{\pi}{2}\right)$, $C(-2, 0)$, $D\left(-3, \frac{\pi}{4}\right)$, $E\left(-1, \frac{\pi}{2}\right)$, $F\left(-3, \frac{\pi}{2}\right)$, $G\left(2, -\frac{\pi}{4}\right)$, $H(2, 0)$.

17. $\left(\frac{\sqrt{3}}{2}, -\frac{1}{2}\right)$ **18.** $(-3, -3)$ **19.** $\left(\frac{\sqrt{6}}{2}, \frac{-\sqrt{2}}{2}\right)$

21. $(-\sqrt{2}, \sqrt{2})$ **22.** $\left(5, \frac{\pi}{2}\right)$ **23.** $(5, 0)$

25. $\left(2, \frac{11\pi}{6}\right)$ **26.** $(5, 3.785)$ **27.** $(13, 4.318)$

EXERCISE 11.3 PAGE 199

1. $r = 3$

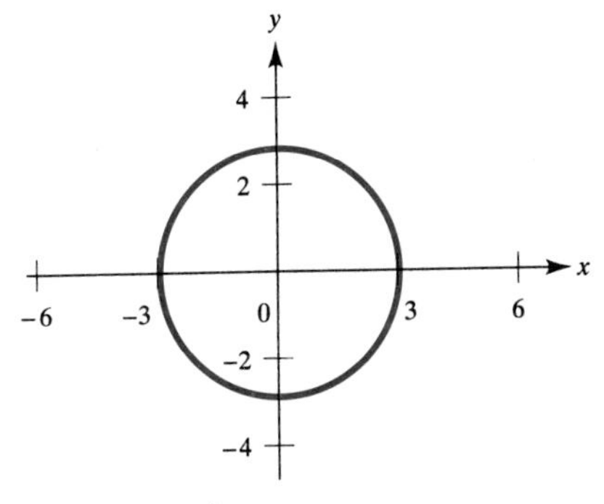

2. $r = -2$

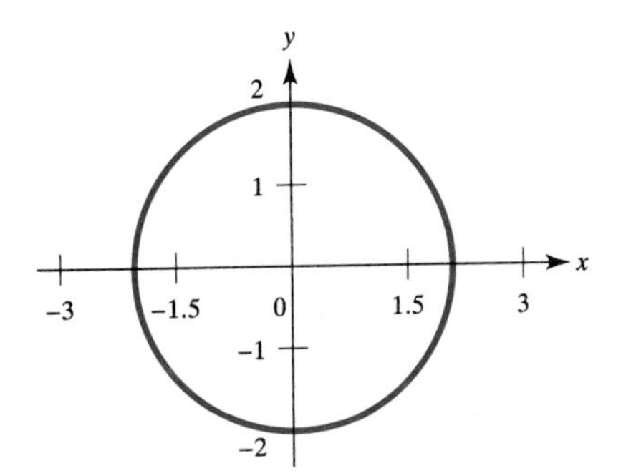

3. $r = \frac{4}{\sin\theta}$

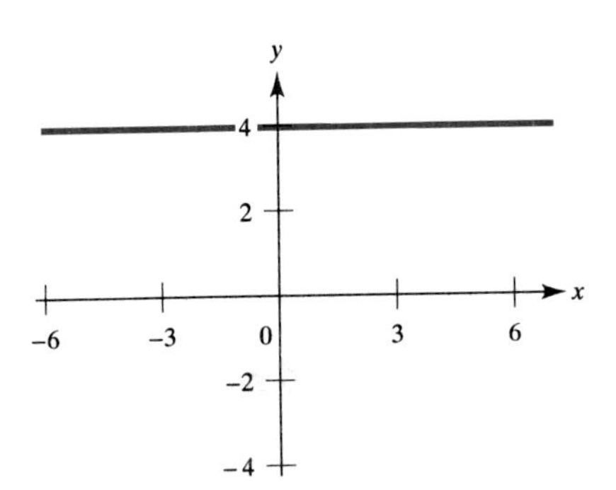

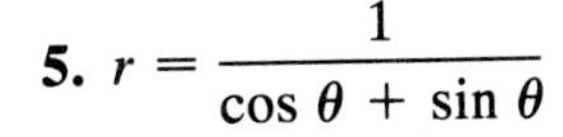

5. $r = \frac{1}{\cos\theta + \sin\theta}$

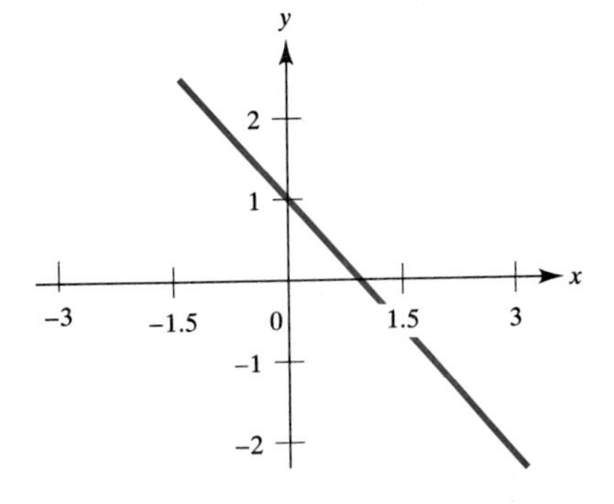

6. $r = \cos \theta$

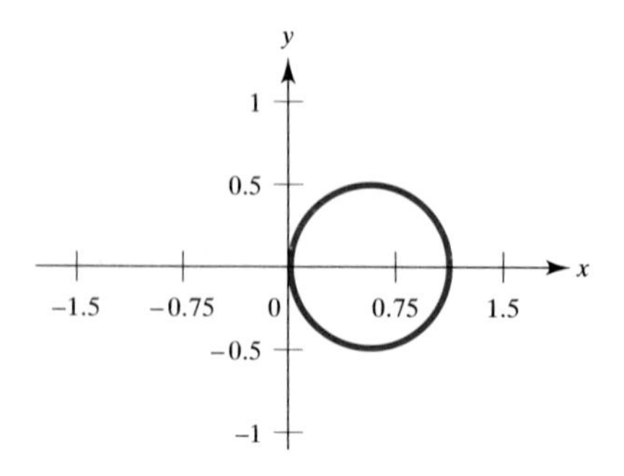

7. $r = 2 \sin \theta$

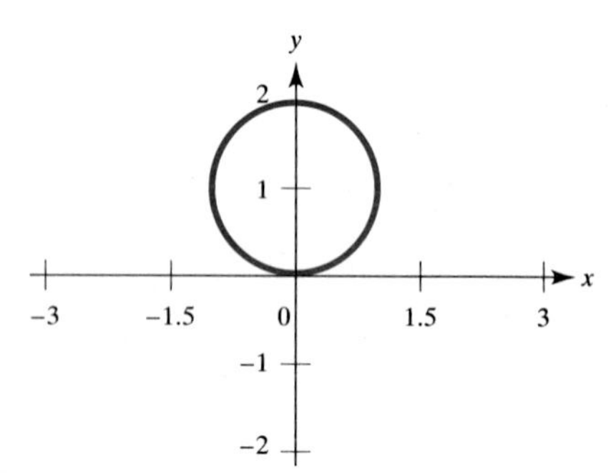

9. $r = \cos 2\theta$

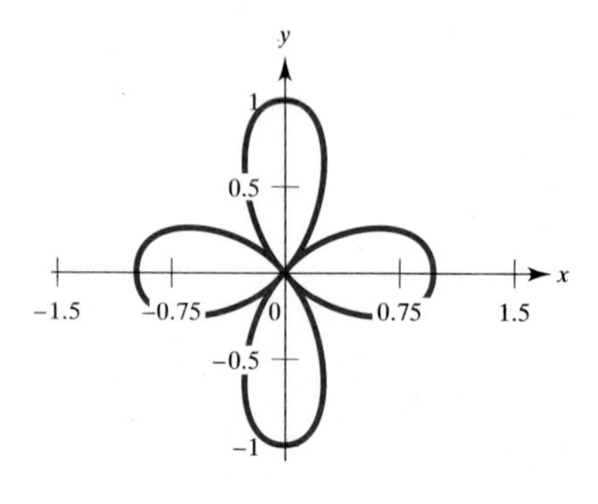

10. $r = \sin 3\theta$

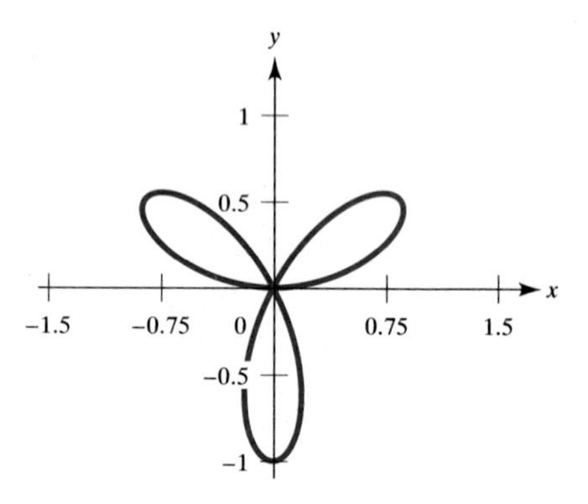

11. $r = \sin \dfrac{\theta}{2}$

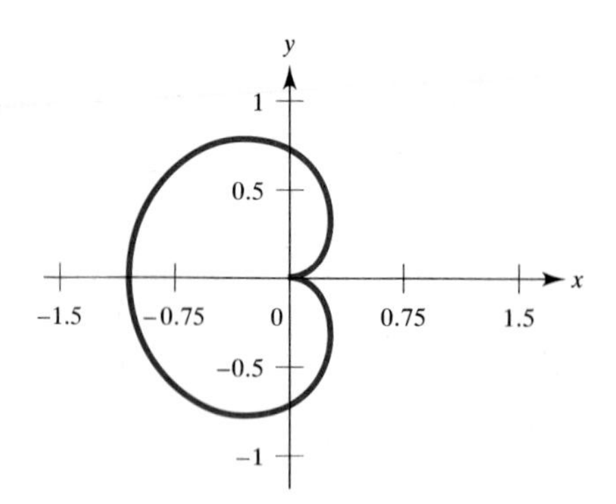

13. $r = \sin \dfrac{3\theta}{2}$

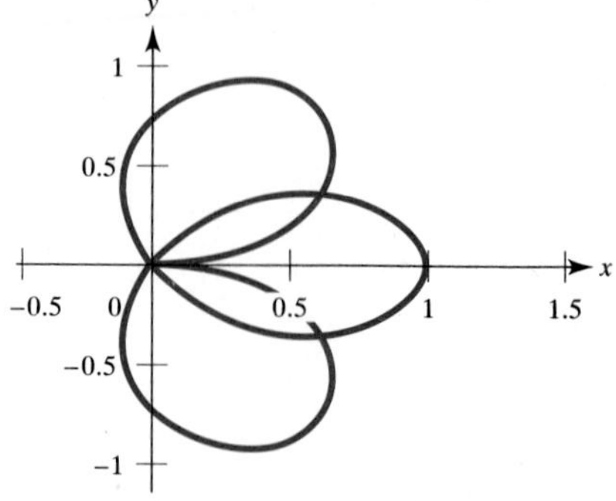

14. $r = 2 + 3 \cos \theta$

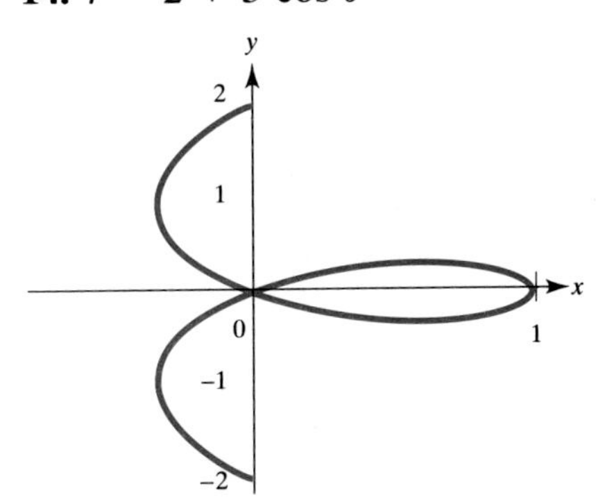

15. $r = \sin^2 \theta + \cos^2 \theta$

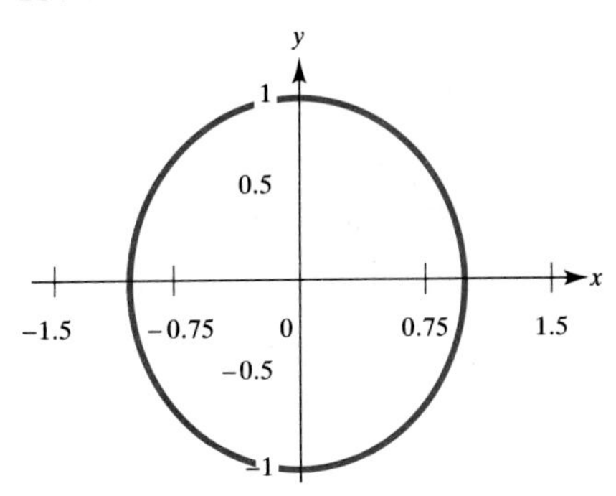

REVIEW EXERCISES PAGE 199

1. $x = 2 \sin \theta \cos \theta$
$y = \cos^2 \theta - \sin^2 \theta$

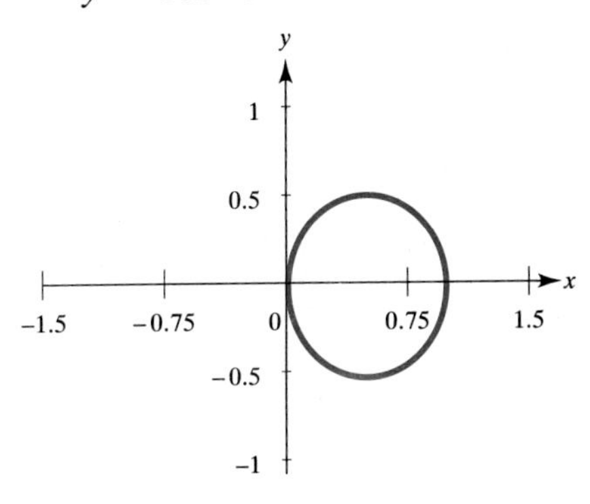

2. $x = \cos^2 \theta$
$y = \cos \theta \sin \theta$

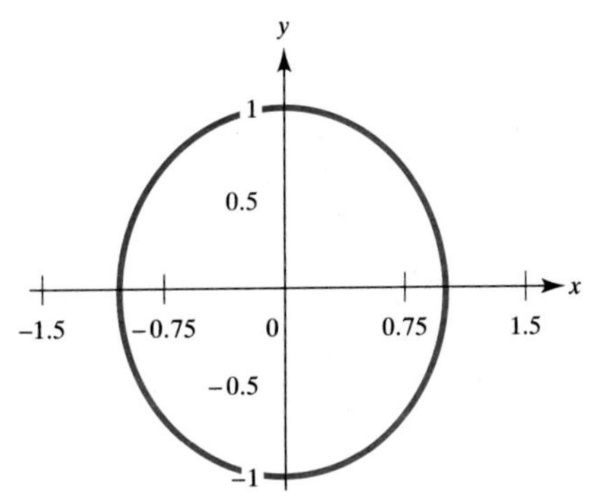

3. $x = \tan \theta$
$y = \sec \theta$
$x^2 - y^2 = 1$ (hyperbola)

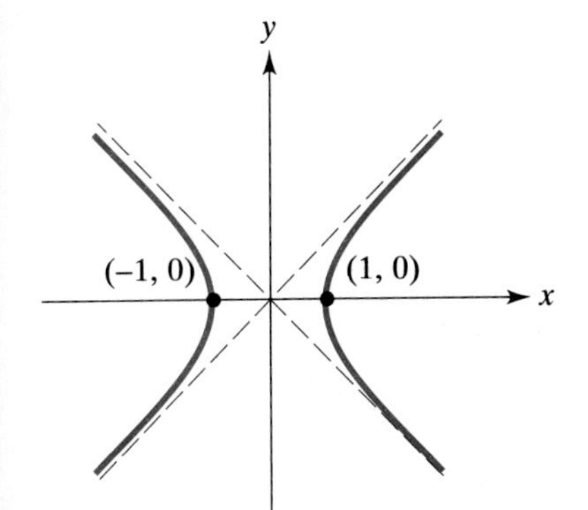

5. $\left(2\sqrt{2}, \frac{5\pi}{4}\right)$

6. $\left(4, \frac{\pi}{2}\right)$

7. $\left(3, \frac{3\pi}{2}\right)$

9. $(\sqrt{2}, \pi)$

10. $\left(\sqrt{6}, \frac{3\pi}{4}\right)$

11. $r = 2 \cos \theta$

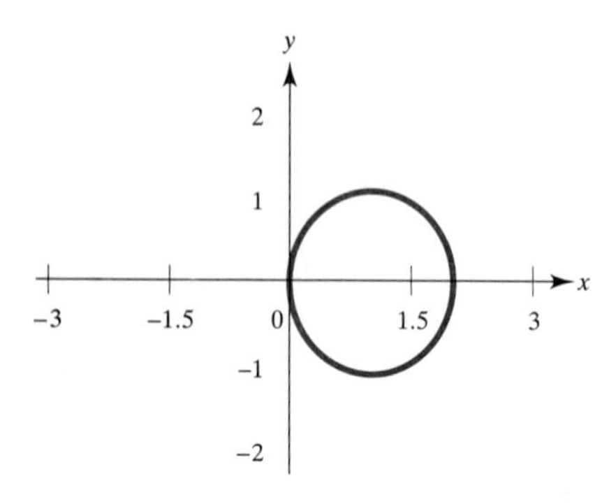

13. $r = \theta$

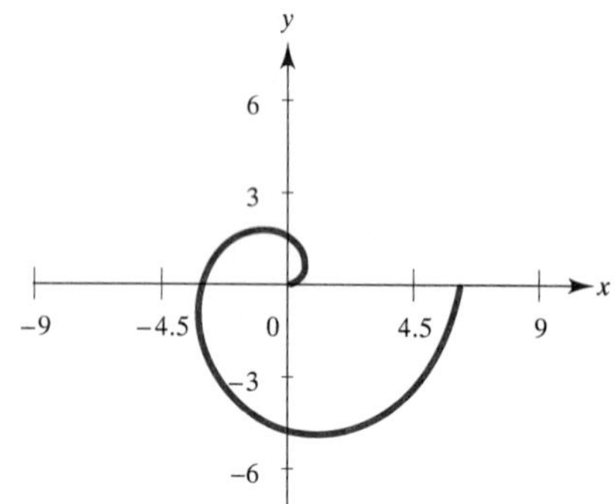

14. $r = \sin 2\theta$

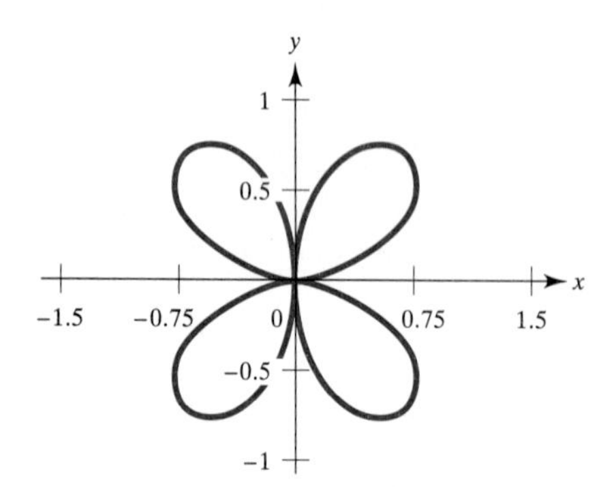

15. $r = \cos 2\theta$

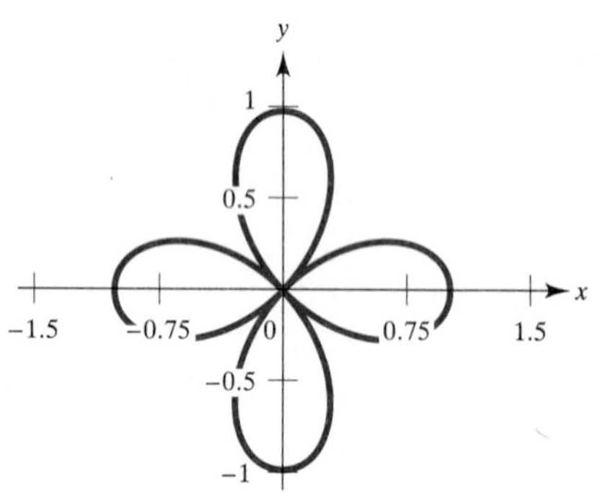

17. The ball will be about 3 ft 8 in. above the ground.

EXERCISE 12.1 PAGES 202–203

1. $8 - 2i$ **2.** $8 - 7i$ **3.** -6
5. $26 + 19i$ **6.** $-12 + 31i$ **7.** $44 + 21i\sqrt{5}$
9. $-55 - 48i$ **10.** $35 + 12i$ **11.** 1
13. 0 **14.** $-2i$ **15.** $-\frac{26}{37} + \frac{29}{37}i$
17. $\frac{31}{13} + \frac{14}{13}i$ **18.** $\frac{4}{53} - \frac{67}{53}i$ **19.** $3 + \frac{5}{2}i$
21. $x = \frac{1}{2}, y = -3$ **22.** $x = 2, y = 8$
23. $x = -4, y = 1; x = -\frac{1}{5}, y = 20$ **25.** 0

EXERCISE 12.2 PAGES 207–208

1. $2 + 6i$ **2.** $2 - 4i$ **3.** $-7 - 2i$
5. $7 + 3i$ **6.** $-4 - 5i$ **7.** $6 + 5i$

9. $2(\cos 3\pi/4 + i \sin 3\pi/4)$
10. $3\sqrt{2}(\cos 5\pi/4 + i \sin 5\pi/4)$
11. $4\sqrt{2}(\cos \pi/4 + i \sin \pi/4)$
13. $7(\cos 0 + i \sin 0)$
14. $6(\cos 3\pi/2 + i \sin 3\pi/2)$
15. $9(\cos \pi/2 + i \sin \pi/2)$
17. $10(\cos 3.785 + i \sin 3.785)$
18. $17(\cos 0.490 + i \sin 0.490)$
19. $2(\cos 11\pi/6 + i \sin 11\pi/6)$
21. $3 - 3i\sqrt{3}$
22. $-2\sqrt{3} + 2i$
23. $-9.397 - 3.420i$
25. (a) Equation (3) demands a positive sign in front of $i \sin \theta$.
(b) $2(\cos \pi/6 - i \sin \pi/6) = 2(\cos 11\pi/6 + i \sin 11\pi/6)$.
27. (a) 0, (b) π, (c) $\pi/2$, (d) $3\pi/2$

EXERCISE 12.3 PAGES 213–214

1. $-15 + 15i\sqrt{3}$
2. $-4\sqrt{2} - 4i\sqrt{2}$
3. $3\sqrt{3} + 3i$
5. $5 + 5i\sqrt{3}$
6. $24 - 8i\sqrt{3}$
7. $36\sqrt{2}$
9. -48
10. $20 - 20i\sqrt{3}$
11. $-2i$
13. -32
14. $-4\sqrt{3} + 4i$
15. -324
17. $128 - 128i$
18. $1024i$
19. $-\dfrac{1}{2} + \dfrac{i\sqrt{3}}{2}$
21. $-64\sqrt{3} - 64i$
22. $-8432 + 5376i$
23. $6.549 + 7.557i, -9.819 + 1.893i, 3.271 - 9.450i$
25. $2 + 2i, -2 - 2i$
26. $1 - i, -1 + i$
27. $-\sqrt{3} + i, \sqrt{3} - i$
29. $2\sqrt{3} + 2i, -2\sqrt{3} + 2i, -4i$
30. $\sqrt{2} + i\sqrt{2}, -1.932 + 0.518i, 0.518 - 1.932i$
31. $2\sqrt{2} + 2i\sqrt{2}, -2\sqrt{2} + 2i\sqrt{2}, -2\sqrt{2} - 2i\sqrt{2}, 2\sqrt{2} - 2i\sqrt{2}$
33. $1, 0.309 + 0.951i, -0.809 + 0.588i, -0.809 - 0.588i,$
$0.309 - 0.951i$
34. $1 + i\sqrt{3}, -\sqrt{3} + i, -1 - i\sqrt{3}, \sqrt{3} - i$
35. $10, 5 + 5i\sqrt{3}, -5 + 5i\sqrt{3}, -10, -5 - 5i\sqrt{3}, 5 - 5i\sqrt{3}$
37. $0.951 + 0.309i, 0.588 + 0.809i, i, -0.588 + 0.809i,$
$-0.951 + 0.309i, -0.951 - 0.309i, -0.588 - 0.809i, -i,$
$0.588 - 0.809i, 0.951 - 0.309i$
38. $1.618 + 1.176i, -0.618 + 1.902i, -2, -0.618 - 1.902i,$
$1.618 - 1.176i$

REVIEW EXERCISES PAGE 215

1. $(5 - \pi) + (1 - 2\sqrt{3})i$
2. $2\sqrt{3}$
3. 5
5. 0
6. i
7. -15

9. $16 + 16i\sqrt{3}$

10. $2\sqrt{2}(\cos 7\pi/4 + i \sin 7\pi/4)$

11. $6(\cos 3\pi/2 + i \sin 3\pi/2)$

13. $\sqrt{2}(\cos \pi/4 + i \sin \pi/4)$

14. $1.070 + 0.213i$
$-0.213 + 1.070i$
$-1.070 - 0.213i$
$0.213 - 1.070i$

15. $-0.174 + 0.985i$
$-0.766 - 0.643i$
$0.940 - 0.342i$

EXERCISE 13.1 PAGES 219–220

2. $y - b = \left(\dfrac{d - b}{c - a}\right)(x - a)$

3. $x = e, y = \pi$

5. $y + 6 - \pi = -2(x + e - 1)$

6. $y = -\pi x - 1$

7. $y - 6 = \frac{3}{7}(x - 5)$

9. $x = -12, x = 4$

10. $y + 6 = -7(x - 5)$

11. $y + 6 = \frac{1}{7}(x - 5)$

13. $88.4°$

EXERCISE 13.2 PAGES 223–224

1. $(y - 1)^2 = 8(x - 4)$
2. $(x - 3)^2 = 8(y - 2)$
3. $(y + 3)^2 = -8(x - 4)$
5. $(x + 1)^2 = 2(y + 2)$
6. $y = x^2$
7. $-x = y^2$
9. $x = y^2$
10. $y^2 = -8(x + 1), V(-1, 0), F(-3, 0)$

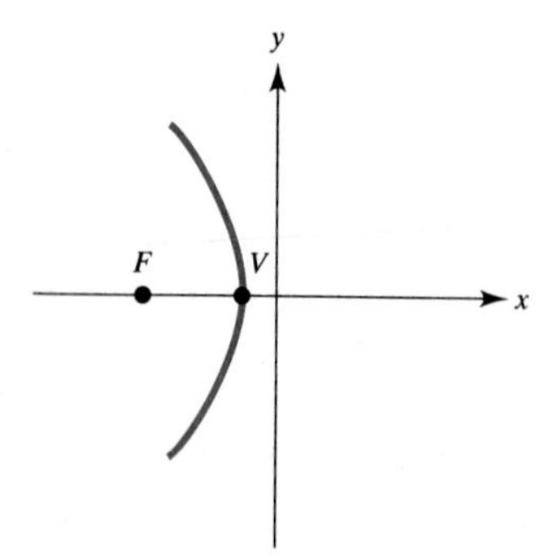

11. $y^2 = 12(x + 4)$, $V(-4, 0)$, $F(-1, 0)$

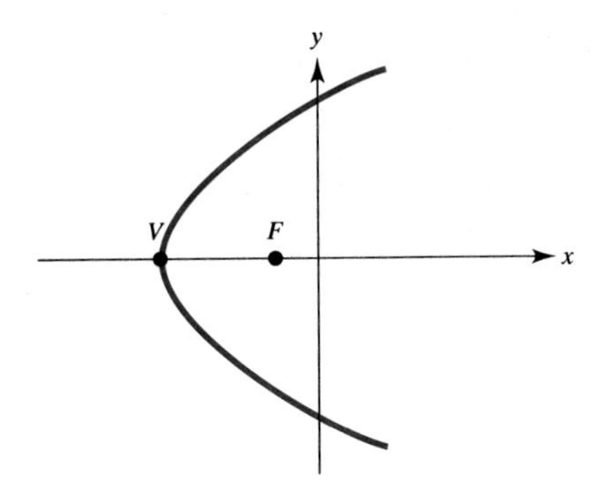

13. $(y - 2)^2 = -8(x - 4)$, $V(4, 2)$, $F(2, 2)$

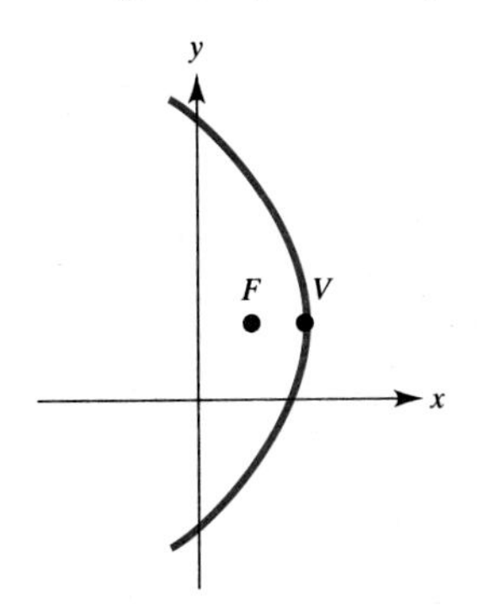

14. $(x - 4)^2 = 6(y + 4)$, $V(4, -4)$, $F(4, -\frac{5}{2})$

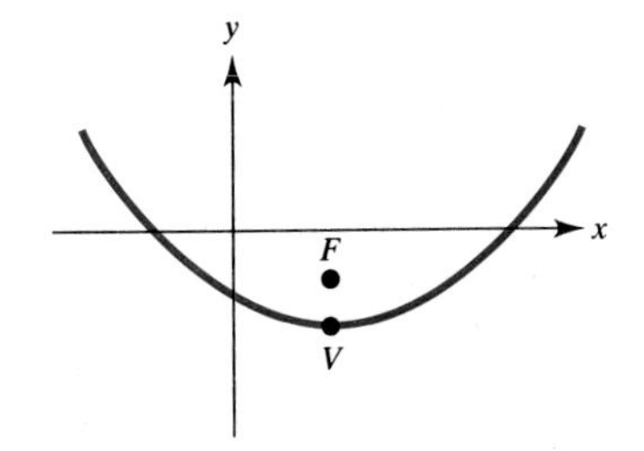

15. At 10:00 P.M. they are 18 miles apart. The southbound ship is N 33.7° W of the eastbound ship at that time.

EXERCISE 13.3 PAGES 227–228

1. $F(\pm\sqrt{5}, 0)$, $C(0, 0)$, $V(\pm 3, 0)$

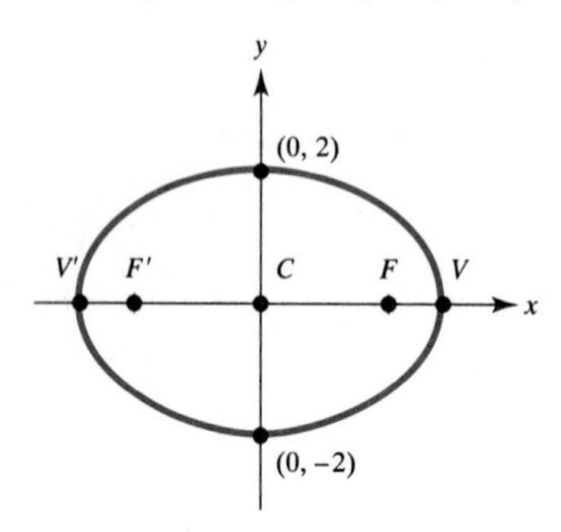

2. $F(0, \pm\sqrt{5})$, $C(0, 0)$, $V(0, \pm 3)$

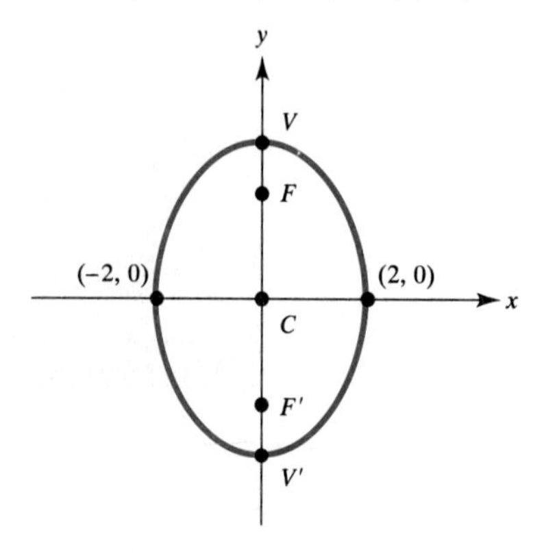

3. $F(3 \pm \sqrt{7}, 2)$, $C(3, 2)$, $V(3 \pm 4, 2)$

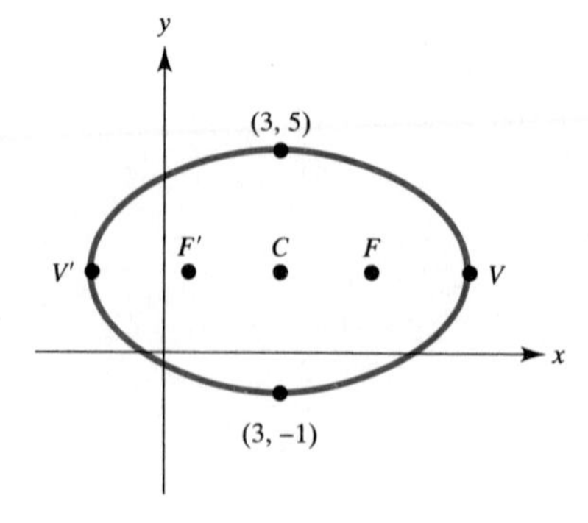

5. $F(1, -3 \pm \sqrt{7}),\ C(1, -3),\ V(1, -3 \pm 4)$

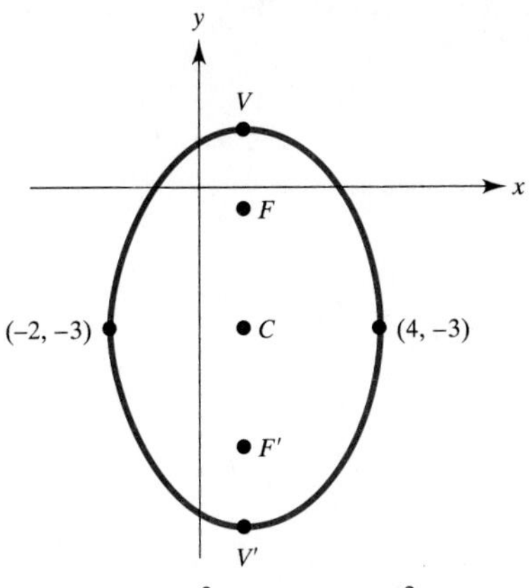

6. $\dfrac{(x+1)^2}{4} + \dfrac{(y-2)^2}{16} = 1,\ F(-1, 2 \pm \sqrt{12}),\ C(-1, 2),\ V(-1, 2 \pm 4)$

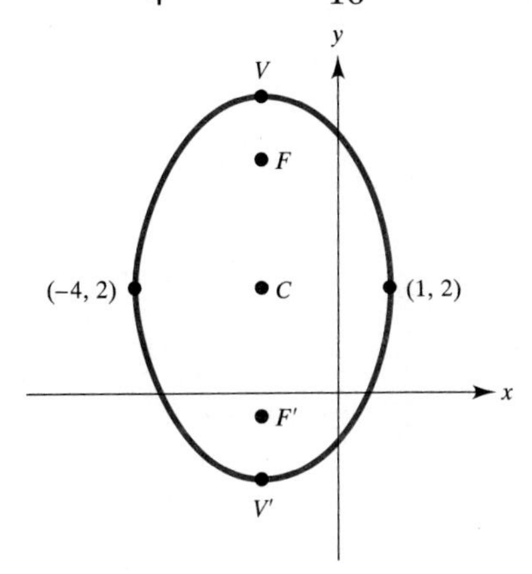

7. $\dfrac{(y-3)^2}{9} + \dfrac{(x-4)^2}{4} = 1,\ F(4, 3 \pm \sqrt{5}),\ C(4, 3),\ V(4, 3 \pm 3)$

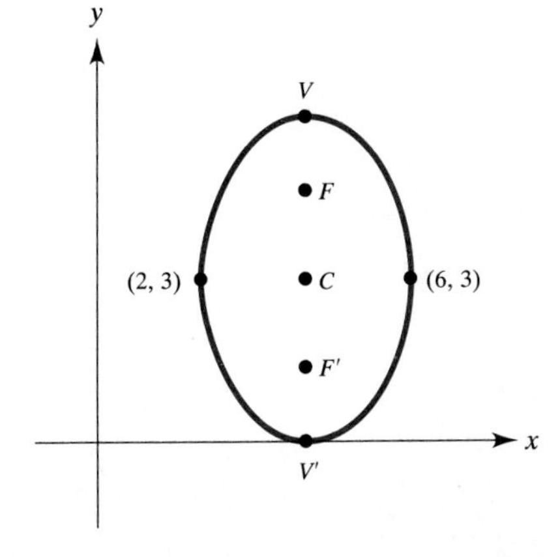

9. $\dfrac{(x-5)^2}{4}+\dfrac{(y-1)^2}{9}=1$

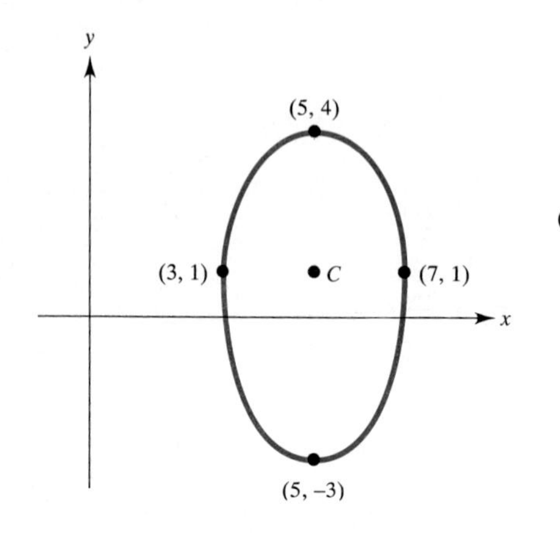

10. $\dfrac{x^2}{36}+\dfrac{(y-3)^2}{20}=1$

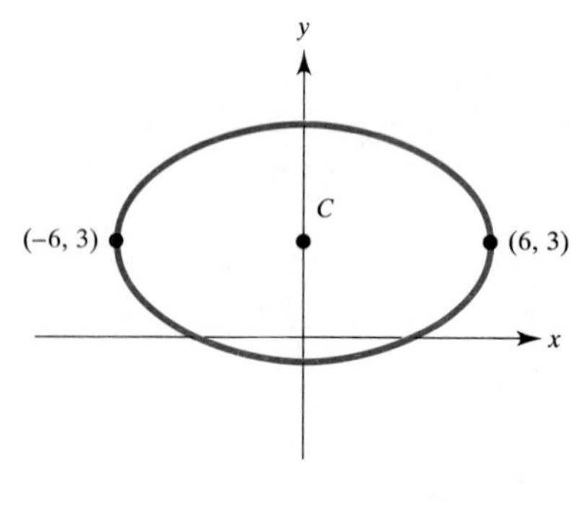

11. $\dfrac{(x+1)^2}{13}+\dfrac{(y+1)^2}{9}=1$

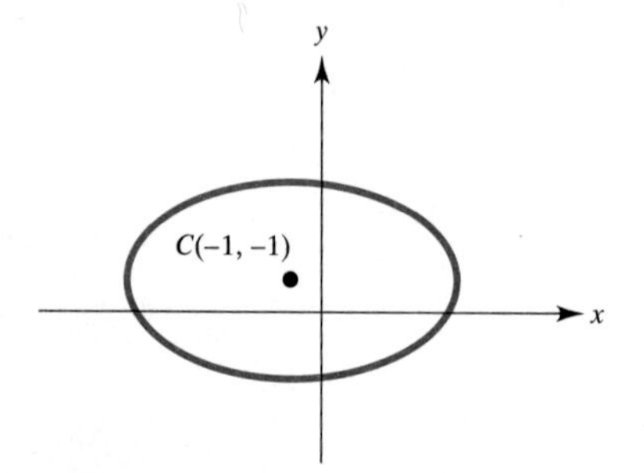

13. $\dfrac{(x-5)^2}{64}+\dfrac{(y-4)^2}{9}=1$

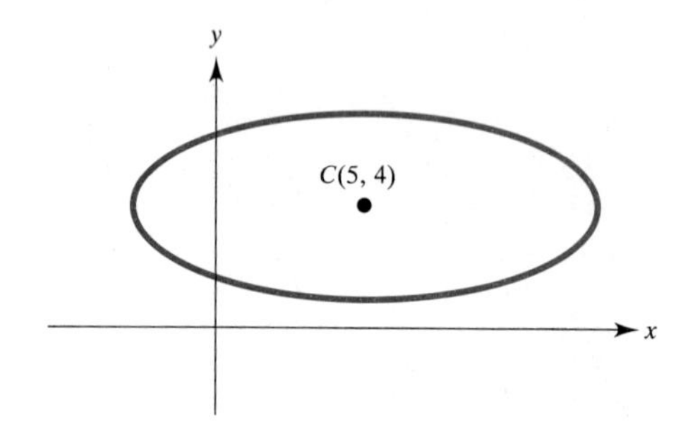

14. $\dfrac{x^2}{81}+\dfrac{y^2}{56}=1$

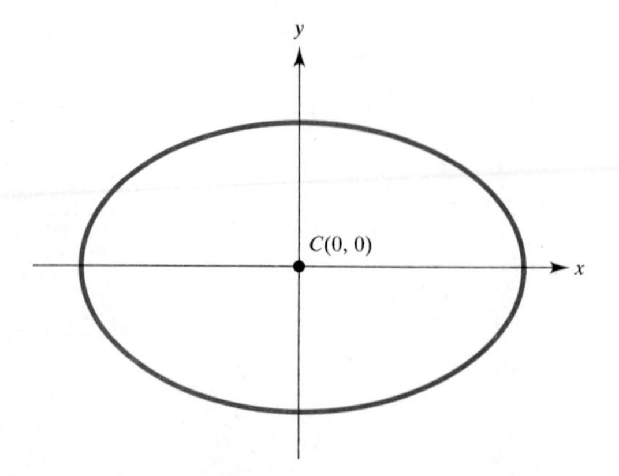

15. $\dfrac{(x-3)^2}{24} + \dfrac{(y-2)^2}{49} = 1$

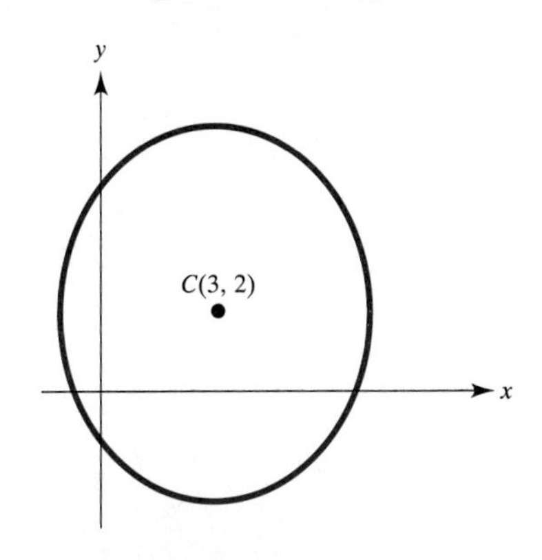

18. 

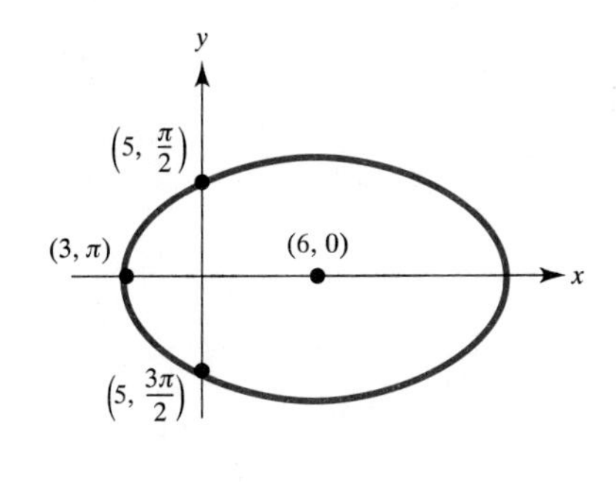

EXERCISE 13.4 PAGE 231

1. $F(\pm\sqrt{20}, 0)$, $V(\pm 4, 0)$, $C(0, 0)$, $y = \pm 2x$

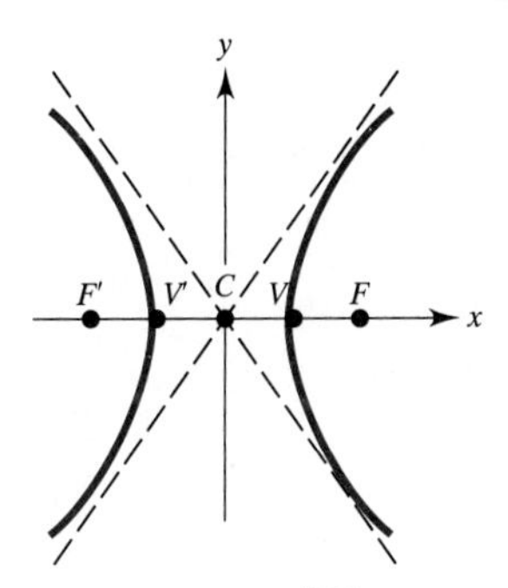

2. $F(0, \pm\sqrt{20})$, $V(0, \pm 4)$, $C(0, 0)$, $y = \pm 2x$

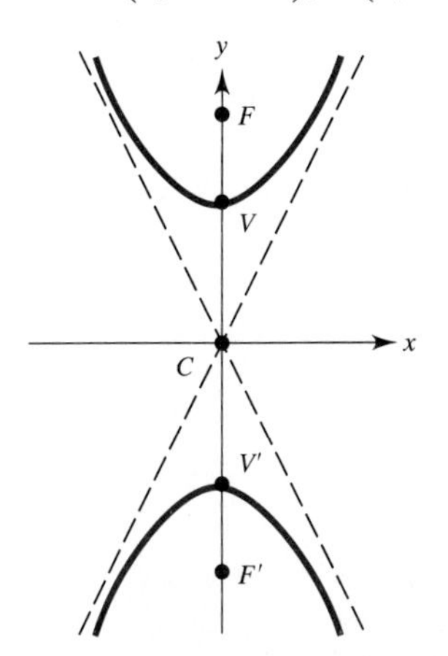

3. $F(-1, 2 \pm \sqrt{14})$, $V(-1, 2 \pm \sqrt{5})$, $C(-1, 2)$, $y - 2 = \pm \frac{\sqrt{5}}{3}(x + 1)$

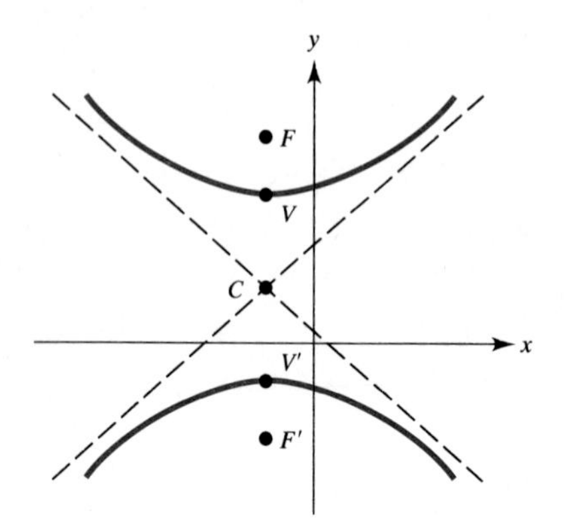

5. $\dfrac{(x+5)^2}{4} - \dfrac{y^2}{9} = 1$, $F(-5 \pm \sqrt{13}, 0)$, $V(-5 \pm 2, 0)$, $C(-5, 0)$,
$y = \pm\frac{3}{2}(x + 5)$

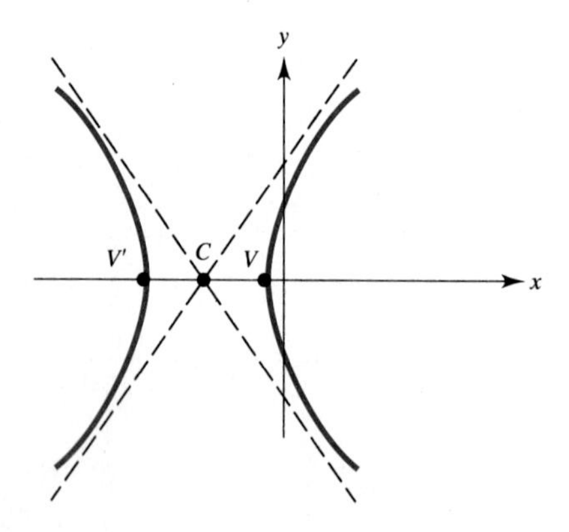

6. $F(\pm 10, 0)$, $V(\pm 8, 0)$, $C(0, 0)$, $3x \pm 4y = 0$

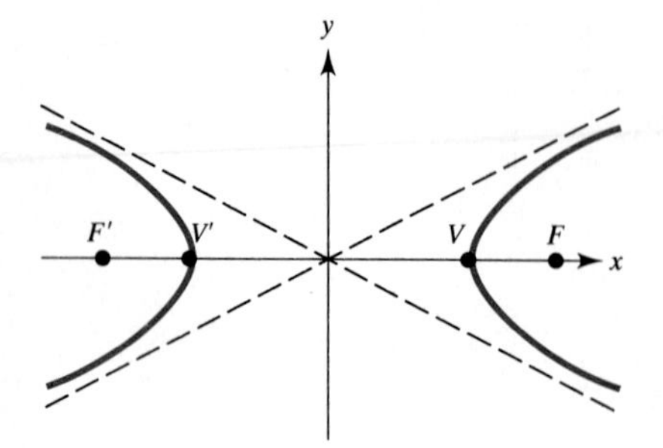

7. $F\left(-3, -1 \pm \frac{\sqrt{13}}{3}\right)$, $V(-3, -1 \pm 1)$, $C(-3, -1)$,

$y + 1 = \pm \frac{3}{2}(x + 3)$

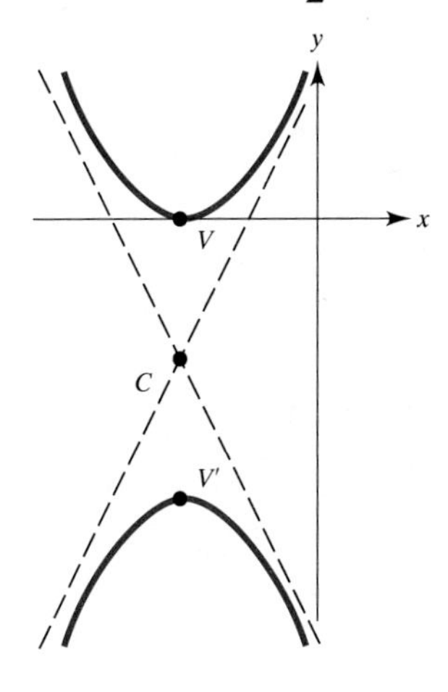

9. $x^2 - y^2 = 1$

10. $\frac{(x + 2)^2}{36} - \frac{(y - 2)^2}{28} = 1$

11. $\frac{(y + 1)^2}{9} - \frac{(x - 3)^2}{7} = 1$

EXERCISE 13.5 PAGE 236

1. $\theta = 45°$, $x'^2 - y'^2 = 14$

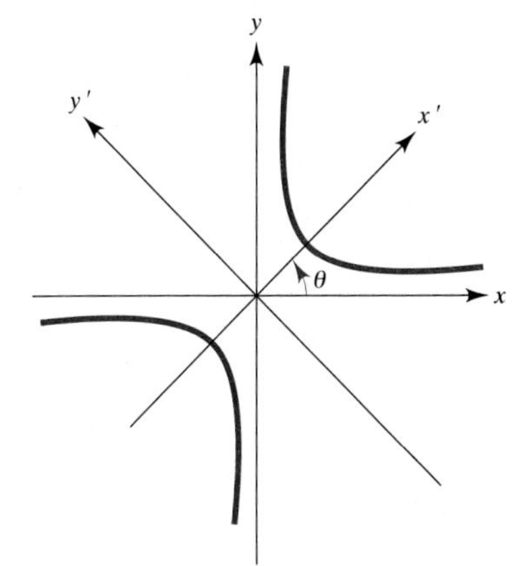

2. $\theta = 30°, 2\left(x' - \frac{\sqrt{3}}{4}\right)^2 = y' + \frac{11}{8}$

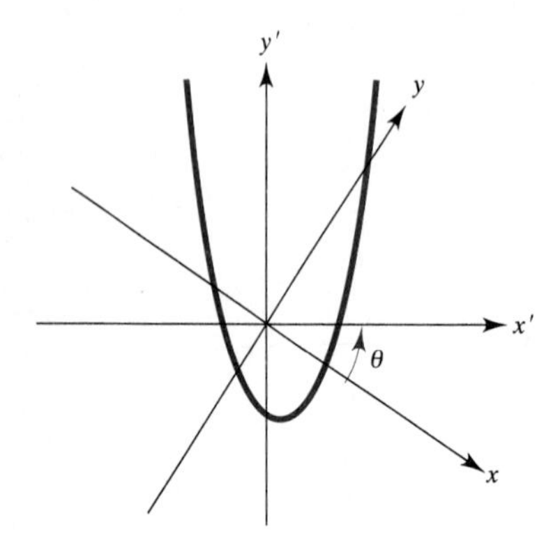

3. $\theta = 45°, \frac{(y' - \sqrt{2})^2}{2} - \frac{x'^2}{2} = 1.$

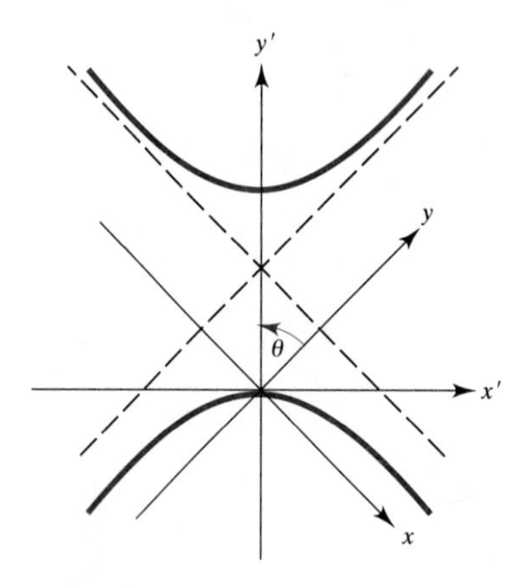

5. $\theta = 30°, x'^2 + 5y'^2 = 4$

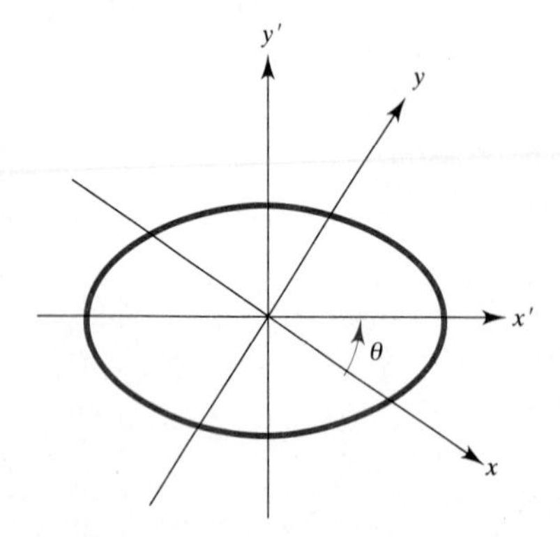

6. $\theta = 45°, 3x'^2 + y'^2 = 2$

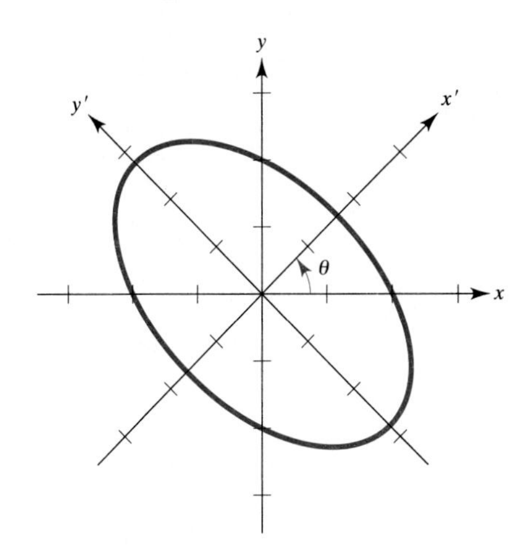

7. $\theta = \frac{1}{2}\tan^{-1}\left(\frac{-24}{7}\right), \frac{(x'-2)^2}{4} + (y'-1)^2 = 1$

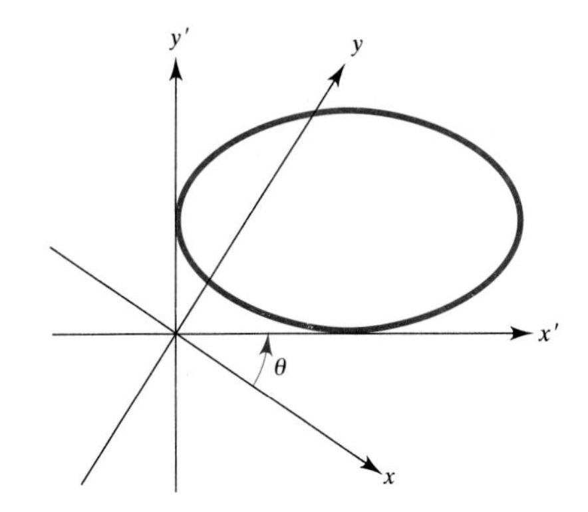

REVIEW EXERCISES PAGE 236

1. $2x + 3y = 28$
2. $x'^2 + y'^2 = 5$
3. $(y + 3)^2 = -6(x - \frac{7}{2})$

5. $\dfrac{(y+3)^2}{4} - \dfrac{(x-2)^2}{12} = 1$

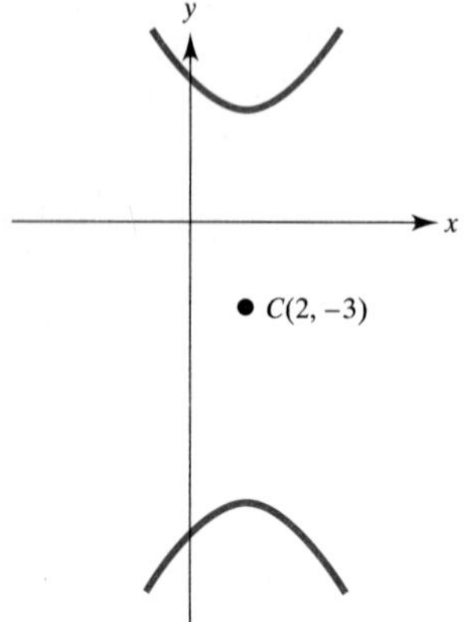

6. $\theta = 45°, \dfrac{x'^2}{9} + \dfrac{y'^2}{4} = 1$

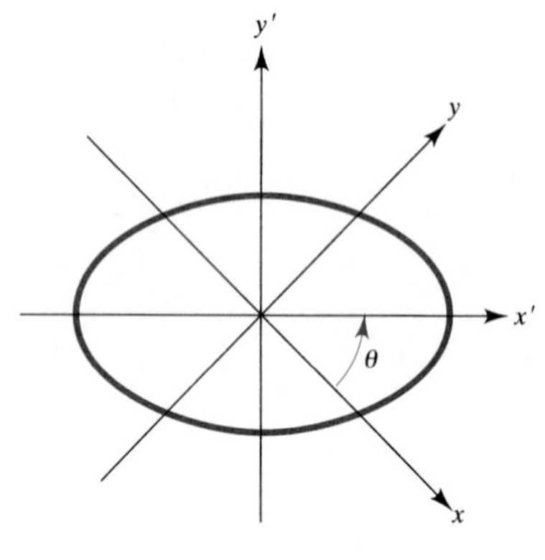

EXERCISE A.1 PAGE 239

3. 0.09983342

EXERCISE A.2 PAGE 241

1. Odd
2. Even
3.

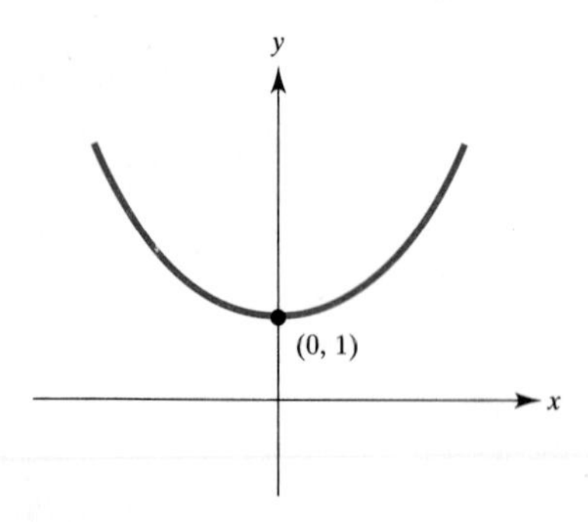

EXERCISE A.3 PAGE 243

1. 0.86858
2. −0.43429
3. 4.6052
5. −4.6052
6. −23.026
7. 6.9078
9. 0.1448
10. −0.2171

EXERCISE A.4 PAGES 244–245

1. 0.9573
2. 0.6569
3. 0.9232
5. 2.989
6. 0.6468
7. 8.144
9. 0.6455
10. 2.257
11. 0.6129
13. 33.9°
14. 44.1°
15. 60.5°
17. 78.9°
18. 84.2°
19. 2.5°
21. 0.7°
22. 15.6°
23. 52.6°

EXERCISE A.5 PAGE 247

1. 0.9387
2. 0.7824
3. 4.131
5. 4.300
6. 0.9530
7. 0.9147
9. 0.3853
10. 0.6761
11. 0.2161
13. 0.4209
14. 0.6779
15. 0.6310
17. 57.88°
18. 8.70°
19. 79.37°
21. 11.37°
22. 73.74°
23. 27.36°
25. 65.43°
26. 37.29°
27. 88.62°
29. 21.98°
30. 61.28°
31. 41.63°

TABLE

Values of the Trigonometric Functions to Four Places

Table

Radians	Degrees	sin	tan	cot	cos		
.0000	0.0°	.0000	.0000	–	1.0000	90.0°	1.5708
.0017	0.1°	.0017	.0017	573.0	1.0000	89.9°	1.5691
.0035	0.2°	.0035	.0035	286.5	1.0000	89.8°	1.5673
.0052	0.3°	.0052	.0052	191.0	1.0000	89.7°	1.5656
.0070	0.4°	.0070	.0070	143.2	1.0000	89.6°	1.5638
.0087	0.5°	.0087	.0087	114.6	1.0000	89.5°	1.5621
.0105	0.6°	.0105	.0105	95.49	.9999	89.4°	1.5603
.0122	0.7°	.0122	.0122	81.85	.9999	89.3°	1.5586
.0140	0.8°	.0140	.0140	71.62	.9999	89.2°	1.5568
.0157	0.9°	.0157	.0157	63.66	.9999	89.1°	1.5551
.0175	1.0°	.0175	.0175	57.29	.9998	89.0°	1.5533
.0192	1.1°	.0192	.0192	52.08	.9998	88.9°	1.5516
.0209	1.2°	.0209	.0209	47.74	.9998	88.8°	1.5499
.0227	1.3°	.0227	.0227	44.07	.9997	88.7°	1.5481
.0244	1.4°	.0244	.0244	40.92	.9997	88.6°	1.5464
.0262	1.5°	.0262	.0262	38.19	.9997	88.5°	1.5446
.0279	1.6°	.0279	.0279	35.80	.9996	88.4°	1.5429
.0297	1.7°	.0297	.0297	33.69	.9996	88.3°	1.5411
.0314	1.8°	.0314	.0314	31.82	.9995	88.2°	1.5394
.0332	1.9°	.0332	.0332	30.14	.9995	88.1°	1.5376
.0349	2.0°	.0349	.0349	28.64	.9994	88.0°	1.5359
.0367	2.1°	.0366	.0367	27.27	.9993	87.9°	1.5341
.0384	2.2°	.0384	.0384	26.03	.9993	87.8°	1.5324
.0401	2.3°	.0401	.0402	24.90	.9992	87.7°	1.5307
.0419	2.4°	.0419	.0419	23.86	.9991	87.6°	1.5289
.0436	2.5°	.0436	.0437	22.90	.9990	87.5°	1.5272
.0454	2.6°	.0454	.0454	22.02	.9990	87.4°	1.5254
.0471	2.7°	.0471	.0472	21.20	.9989	87.3°	1.5237
.0489	2.8°	.0488	.0489	20.45	.9988	87.2°	1.5219
.0506	2.9°	.0506	.0507	19.74	.9987	87.1°	1.5202
.0524	3.0°	.0523	.0524	19.08	.9986	87.0°	1.5184
.0541	3.1°	.0541	.0542	18.46	.9985	86.9°	1.5167
.0559	3.2°	.0558	.0559	17.89	.9984	86.8°	1.5149
.0576	3.3°	.0576	.0577	17.34	.9983	86.7°	1.5132
.0593	3.4°	.0593	.0594	16.83	.9982	86.6°	1.5115
.0611	3.5°	.0610	.0612	16.35	.9981	86.5°	1.5097
.0628	3.6°	.0628	.0629	15.89	.9980	86.4°	1.5080
.0646	3.7°	.0645	.0647	15.46	.9979	86.3°	1.5062
.0663	3.8°	.0663	.0664	15.06	.9978	86.2°	1.5045
.0681	3.9°	.0680	.0682	14.67	.9977	86.1°	1.5027
.0698	4.0°	.0698	.0699	14.30	.9976	86.0°	1.5010
.0716	4.1°	.0715	.0717	13.95	.9974	85.9°	1.4992
.0733	4.2°	.0732	.0734	13.62	.9973	85.8°	1.4975
.0750	4.3°	.0750	.0752	13.30	.9972	85.7°	1.4957
.0768	4.4°	.0767	.0769	13.00	.9971	85.6°	1.4940
.0785	4.5°	.0785	.0787	12.71	.9969	85.5°	1.4923
.0803	4.6°	.0802	.0805	12.43	.9968	85.4°	1.4905
.0820	4.7°	.0819	.0822	12.16	.9966	85.3°	1.4888
.0838	4.8°	.0837	.0840	11.91	.9965	85.2°	1.4870
.0855	4.9°	.0854	.0857	11.66	.9963	85.1°	1.4853
.0873	5.0°	.0872	.0875	11.43	.9962	85.0°	1.4835
		cos	cot	tan	sin	Degrees	Radians

Table (*Continued*)

Radians	Degrees	sin	tan	cot	cos		
.0873	5.0°	.0872	.0875	11.43	.9962	85.0°	1.4835
.0890	5.1°	.0889	.0892	11.20	.9960	84.9°	1.4818
.0908	5.2°	.0906	.0910	10.99	.9959	84.8°	1.4800
.0925	5.3°	.0924	.0928	10.78	.9957	84.7°	1.4783
.0942	5.4°	.0941	.0945	10.58	.9956	84.6°	1.4765
.0960	5.5°	.0958	.0963	10.39	.9954	84.5°	1.4748
.0977	5.6°	.0976	.0981	10.20	.9952	84.4°	1.4731
.0995	5.7°	.0993	.0998	10.02	.9951	84.3°	1.4713
.1012	5.8°	.1011	.1016	9.845	.9949	84.2°	1.4696
.1030	5.9°	.1028	.1033	9.677	.9947	84.1°	1.4678
.1047	6.0°	.1045	.1051	9.514	.9945	84.0°	1.4661
.1065	6.1°	.1063	.1069	9.357	.9943	83.9°	1.4643
.1082	6.2°	.1080	.1086	9.205	.9942	83.8°	1.4626
.1100	6.3°	.1097	.1104	9.058	.9940	83.7°	1.4608
.1117	6.4°	.1115	.1122	8.915	.9938	83.6°	1.4591
.1134	6.5°	.1132	.1139	8.777	.9936	83.5°	1.4573
.1152	6.6°	.1149	.1157	8.643	.9934	83.4°	1.4556
.1169	6.7°	.1167	.1175	8.513	.9932	83.3°	1.4539
.1187	6.8°	.1184	.1192	8.386	.9930	83.2°	1.4521
.1204	6.9°	.1201	.1210	8.264	.9928	83.1°	1.4504
.1222	7.0°	.1219	.1228	8.144	.9925	83.0°	1.4486
.1239	7.1°	.1236	.1246	8.028	.9923	82.9°	1.4469
.1257	7.2°	.1253	.1263	7.916	.9921	82.8°	1.4451
.1274	7.3°	.1271	.1281	7.806	.9919	82.7°	1.4434
.1292	7.4°	.1288	.1299	7.700	.9917	82.6°	1.4416
.1309	7.5°	.1305	.1317	7.596	.9914	82.5°	1.4399
.1326	7.6°	.1323	.1334	7.495	.9912	82.4°	1.4382
.1344	7.7°	.1340	.1352	7.396	.9910	82.3°	1.4364
.1361	7.8°	.1357	.1370	7.300	.9907	82.2°	1.4347
.1379	7.9°	.1374	.1388	7.207	.9905	82.1°	1.4329
.1396	8.0°	.1392	.1405	7.115	.9903	82.0°	1.4312
.1414	8.1°	.1409	.1423	7.026	.9900	81.9°	1.4294
.1431	8.2°	.1426	.1441	6.940	.9898	81.8°	1.4277
.1449	8.3°	.1444	.1459	6.855	.9895	81.7°	1.4259
.1466	8.4°	.1461	.1477	6.772	.9893	81.6°	1.4242
.1484	8.5°	.1478	.1495	6.691	.9890	81.5°	1.4224
.1501	8.6°	.1495	.1512	6.612	.9888	81.4°	1.4207
.1518	8.7°	.1513	.1530	6.535	.9885	81.3°	1.4190
.1536	8.8°	.1530	.1548	6.460	.9882	81.2°	1.4172
.1553	8.9°	.1547	.1566	6.386	.9880	81.1°	1.4155
.1571	9.0°	.1564	.1584	6.314	.9877	81.0°	1.4137
.1588	9.1°	.1582	.1602	6.243	.9874	80.9°	1.4120
.1606	9.2°	.1599	.1620	6.174	.9871	80.8°	1.4102
.1623	9.3°	.1616	.1638	6.107	.9869	80.7°	1.4085
.1641	9.4°	.1633	.1655	6.041	.9866	80.6°	1.4067
.1658	9.5°	.1650	.1673	5.976	.9863	80.5°	1.4050
.1676	9.6°	.1668	.1691	5.912	.9860	80.4°	1.4032
.1693	9.7°	.1685	.1709	5.850	.9857	80.3°	1.4015
.1710	9.8°	.1702	.1727	5.789	.9854	80.2°	1.3998
.1728	9.9°	.1719	.1745	5.730	.9851	80.1°	1.3980
.1745	10.0°	.1736	.1763	5.671	.9848	80.0°	1.3963
		cos	cot	tan	sin	Degrees	Radians

Table *(Continued)*

Radians	Degrees	sin	tan	cot	cos		
.1745	10.0°	.1736	.1763	5.671	.9848	80.0°	1.3963
.1763	10.1°	.1754	.1781	5.614	.9845	79.9°	1.3945
.1780	10.2°	.1771	.1799	5.558	.9842	79.8°	1.3928
.1798	10.3°	.1788	.1817	5.503	.9839	79.7°	1.3910
.1815	10.4°	.1805	.1835	5.449	.9836	79.6°	1.3893
.1833	10.5°	.1822	.1853	5.396	.9833	79.5°	1.3875
.1850	10.6°	.1840	.1871	5.343	.9829	79.4°	1.3858
.1868	10.7°	.1857	.1890	5.292	.9826	79.3°	1.3840
.1885	10.8°	.1874	.1908	5.242	.9823	79.2°	1.3823
.1902	10.9°	.1891	.1926	5.193	.9820	79.1°	1.3806
.1920	11.0°	.1908	.1944	5.145	.9816	79.0°	1.3788
.1937	11.1°	.1925	.1962	5.097	.9813	78.9°	1.3771
.1955	11.2°	.1942	.1980	5.050	.9810	78.8°	1.3753
.1972	11.3°	.1959	.1998	5.005	.9806	78.7°	1.3736
.1990	11.4°	.1977	.2016	4.959	.9803	78.6°	1.3718
.2007	11.5°	.1994	.2035	4.915	.9799	78.5°	1.3701
.2025	11.6°	.2011	.2053	4.872	.9796	78.4°	1.3683
.2042	11.7°	.2028	.2071	4.829	.9792	78.3°	1.3666
.2059	11.8°	.2045	.2089	4.787	.9789	78.2°	1.3648
.2077	11.9°	.2062	.2107	4.745	.9785	78.1°	1.3631
.2094	12.0°	.2079	.2126	4.705	.9781	78.0°	1.3614
.2112	12.1°	.2096	.2144	4.665	.9778	77.9°	1.3596
.2129	12.2°	.2113	.2162	4.625	.9774	77.8°	1.3579
.2147	12.3°	.2130	.2180	4.586	.9770	77.7°	1.3561
.2164	12.4°	.2147	.2199	4.548	.9767	77.6°	1.3544
.2182	12.5°	.2164	.2217	4.511	.9763	77.5°	1.3526
.2199	12.6°	.2181	.2235	4.474	.9759	77.4°	1.3509
.2217	12.7°	.2198	.2254	4.437	.9755	77.3°	1.3491
.2234	12.8°	.2215	.2272	4.402	.9751	77.2°	1.3474
.2251	12.9°	.2233	.2290	4.366	.9748	77.1°	1.3456
.2269	13.0°	.2250	.2309	4.331	.9744	77.0°	1.3439
.2286	13.1°	.2267	.2327	4.297	.9740	76.9°	1.3422
.2304	13.2°	.2284	.2345	4.264	.9736	76.8°	1.3404
.2321	13.3°	.2300	.2364	4.230	.9732	76.7°	1.3387
.2339	13.4°	.2317	.2382	4.198	.9728	76.6°	1.3369
.2356	13.5°	.2334	.2401	4.165	.9724	76.5°	1.3352
.2374	13.6°	.2351	.2419	4.134	.9720	76.4°	1.3334
.2391	13.7°	.2368	.2438	4.102	.9715	76.3°	1.3317
.2409	13.8°	.2385	.2456	4.071	.9711	76.2°	1.3299
.2426	13.9°	.2402	.2475	4.041	.9707	76.1°	1.3282
.2443	14.0°	.2419	.2493	4.011	.9703	76.0°	1.3265
.2461	14.1°	.2436	.2512	3.981	.9699	75.9°	1.3247
.2478	14.2°	.2453	.2530	3.952	.9694	75.8°	1.3230
.2496	14.3°	.2470	.2549	3.923	.9690	75.7°	1.3212
.2513	14.4°	.2487	.2568	3.895	.9686	75.6°	1.3195
.2531	14.5°	.2504	.2586	3.867	.9681	75.5°	1.3177
.2548	14.6°	.2521	.2605	3.839	.9677	75.4°	1.3160
.2566	14.7°	.2538	.2623	3.812	.9673	75.3°	1.3142
.2583	14.8°	.2554	.2642	3.785	.9668	75.2°	1.3125
.2601	14.9°	.2571	.2661	3.758	.9664	75.1°	1.3107
.2618	15.0°	.2588	.2679	3.732	.9659	75.0°	1.3090
		cos	cot	tan	sin	Degrees	Radians

Table (*Continued*)

Radians	Degrees	sin	tan	cot	cos		
.2618	15.0°	.2588	.2679	3.732	.9659	75.0°	1.3090
.2635	15.1°	.2605	.2698	3.706	.9655	74.9°	1.3073
.2653	15.2°	.2622	.2717	3.681	.9650	74.8°	1.3055
.2670	15.3°	.2639	.2736	3.655	.9646	74.7°	1.3038
.2688	15.4°	.2656	.2754	3.630	.9641	74.6°	1.3020
.2705	15.5°	.2672	.2773	3.606	.9636	74.5°	1.3003
.2723	15.6°	.2689	.2792	3.582	.9632	74.4°	1.2985
.2740	15.7°	.2706	.2811	3.558	.9627	74.3°	1.2968
.2758	15.8°	.2723	.2830	3.534	.9622	74.2°	1.2950
.2775	15.9°	.2740	.2849	3.511	.9617	74.1°	1.2933
.2793	16.0°	.2756	.2867	3.487	.9613	74.0°	1.2915
.2810	16.1°	.2773	.2886	3.465	.9608	73.9°	1.2898
.2827	16.2°	.2790	.2905	3.442	.9603	73.8°	1.2881
.2845	16.3°	.2807	.2924	3.420	.9598	73.7°	1.2863
.2862	16.4°	.2823	.2943	3.398	.9593	73.6°	1.2846
.2880	16.5°	.2840	.2962	3.376	.9588	73.5°	1.2828
.2897	16.6°	.2857	.2981	3.354	.9583	73.4°	1.2811
.2915	16.7°	.2874	.3000	3.333	.9578	73.3°	1.2793
.2932	16.8°	.2890	.3019	3.312	.9573	73.2°	1.2776
.2950	16.9°	.2907	.3038	3.291	.9568	73.1°	1.2758
.2967	17.0°	.2924	.3057	3.271	.9563	73.0°	1.2741
.2985	17.1°	.2940	.3076	3.251	.9558	72.9°	1.2723
.3002	17.2°	.2957	.3096	3.230	.9553	72.8°	1.2706
.3019	17.3°	.2974	.3115	3.211	.9548	72.7°	1.2689
.3037	17.4°	.2990	.3134	3.191	.9542	72.6°	1.2671
.3054	17.5°	.3007	.3153	3.172	.9537	72.5°	1.2654
.3072	17.6°	.3024	.3172	3.152	.9532	72.4°	1.2636
.3089	17.7°	.3040	.3191	3.133	.9527	72.3°	1.2619
.3107	17.8°	.3057	.3211	3.115	.9521	72.2°	1.2601
.3124	17.9°	.3074	.3230	3.096	.9516	72.1°	1.2584
.3142	18.0°	.3090	.3249	3.078	.9511	72.0°	1.2566
.3159	18.1°	.3107	.3269	3.060	.9505	71.9°	1.2549
.3176	18.2°	.3123	.3288	3.042	.9500	71.8°	1.2531
.3194	18.3°	.3140	.3307	3.024	.9494	71.7°	1.2514
.3211	18.4°	.3156	.3327	3.006	.9489	71.6°	1.2497
.3229	18.5°	.3173	.3346	2.989	.9483	71.5°	1.2479
.3246	18.6°	.3190	.3365	2.971	.9478	71.4°	1.2462
.3264	18.7°	.3206	.3385	2.954	.9472	71.3°	1.2444
.3281	18.8°	.3223	.3404	2.937	.9466	71.2°	1.2427
.3299	18.9°	.3239	.3424	2.921	.9461	71.1°	1.2409
.3316	19.0°	.3256	.3443	2.904	.9455	71.0°	1.2392
.3334	19.1°	.3272	.3463	2.888	.9449	70.9°	1.2374
.3351	19.2°	.3289	.3482	2.872	.9444	70.8°	1.2357
.3368	19.3°	.3305	.3502	2.856	.9438	70.7°	1.2339
.3386	19.4°	.3322	.3522	2.840	.9432	70.6°	1.2322
.3403	19.5°	.3338	.3541	2.824	.9426	70.5°	1.2305
.3421	19.6°	.3355	.3561	2.808	.9421	70.4°	1.2287
.3438	19.7°	.3371	.3581	2.793	.9415	70.3°	1.2270
.3456	19.8°	.3387	.3600	2.778	.9409	70.2°	1.2252
.3473	19.9°	.3404	.3620	2.762	.9403	70.1°	1.2235
.3491	20.0°	.3420	.3640	2.747	.9397	70.0°	1.2217
		cos	cot	tan	sin	Degrees	Radians

Table (*Continued*)

Radians	Degrees	sin	tan	cot	cos		
.3491	20.0°	.3420	.3640	2.747	.9397	70.0°	1.2217
.3508	20.1°	.3437	.3659	2.733	.9391	69.9°	1.2200
.3526	20.2°	.3453	.3679	2.718	.9385	69.8°	1.2182
.3543	20.3°	.3469	.3699	2.703	.9379	69.7°	1.2165
.3560	20.4°	.3486	.3719	2.689	.9373	69.6°	1.2147
.3578	20.5°	.3502	.3739	2.675	.9367	69.5°	1.2130
.3595	20.6°	.3518	.3759	2.660	.9361	69.4°	1.2113
.3613	20.7°	.3535	.3779	2.646	.9354	69.3°	1.2095
.3630	20.8°	.3551	.3799	2.633	.9348	69.2°	1.2078
.3648	20.9°	.3567	.3819	2.619	.9342	69.1°	1.2060
.3665	21.0°	.3584	.3839	2.605	.9336	69.0°	1.2043
.3683	21.1°	.3600	.3859	2.592	.9330	68.9°	1.2025
.3700	21.2°	.3616	.3879	2.578	.9323	68.8°	1.2008
.3718	21.3°	.3633	.3899	2.565	.9317	68.7°	1.1990
.3735	21.4°	.3649	.3919	2.552	.9311	68.6°	1.1973
.3752	21.5°	.3665	.3939	2.539	.9304	68.5°	1.1956
.3770	21.6°	.3681	.3959	2.526	.9298	68.4°	1.1938
.3787	21.7°	.3697	.3979	2.513	.9291	68.3°	1.1921
.3805	21.8°	.3714	.4000	2.500	.9285	68.2°	1.1903
.3822	21.9°	.3730	.4020	2.488	.9278	68.1°	1.1886
.3840	22.0°	.3746	.4040	2.475	.9272	68.0°	1.1868
.3857	22.1°	.3762	.4061	2.463	.9265	67.9°	1.1851
.3875	22.2°	.3778	.4081	2.450	.9259	67.8°	1.1833
.3892	22.3°	.3795	.4101	2.438	.9252	67.7°	1.1816
.3910	22.4°	.3811	.4122	2.426	.9245	67.6°	1.1798
.3927	22.5°	.3827	.4142	2.414	.9239	67.5°	1.1781
.3944	22.6°	.3843	.4163	2.402	.9232	67.4°	1.1764
.3962	22.7°	.3859	.4183	2.391	.9225	67.3°	1.1746
.3979	22.8°	.3875	.4204	2.379	.9219	67.2°	1.1729
.3997	22.9°	.3891	.4224	2.367	.9212	67.1°	1.1711
.4014	23.0°	.3907	.4245	2.356	.9205	67.0°	1.1694
.4032	23.1°	.3923	.4265	2.344	.9198	66.9°	1.1676
.4049	23.2°	.3939	.4286	2.333	.9191	66.8°	1.1659
.4067	23.3°	.3955	.4307	2.322	.9184	66.7°	1.1641
.4084	23.4°	.3971	.4327	2.311	.9178	66.6°	1.1624
.4102	23.5°	.3987	.4348	2.300	.9171	66.5°	1.1606
.4119	23.6°	.4003	.4369	2.289	.9164	66.4°	1.1589
.4136	23.7°	.4019	.4390	2.278	.9157	66.3°	1.1572
.4154	23.8°	.4035	.4411	2.267	.9150	66.2°	1.1554
.4171	23.9°	.4051	.4431	2.257	.9143	66.1°	1.1537
.4189	24.0°	.4067	.4452	2.246	.9135	66.0°	1.1519
.4206	24.1°	.4083	.4473	2.236	.9128	65.9°	1.1502
.4224	24.2°	.4099	.4494	2.225	.9121	65.8°	1.1484
.4241	24.3°	.4115	.4515	2.215	.9114	65.7°	1.1467
.4259	24.4°	.4131	.4536	2.204	.9107	65.6°	1.1449
.4276	24.5°	.4147	.4557	2.194	.9100	65.5°	1.1432
.4294	24.6°	.4163	.4578	2.184	.9092	65.4°	1.1414
.4311	24.7°	.4179	.4599	2.174	.9085	65.3°	1.1397
.4328	24.8°	.4195	.4621	2.164	.9078	65.2°	1.1380
.4346	24.9°	.4210	.4642	2.154	.9070	65.1°	1.1362
.4363	25.0°	.4226	.4663	2.145	.9063	65.0°	1.1345
		cos	cot	tan	sin	Degrees	Radians

Table (*Continued*)

Radians	Degrees	sin	tan	cot	cos		
.4363	25.0°	.4226	.4663	2.145	.9063	65.0°	1.1345
.4381	25.1°	.4242	.4684	2.135	.9056	64.9°	1.1327
.4398	25.2°	.4258	.4706	2.125	.9048	64.8°	1.1310
.4416	25.3°	.4274	.4727	2.116	.9041	64.7°	1.1292
.4433	25.4°	.4289	.4748	2.106	.9033	64.6°	1.1275
.4451	25.5°	.4305	.4770	2.097	.9026	64.5°	1.1257
.4468	25.6°	.4321	.4791	2.087	.9018	64.4°	1.1240
.4485	25.7°	.4337	.4813	2.078	.9011	64.3°	1.1222
.4503	25.8°	.4352	.4834	2.069	.9003	64.2°	1.1205
.4520	25.9°	.4368	.4856	2.059	.8996	64.1°	1.1188
.4538	26.0°	.4384	.4877	2.050	.8988	64.0°	1.1170
.4555	26.1°	.4399	.4899	2.041	.8980	63.9°	1.1153
.4573	26.2°	.4415	.4921	2.032	.8973	63.8°	1.1135
.4590	26.3°	.4431	.4942	2.023	.8965	63.7°	1.1118
.4608	26.4°	.4446	.4964	2.014	.8957	63.6°	1.1100
.4625	26.5°	.4462	.4986	2.006	.8949	63.5°	1.1083
.4643	26.6°	.4478	.5008	1.997	.8942	63.4°	1.1065
.4660	26.7°	.4493	.5029	1.988	.8934	63.3°	1.1048
.4677	26.8°	.4509	.5051	1.980	.8926	63.2°	1.1030
.4695	26.9°	.4524	.5073	1.971	.8918	63.1°	1.1013
.4712	27.0°	.4540	.5095	1.963	.8910	63.0°	1.0996
.4730	27.1°	.4555	.5117	1.954	.8902	62.9°	1.0978
.4747	27.2°	.4571	.5139	1.946	.8894	62.8°	1.0961
.4765	27.3°	.4586	.5161	1.937	.8886	62.7°	1.0943
.4782	27.4°	.4602	.5184	1.929	.8878	62.6°	1.0926
.4800	27.5°	.4617	.5206	1.921	.8870	62.5°	1.0908
.4817	27.6°	.4633	.5228	1.913	.8862	62.4°	1.0891
.4835	27.7°	.4648	.5250	1.905	.8854	62.3°	1.0873
.4852	27.8°	.4664	.5272	1.897	.8846	62.2°	1.0856
.4869	27.9°	.4679	.5295	1.889	.8838	62.1°	1.0838
.4887	28.0°	.4695	.5317	1.881	.8829	62.0°	1.0821
.4904	28.1°	.4710	.5340	1.873	.8821	61.9°	1.0804
.4922	28.2°	.4726	.5362	1.865	.8813	61.8°	1.0786
.4939	28.3°	.4741	.5384	1.857	.8805	61.7°	1.0769
.4957	28.4°	.4756	.5407	1.849	.8796	61.6°	1.0751
.4974	28.5°	.4772	.5430	1.842	.8788	61.5°	1.0734
.4992	28.6°	.4787	.5452	1.834	.8780	61.4°	1.0716
.5009	28.7°	.4802	.5475	1.827	.8771	61.3°	1.0699
.5027	28.8°	.4818	.5498	1.819	.8763	61.2°	1.0681
.5044	28.9°	.4833	.5520	1.811	.8755	61.1°	1.0664
.5061	29.0°	.4848	.5543	1.804	.8746	61.0°	1.0647
.5079	29.1°	.4863	.5566	1.797	.8738	60.9°	1.0629
.5096	29.2°	.4879	.5589	1.789	.8729	60.8°	1.0612
.5114	29.3°	.4894	.5612	1.782	.8721	60.7°	1.0594
.5131	29.4°	.4909	.5635	1.775	.8712	60.6°	1.0577
.5149	29.5°	.4924	.5658	1.767	.8704	60.5°	1.0559
.5166	29.6°	.4939	.5681	1.760	.8695	60.4°	1.0542
.5184	29.7°	.4955	.5704	1.753	.8686	60.3°	1.0524
.5201	29.8°	.4970	.5727	1.746	.8678	60.2°	1.0507
.5219	29.9°	.4985	.5750	1.739	.8669	60.1°	1.0489
.5236	30.0°	.5000	.5774	1.732	.8660	60.0°	1.0472
		cos	cot	tan	sin	Degrees	Radians

Table *(Continued)*

Radians	Degrees	sin	tan	cot	cos		
.5236	30.0°	.5000	.5774	1.732	.8660	60.0°	1.0472
.5253	30.1°	.5015	.5797	1.725	.8652	59.9°	1.0455
.5271	30.2°	.5030	.5820	1.718	.8643	59.8°	1.0437
.5288	30.3°	.5045	.5844	1.711	.8634	59.7°	1.0420
.5306	30.4°	.5060	.5867	1.704	.8625	59.6°	1.0402
.5323	30.5°	.5075	.5890	1.698	.8616	59.5°	1.0385
.5341	30.6°	.5090	.5914	1.691	.8607	59.4°	1.0367
.5358	30.7°	.5105	.5938	1.684	.8599	59.3°	1.0350
.5376	30.8°	.5120	.5961	1.678	.8590	59.2°	1.0332
.5393	30.9°	.5135	.5985	1.671	.8581	59.1°	1.0315
.5411	31.0°	.5150	.6009	1.664	.8572	59.0°	1.0297
.5428	31.1°	.5165	.6032	1.658	.8563	58.9°	1.0280
.5445	31.2°	.5180	.6056	1.651	.8554	58.8°	1.0263
.5463	31.3°	.5195	.6080	1.645	.8545	58.7°	1.0245
.5480	31.4°	.5210	.6104	1.638	.8536	58.6°	1.0228
.5498	31.5°	.5225	.6128	1.632	.8526	58.5°	1.0210
.5515	31.6°	.5240	.6152	1.625	.8517	58.4°	1.0193
.5533	31.7°	.5255	.6176	1.619	.8508	58.3°	1.0175
.5550	31.8°	.5270	.6200	1.613	.8499	58.2°	1.0158
.5568	31.9°	.5284	.6224	1.607	.8490	58.1°	1.0140
.5585	32.0°	.5299	.6249	1.600	.8480	58.0°	1.0123
.5603	32.1°	.5314	.6273	1.594	.8471	57.9°	1.0105
.5620	32.2°	.5329	.6297	1.588	.8462	57.8°	1.0088
.5637	32.3°	.5344	.6322	1.582	.8453	57.7°	1.0071
.5655	32.4°	.5358	.6346	1.576	.8443	57.6°	1.0053
.5672	32.5°	.5373	.6371	1.570	.8434	57.5°	1.0036
.5690	32.6°	.5388	.6395	1.564	.8425	57.4°	1.0018
.5707	32.7°	.5402	.6420	1.558	.8415	57.3°	1.0001
.5725	32.8°	.5417	.6445	1.552	.8406	57.2°	.9983
.5742	32.9°	.5432	.6469	1.546	.8396	57.1°	.9966
.5760	33.0°	.5446	.6494	1.540	.8387	57.0°	.9948
.5777	33.1°	.5461	.6519	1.534	.8377	56.9°	.9931
.5794	33.2°	.5476	.6544	1.528	.8368	56.8°	.9913
.5812	33.3°	.5490	.6569	1.522	.8358	56.7°	.9896
.5829	33.4°	.5505	.6594	1.517	.8348	56.6°	.9879
.5847	33.5°	.5519	.6619	1.511	.8339	56.5°	.9861
.5864	33.6°	.5534	.6644	1.505	.8329	56.4°	.9844
.5882	33.7°	.5548	.6669	1.499	.8320	56.3°	.9826
.5899	33.8°	.5563	.6694	1.494	.8310	56.2°	.9809
.5917	33.9°	.5577	.6720	.1488	.8300	56.1°	.9791
.5934	34.0°	.5592	.6745	1.483	.8290	56.0°	.9774
.5952	34.1°	.5606	.6771	1.477	.8281	55.9°	.9756
.5969	34.2°	.5621	.6796	1.471	.8271	55.8°	.9739
.5986	34.3°	.5635	.6822	1.466	.8261	55.7°	.9721
.6004	34.4°	.5650	.6847	1.460	.8251	55.6°	.9704
.6021	34.5°	.5664	.6873	1.455	.8241	55.5°	.9687
.6039	34.6°	.5678	.6899	1.450	.8231	55.4°	.9669
.6056	34.7°	.5693	.6924	1.444	.8221	55.3°	.9652
.6074	34.8°	.5707	.6950	1.439	.8211	55.2°	.9634
.6091	34.9°	.5721	.6976	1.433	.8202	55.1°	.9617
.6109	35.0°	.5736	.7002	1.428	.8192	55.0°	.9599
		cos	cot	tan	sin	Degrees	Radians

Table (*Continued*)

Radians	Degrees	sin	tan	cot	cos		
.6109	35.0°	.5736	.7002	1.428	.8192	55.0°	.9599
.6126	35.1°	.5750	.7028	1.423	.8181	54.9°	.9582
.6144	35.2°	.5764	.7054	1.418	.8171	54.8°	.9564
.6161	35.3°	.5779	.7080	1.412	.8161	54.7°	.9547
.6178	35.4°	.5793	.7107	1.407	.8151	54.6°	.9530
.6196	35.5°	.5807	.7133	1.402	.8141	54.5°	.9512
.6213	35.6°	.5821	.7159	1.397	.8131	54.4°	.9495
.6231	35.7°	.5835	.7186	1.392	.8121	54.3°	.9477
.6248	35.8°	.5850	.7212	1.387	.8111	54.2°	.9460
.6266	35.9°	.5864	.7239	1.381	.8100	54.1°	.9442
.6283	36.0°	.5878	.7265	1.376	.8090	54.0°	.9425
.6301	36.1°	.5892	.7292	1.371	.8080	53.9°	.9407
.6318	36.2°	.5906	.7319	1.366	.8070	53.8°	.9390
.6336	36.3°	.5920	.7346	1.361	.8059	53.7°	.9372
.6353	36.4°	.5934	.7373	1.356	.8049	53.6°	.9355
.6370	36.5°	.5948	.7400	1.351	.8039	53.5°	.9338
.6388	36.6°	.5962	.7427	1.347	.8028	53.4°	.9320
.6405	36.7°	.5976	.7454	1.342	.8018	53.3°	.9303
.6423	36.8°	.5990	.7481	1.337	.8007	53.2°	.9285
.6440	36.9°	.6004	.7508	1.332	.7997	53.1°	.9268
.6458	37.0°	.6018	.7536	1.327	.7986	53.0°	.9250
.6475	37.1°	.6032	.7563	1.322	.7976	52.9°	.9233
.6493	37.2°	.6046	.7590	1.317	.7965	52.8°	.9215
.6510	37.3°	.6060	.7618	1.313	.7955	52.7°	.9198
.6528	37.4°	.6074	.7646	1.308	.7944	52.6°	.9180
.6545	37.5°	.6088	.7673	1.303	.7934	52.5°	.9163
.6562	37.6°	.6101	.7701	1.299	.7923	52.4°	.9146
.6580	37.7°	.6115	.7729	1.294	.7912	52.3°	.9128
.6597	37.8°	.6129	.7757	1.289	.7902	52.2°	.9111
.6615	37.9°	.6143	.7785	1.285	.7891	52.1°	.9093
.6632	38.0°	.6157	.7813	1.280	.7880	52.0°	.9076
.6650	38.1°	.6170	.7841	1.275	.7869	51.9°	.9058
.6667	38.2°	.6184	.7869	1.271	.7859	51.8°	.9041
.6685	38.3°	.6198	.7898	1.266	.7848	51.7°	.9023
.6702	38.4°	.6211	.7926	1.262	.7837	51.6°	.9006
.6720	38.5°	.6225	.7954	1.257	.7826	51.5°	.8988
.6737	38.6°	.6239	.7983	1.253	.7815	51.4°	.8971
.6754	38.7°	.6252	.8012	1.248	.7804	51.3°	.8954
.6772	38.8°	.6266	.8040	1.244	.7793	51.2°	.8936
.6789	38.9°	.6280	.8069	1.239	.7782	51.1°	.8919
.6807	39.0°	.6293	.8098	1.235	.7771	51.0°	.8901
.6824	39.1°	.6307	.8127	1.231	.7760	50.9°	.8884
.6842	39.2°	.6320	.8156	1.226	.7749	50.8°	.8866
.6859	39.3°	.6334	.8185	1.222	.7738	50.7°	.8849
.6877	39.4°	.6347	.8214	1.217	.7727	50.6°	.8831
.6894	39.5°	.6361	.8243	1.213	.7716	50.5°	.8814
.6912	39.6°	.6374	.8273	1.209	.7705	50.4°	.8796
.6929	39.7°	.6388	.8302	1.205	.7694	50.3°	.8779
.6946	39.8°	.6401	.8332	1.200	.7683	50.2°	.8762
.6964	39.9°	.6414	.8361	1.196	.7672	50.1°	.8744
.6981	40.0°	.6428	.8391	1.192	.7660	50.0°	.8727
		cos	cot	tan	sin	Degrees	Radians

Table *(Continued)*

Radians	Degrees	sin	tan	cot	cos		
.6981	40.0°	.6428	.8391	1.192	.7660	50.0°	.8727
.6999	40.1°	.6441	.8421	1.188	.7649	49.9°	.8709
.7016	40.2°	.6455	.8451	1.183	.7638	49.8°	.8692
.7034	40.3°	.6468	.8481	1.179	.7627	49.7°	.8674
.7051	40.4°	.6481	.8511	1.175	.7615	49.6°	.8657
.7069	40.5°	.6494	.8541	1.171	.7604	49.5°	.8639
.7086	40.6°	.6508	.8571	1.167	.7593	49.4°	.8622
.7103	40.7°	.6521	.8601	1.163	.7581	49.3°	.8604
.7121	40.8°	.6534	.8632	1.159	.7570	49.2°	.8587
.7138	40.9°	.6547	.8662	1.154	.7559	49.1°	.8570
.7156	41.0°	.6561	.8693	1.150	.7547	49.0°	.8552
.7173	41.1°	.6574	.8724	1.146	.7536	48.9°	.8535
.7191	41.2°	.6587	.8754	1.142	.7524	48.8°	.8517
.7208	41.3°	.6600	.8785	1.138	.7513	48.7°	.8500
.7226	41.4°	.6613	.8816	1.134	.7501	48.6°	.8482
.7243	41.5°	.6626	.8847	1.130	.7490	48.5°	.8465
.7261	41.6°	.6639	.8878	1.126	.7478	48.4°	.8447
.7278	41.7°	.6652	.8910	1.122	.7466	48.3°	.8430
.7295	41.8°	.6665	.8941	1.118	.7455	48.2°	.8412
.7313	41.9°	.6678	.8972	1.115	.7443	48.1°	.8395
.7330	42.0°	.6691	.9004	1.111	.7431	48.0°	.8378
.7348	42.1°	.6704	.9036	1.107	.7420	47.9°	.8360
.7365	42.2°	.6717	.9067	1.103	.7408	47.8°	.8343
.7383	42.3°	.6730	.9099	1.099	.7396	47.7°	.8325
.7400	42.4°	.6743	.9131	1.095	.7385	47.6°	.8308
.7418	42.5°	.6756	.9163	1.091	.7373	47.5°	.8290
.7435	42.6°	.6769	.9195	1.087	.7361	47.4°	.8273
.7453	42.7°	.6782	.9228	1.084	.7349	47.3°	.8255
.7470	42.8°	.6794	.9260	1.080	.7337	47.2°	.8238
.7487	42.9°	.6807	.9293	1.076	.7325	47.1°	.8221
.7505	43.0°	.6820	.9325	1.072	.7314	47.0°	.8203
.7522	43.1°	.6833	.9358	1.069	.7302	46.9°	.8186
.7540	43.2°	.6845	.9391	1.065	.7290	46.8°	.8168
.7557	43.3°	.6858	.9424	1.061	.7278	46.7°	.8151
.7575	43.4°	.6871	.9457	1.057	.7266	46.6°	.8133
.7592	43.5°	.6884	.9490	1.054	.7254	46.5°	.8116
.7610	43.6°	.6896	.9523	1.050	.7242	46.4°	.8098
.7627	43.7°	.6909	.9556	1.046	.7230	46.3°	.8081
.7645	43.8°	.6921	.9590	1.043	.7218	46.2°	.8063
.7662	43.9°	.6934	.9623	1.039	.7206	46.1°	.8046
.7679	44.0°	.6947	.9657	1.036	.7193	46.0°	.8029
.7697	44.1°	.6959	.9691	1.032	.7181	45.9°	.8011
.7714	44.2°	.6972	.9725	1.028	.7169	45.8°	.7994
.7732	44.3°	.6984	.9759	1.025	.7157	45.7°	.7976
.7749	44.4°	.6997	.9793	1.021	.7145	45.6°	.7959
.7767	44.5°	.7009	.9827	1.018	.7133	45.5°	.7941
.7784	44.6°	.7022	.9861	1.014	.7120	45.4°	.7924
.7802	44.7°	.7034	.9896	1.011	.7108	45.3°	.7906
.7819	44.8°	.7046	.9930	1.007	.7096	45.2°	.7889
.7837	44.9°	.7059	.9965	1.003	.7083	45.1°	.7871
.7854	45.0°	.7071	1.0000	1.000	.7071	45.0°	.7854
		cos	cot	tan	sin	Degrees	Radians

Index

H

I

L

M

N

O

P

Q

R

S

T

V

W

X

Y